Logodaedalus

logodaedalus
WORD HISTORIES OF INGENUITY IN EARLY MODERN EUROPE

Alexander Marr, Raphaële Garrod,
José Ramón Marcaida, and Richard J. Oosterhoff

UNIVERSITY OF PITTSBURGH PRESS

The research leading to these findings has received funding from the European Research Council under the European Union's Seventh Framework Programme (FP7/2007-2013)/ ERC grant agreement no 617391.

Published by the University of Pittsburgh Press, Pittsburgh, Pa., 15260
Copyright © 2018, University of Pittsburgh Press
All rights reserved
Manufactured in the United States of America
Printed on acid-free paper

Cataloging-in-Publication data is available from the Library of Congress

ISBN 13: 978-0-8229-4541-3
ISBN 10: 0-8229-4541-x

Cover art: Johann Neudörffer, from *Ein gute Ordnung, vnd kurtze vnterricht, der fürnemsten grunde aus denen die Jungen, Zierlichs schreybens begirlich, mit besonderer kunst vnd behendigkeyt vnterricht vnd geübt möge[n] werden* (1538), 131.
Cover design: Joel W. Coggins

Ein artlicher redner / logodaedalus. Ein artlicher und kunstreicher handwercksman / mechanicus, mechanarius ingeniosus, solers fabricator operum, quae ingenio pariter et manu perficiuntur.

• • •

An elegant author: logodaedalus. An elegant and crafty handworker: a mechanic, ingenious mechanician, skillful maker of works done by wit and hand together.

GEORG HENISCH, TEÜTSCHE SPRACH UND WEISSHEIT (1616)

Contents

List of Illustrations
 IX

Acknowledgments
 XIII

Note on Conventions
 XV

Introduction
 1

1. LATIN
    Genius · Ingenium
    19

2. ITALIAN
    Genio · Ingegno
    53

3. SPANISH
    Ingenio · Agudeza
    87

4. FRENCH
    Engin · Esprit · Naturel · Génie
    121

5. GERMAN AND DUTCH
    Art/Aard · Sinnlichkeit/sinrijk · Geest · Gemüt
    153

6. ENGLISH
    Genius · Ingenuity · Wit · Cunning
    193

Conclusion
 235

Notes
 241

Bibliography
 307

Index
 347

List of Illustrations

Fig. 1.1.
Otto van Veen, *Q. Horati Flacci Emblemata: imaginibus in aes incisis, notisque illustrata* (Antwerp: Jerome Verdussen, 1607), 41. Reproduced by kind permission of the Syndics of Cambridge University Library (Rare Books X.8.15) — 26

Fig. 1.2.
Hadrianus Junius, *Emblemata* (Antwerp: Christoph Plantin, 1565), 40. Reproduced by kind permission of the Syndics of Cambridge University Library (Rare Books O*.6.1[E]) — 27

Fig. 1.3.
Ambrogio Calepino, *Dictionarium undecim linguarum : iam postremo accurata emendatione, atque infinitorum locorum augmentatione, collectis ex bonorum authorum monumentis . . . respondent autem Latinis vocabulis, Hebraica, Græca, Gallica, Italica, Germanica, Belgica, Hispanica, Polonica, Ungarica, Anglica. Onomasticum verò : hoc est, propriorum nominum . . . maxima etiam accessione locupletatum, & praecipuarum rerum Germanica explanatione illustratum, seorsim adjunximus* (Basel: Sebastian Henricpetri, 1590). Reproduced by kind permission of the Syndics of Cambridge University Library (Gonville and Caius College Lower Library M.20.16) — 31

Fig. 1.4.
Robert Estienne, *Dictionarium, seu latinae linguae thesaurus* (Paris: Robert Estienne, 1531), 411v-412r. Reproduced by kind permission of the Syndics of Cambridge University Library (Rare Books Aa*.8.28[C]) — 42–43

Fig. 2.1.
Accademici della Crusca, *Vocabolario degli Accademici della Crusca* (Venice: Giovanni Alberti, 1612). Private collection, Cambridge — 61

Fig. 2.2.
Anonymous engraver after Domenico de Michelino, *Dante as Poet of the Divine Comedy*, ca. 1460. Copperplate engraving. The Albertina Museum, Vienna   77

Fig. 2.3.
Italian school, *stuccio* of drawing instruments, early seventeenth century. Mixed metals. Istituto e Museo di Storia della Scienza, Florence. Inventory number: 671. Museo Galileo, Florence—Photographic Archives   81

Fig. 3.1.
Miguel de Cervantes. *El ingenioso hidalgo Don Quixote de la Mancha*. Valencia: Pedro Patricio Mey, 1605. Reproduced by kind permission of the Syndics of Cambridge University Library (Hisp.8.60.7)   89

Fig. 3.2.
Sebastián de Covarrubias. *Emblemas Morales*. Madrid: Luis Sánchez, 1610. Fol. 285r. Reproduced by kind permission of the Syndics of Cambridge University Library (F161.d.8.2)   107

Fig. 3.3.
Real Academia Española. *Diccionario de la lengua castellana*. Madrid: Herederos de Francisco del Hierro, 1734, p. 270. Reproduced by kind permission of the Syndics of Cambridge University Library (Hisp.3.72.1-)   108

Fig. 4.1.
Académie française, *Dictionnaire de l'Académie françoise* (Paris, 1694). Engraving (36.4 x 23.5 cm), drawn by Jean-Baptiste Corneille, engraved by Jean Mariette and Gérard Edelinck. Collection du château de Versailles. Courtesy of Château de Versailles   123

Fig. 4.2.

"Aiguillier, bonnetier," engraving in *Recueil de planches, sur les sciences, les arts libéraux, et les arts méchaniques, avec leur explication. Deux cens soixante & neuf planches, premiere livraison*, vol. 22 of *Encyclopédie, ou dictionnaire raisonné des arts et des sciences*, ed. Denis Diderot and Jean-Baptiste le Rond d'Alembert. Paris: Briasson, David, Le Breton, Durand, 1762. PL.[1r]. Bibliothèque Mazarine, 2° 3442 Volume 22; http://mazarinum.bibliotheque-mazarine.fr/idurl/1/2131. Creative commons license, CC-BY-NC-ND   130

Fig. 5.1a, b, and c.

Kaspar Stieler, *Der Teutschen Sprache Stammbaum und Fortwachs oder Teutscher Sprachschatz / Teutonicae linguae semina et germina, sive lexicon Germanicum* (Nuremberg: Johann Hofmann, 1691), cols. 2029–2032. Reproduced by kind permission of the Syndics of Cambridge University Library (West Room S950.c.9.366–368)   177–79

Fig. 5.2.

Guillam van Haecht the Younger, *The Cabinet of Cornelis van der Geest* (1628), detail. Oil on panel, 99 x 129.5 cm. Rubenhuis, Antwerp. Collectie Rubenshuis (Antwerpen) RH.S.171   183

Fig. 6.1.

Elephant salt-cellar (ca. 1550). Rock crystal, gold and enamel. The elephant (made in India, ca. 1400) was combined with a later receptacle in Lisbon by Francisco Lopéz. Kunsthistorisches Museum, Vienna, Inv.-Nr. KK 2320. Manfred Werner/Tsui. Creative commons license, CC BY-SA 4.0   196

Fig. 6.2.

Nicholas Hilliard, *Self-portrait, aged 30* (1577), watercolor on vellum, laid on to card. Victoria and Albert Museum, London, Inv. P.155-1910. © Victoria and Albert Museum, London   217

Acknowledgments

We researched and wrote this book under the auspices of the project *Genius before Romanticism: Ingenuity in Early Modern Art & Science*, at the University of Cambridge's Centre for Research in the Arts, Social Sciences and Humanities (CRASSH). The project was generously funded by a European Research Council Consolidator Grant, the support of which we thankfully acknowledge. CRASSH provided a truly congenial environment for our research and we are especially grateful to its director, Simon Goldhill, and to Catherine Hurley and the administrative staff. We owe much to our project's administrators, Gaenor Moore and Rachael Taylor, for their help and forbearance.

This book would probably never have been written had it not been for the encouragement of Richard Scholar and Ita Mac Carthy, who were our project's first visiting fellows. Their advice and friendship have been crucial at every stage. We owe a special debt of thanks to the participants in the project's first colloquium, Word Histories of Ingenuity in Early Modernity (2015), who helped us to map the book's territory. We are very grateful to the following, who generously read chapters and shared their expertise: Monica Azzolini, Paul Binski, Horst Bredekamp, Rodrigo Cacho, Ray Carlson, Irene Gallandra Cooper, Sven Dupré, Puck Fletcher, Sietske Fransen, Tony Grafton, Ann Jefferson, Neil Kenny, Sachiko Kusukawa, Rhodri Lewis, Darin McMahon, Michael Moriarty, Kathryn Murphy, Juan Pimentel, Joan-Pau Rubies, Eileen Reeves, Richard Serjeantson, Deborah Shuger, Jake Soll, Pamela Smith, Peter Stallybrass, Claudia Swan, Paul Taylor, Antonio Urquízar, Andrés Vélez Posada, Joanna Woodall, and David Zagoury.

We are grateful to the anonymous readers for University of Pittsburgh Press, who helped us significantly to improve the book and saved us from errors. Needless to say, any errors that remain are our own. We could not have asked for a better editor than Abby Collier, whom we thank for her support and enthusiasm. Our long-suffering families deserve special thanks for having put up with us and our regular absences. We could not have written this book without them. Tim Chesters has supported this project in innumerable ways. Our best critic and our dear friend, he has read this book many times over. We affectionately dedicate it to him.

Note on Conventions

Original orthography has been retained in most cases, as suits early modern lexicography, with minor exceptions: u/v and i/j have been modernized throughout, as well as some punctuation in Latin. Abbreviations have been silently expanded. Normally we cite dictionary entries by author and date, and by lemma. We only cite page numbers where the relevant lemma is not clear from context, or when discussing paratextual apparatus. Other primary sources are cited by author, date, and page number.

We have italicized keywords in English when referring to them *as words* rather than as notions.

All translations from the original languages are our own, unless otherwise indicated.

In the chapters, synoptic lists of each keyword are presented. Not all instances of every word are included in these lists; rather, we have indicated first uses, or alerted to unusual instances, to provide a concise map. In some cases we have also identified last instances—where a meaning disappeared. The lists are not intended to be exhaustive accounts of these keywords.

Logodaedalus

Introduction

> It is never a waste of time to study the history of a word. Such journeys, whether short or long, monotonous or varied are always instructive.
>
> LUCIEN FEBVRE, "*Civilisation*: Evolution of a Word and a Group of Ideas" (1930)

In the revised edition of *Keywords: A Vocabulary of Culture and Society*, Raymond Williams offered a pithy, twenty-two-line entry for "genius." Noting the English word's origin in the identical Latin term, he observed that "the development towards the dominant modern meaning of [genius as] 'extraordinary ability' is complex; it occurred, interactively, in both English and French and later in German."[1] The present book takes a first step toward charting that complex history by examining the fortunes of words used in early modernity to denote the qualities that would coalesce into something more-or-less recognizable as the modern notion of genius. While this modern notion is as capacious and contested as its earlier counterparts, it conjures up associations with exceptionality that include outstanding creativity and intelligence, inborn brilliance, charisma, and a social and psychological disposition that is outside the norm.[2] By comparison, early modern ingenuity was often described as an inborn power brought out by training and industry, which could be affected by environment. Ingenuity, in its highest form, often manifested as sharp wit, fine skill, quick thinking, and swift execution.

By looking to early modern ingenuity, this book seeks to offer (to borrow Terence Cave's term) a "prehistory" of genius before Romanticism, suspending belief in the teleology of the modern concept by excavating the language used to define and express a set of meanings in flux.[3] In so doing we write against the flow of history and modern disciplinary boundaries, moving backward through time to pursue the ways in which certain words rose to prominence, changed shape, and retreated from view; inflated and deflated semantically; sedimented into settled uses or veered off in unexpected directions. This is a form of historical lexicography, the aim of which is not to assemble dry origin stories characteristic of certain kinds of etymology, but to reveal the particularity and peculiarity of lexical traces as they relate dynamically to culture and society. It is a kind of philology that seeks to show how significant words produced meaning *in* history and *as* history.[4]

This book presumes that without attention to the histories of words we cannot hope to comprehend the complex issues to which their meanings gave rise, nor attend sufficiently to the messy processes by which social, intellectual, and cultural notions are forged and transformed. As such, and unlike those studies that have approached the history of genius through the analeptic retrojection of a predetermined concept, we follow Neil Kenny's lead by beginning with language.[5] In his studies on early modern curiosity, Kenny has offered a robust defense of word history as an alternative to the kind of *Begriffsgeschichte* in which concepts trump language by preceding words in the study of their history.[6] Our book shares Kenny's commitment to word history as a means of avoiding the unhistorical superimposition of inappropriate, fixed meaning on energetic premodern signs.[7] Yet where early modern curiosity clusters around a lexically homogeneous set of terms deriving from a single Latin root (*cura*), the qualities with which we are concerned were signified by a larger constellation of words. This comprised not only *genius* in Latin and its vernacular derivatives, but also *ingenium* and its many translations, along with a range of vernacular terms with Teutonic roots. Indeed, the meanings that started to attach to *genius* around the turn of the eighteenth century, as it journeyed toward "extraordinary ability," came more from *ingenium*-related words than from *genius* per se, such that our study may more accurately be called a history of ingenuity.

Histories of modern genius often cite *ingenium* as its fellow, much as Kant pointed out in the *Critique of Judgment* §46 that "Genius is the inborn predisposition of the mind (*ingenium*) through which nature gives the rule to art."[8] The two etymologies have often overlapped since Latin antiquity, but the cultural dominance of *genius* in English (and *génie* in French, *Genie* in German, and so on) has obscured the fact that, until the late eighteenth century, *ingenium* was by far the richer of the two terms. Since Roman antiquity, lawyers had talked about a cunning trick or evil plot as an *ingenium*. This language borrowed from the notion of *ingenium* as machination; siege machines were known as *ingenia*: the products of human craft. By analogy, every writer knew that Quintilian and Cicero list a swift *ingenium* among the basic requirements of the wordsmith, skilled in constructing arguments as well as in the poetics of literary invention. Indeed classical rhetoric and its reception was one of the most important frameworks in which the culture of ingenuity was articulated. The schools and universities of the Middle Ages and Renaissance proved a fertile seedbed for ingenuity. In grammar teaching, in discussions of oratorical style, and in the study of language as a means of personal cultivation, *ingenium* loomed large.[9] In medical faculties, the *ingenium* was the subject of therapy; in arts faculties, it was the mental capacity for logical invention, the discovery of middle terms. Outside the universities, art theorists turned to *ingenium* as the source of pictorial invention, just as courtiers made it central to questions of social standing and civility. A history of ingenuity could focus on these sites and more.

Yet such an approach would be compromised if we did not first examine the lexical history of ingenuity more generally, which requires the investigation not just of a single word but of a cluster of terms. This comprises, in Latin: *genius* and *ingenium*; in Italian: *genio* and *ingegno*; in Spanish: *ingenio* and *agudeza*; in French: *engin, esprit, naturel*, and *génie*; in German and Dutch: *Art/Aerd, Sinnlichkeit/sinnrijk, Gemüt* (German), and *geest* (Dutch); in English: *genius, ingenuity, wit*, and *cunning*. For reasons practical and methodological, in studying these words we have drawn on a single corpus: printed dictionaries and other lexical works, published between ca. 1470 and ca. 1750.[10] Practically, while this corpus is substantial, it is not so large as to prohibit a thorough, systematic investigation of our keywords. While recent work in digital humanities has made available enormous and diverse corpora that can be searched by keyword, coverage across languages and time periods is sufficiently patchy that such resources are of limited use for comparative studies of early modernity. Moreover, while the kind of quantity-driven approach afforded by big data is useful for identifying broad linguistic (and perhaps conceptual) trends by tracking the frequency and collocations of a given term, such methods can flatten out the microphenomenal twists and turns that are revealed by examining words in close, historical context.[11] By contrast, focusing on dictionaries enables us to tell fine-grained stories through sources that—far from being transparent windows onto pre-existing meanings—reveal the kinds of ideological bias and critical fault-lines that determined how the language of ingenuity was defined, used, and manipulated.[12]

We consider the apparent limitations of this corpus—the "artificiality" of definition compared to the "naturalism" of words in use; conformation to the conventions of genre; time lapse between the currency of a given meaning and its definition by lexicographers—as an opportunity rather than a disadvantage. Early modern dictionaries reveal the struggle to manage and explain the language of ingenuity, not simply in their definitions, but through their absences, repetitions, obfuscations, thefts, and fanciful imaginings. Polyglot lexica, in particular, offer a magnified view of the relationships between local varieties of ingenuity and the translatability (or difficulties therein) between terms, putting the lie to the notion that any of our keywords are simply "untranslatables," as a recent lexicon has implied.[13] Most significantly, this translatability underpins our conviction that a strictly nominalist approach to the history of ingenuity is inadvisable, since early modern dictionary-makers clearly believed that certain synonyms sufficiently (if not fully) captured the meaning of our ingenuity keywords. That is, they maintained a potential distinction between *verba* and *res* by assuming that disparate terms could adequately be used to describe or define the same thing.¹⁴ Thus, our concern here is not with the metaphysics of concepts, their formation and expressibility in language, but with the semantics of culture. In this book we have sought to deploy a procedural tactic that begins with, but need not simply remain at, the level of language. We consider this a necessary first step in identifying a certain something that was frequently expressed in words, but which was equally evident outside and beyond them.

We have selected our terms according to two criteria. First, according to the kinship of words deriving from or translating the Latin *genius* and *ingenium*. Second, according to the prevalence of certain words in this kinship group. By this we mean the collocative frequency with which equivalent terms recur in the dictionaries. To give an example, in *A Worlde of Wordes* (1598), John Florio translates the Italian *ingegnòso* with a string of English synonyms: "wittie, wilie, ingenious, subtile, wise, cunning, craftie, full of invention." Only three of these synonyms recur frequently in close proximity to one another in the English lexica: *witty*, *ingenious*, and *cunning*. According to our criteria, therefore, these three terms are "keywords" in the ingenuity "family" of words, while the remainder—*wilie*, *subtile*, *wise*, and *craftie*—are "neighbors." It is important to emphasize that this method requires judgment rather than simply counting words, not least because in many lexica the synonym strings and definitions are long and include lengthy quotations. As will be apparent from the chapters on individual languages, in applying these criteria we have drawn upon our familiarity with word use in the discursive contexts that shaped early modern ingenuity.¹⁵

In exploring the relationships between the nuclear family members of ingenuity in multiple languages and across several centuries, we are concerned less to identify the history of a single concept with an "ineliminable core" than to reveal the emergence, diffusion, and transformation of a shared culture.¹⁶ This "culture of ingenuity" was capacious, mobile, and frequently idiosyncratic. Yet it was always concerned with specific characteristics, clustering around natural inclination, creative potential, craft talent, or mental agility. These were not new notions. Something comparable is evident in the *engin* so prominent in High Middle Ages romance and in the *métis* that pervaded ancient Greek culture and society.[17] Indeed, it is tempting to define the early modern culture of ingenuity in terms similar to Detienne and Vernant's description of *métis* in their now-classic *Cunning Intelligence in Greek Culture and Society*. For those authors, the semantically polyvalent *métis* is "a type of intelligence and of thought, a way of knowing; it implies a complex but very coherent body of mental attitudes and intellectual behaviour which combine flair, wisdom, forethought, subtlety of mind, deception, resourcefulness, vigilance, opportunism, various skills, and experience acquired over the years."[18] All these properties are observable in early modern ingenuity, which nevertheless differs from its ancient forbear in at least one crucial respect. *Métis*, according to Detienne and Vernant, is "never made manifest for what it is, it is never clearly revealed in a theoretical work that aims to define it. It always appears more or less below the surface, immersed as it were in practical operations . . . ."[19] Yet ingenuity, while equally operative, was subject to explicit theoretical treatment in the period with which we are concerned, be it in Juan Huarte de San Juan's seminal and much translated medical treatise, *Examen de ingenios para las ciencias* (1575), or in Descartes's rules for the direction of the mind, *Regulae ad directionem ingenii* (ca. 1620s, first published in Dutch translation, 1684).[20] Thus, ingenuity not only blossomed in the early modern period; it also attracted numerous attempts at definition and explanation in a host of discursive domains, from natural philosophy to poetics, technical treatises to artistic theory. As one scholar recently put it, by the early modern period "*ingenium* had become a whole empire."[21]

Semantic Axes of Ingenuity

This book cannot hope to chart fully that empire, not least because certain important topics are only dimly apparent in the lexica, such as genial melancholy, inspiration, and the sublime.[22] Rather, we offer an outline map of some of ingenuity's contours, boundaries, and limits, which may be populated, revised, and expanded by future work. Thus the very stable and conservative word histories of *genius* in French and English do not quite do justice to the thriving conceptual history of the notion, which must largely be sought elsewhere.

We have found that the language of ingenuity tends to cluster along four axes of meaning: moral, cognitive, presentational, and artificial (the latter meaning both contrived and artful). These axes are a useful grid against which to plot the changing cultural and epistemological trajectories of ingenuity, revealing what is at stake in this history. They help us compare different emphases and uses of the word family of ingenuity, hinting at the different vocabularies available at different times and places. Not all characteristics were invoked all at once, but varied over the period we examine, across Europe.

First, we mean "moral" in the broadest possible early modern sense, encompassing both nature and identity. For the language of ingenuity addressed the question of what made one distinct, what gifts and abilities one might have—or lack. This grand sense of "moral" ran closely with the narrower question of moral responsibility for one's desires and abilities. Was one's temperament shaped by birth, so that someone born noble (*ingenuus*) was also born gentle? Could trickery or licence be excused by saying one could not determine one's inborn inclinations (*indulgere genium, suivre son génie*)? Unlike Romantic genius, *ingenium* was something everyone had; the challenge, tackled most influentially by Huarte de San Juan, was to classify the specific ingenuity of an individual or a particular group.[23]

Ingenuity's grand moral sense can be distinguished from a second axis of meaning, which turns on its cognitive roles. Theorists remained ambivalent about the source of such capacities. How much was simply what one was born with, and how much could be sharpened, softened, or redirected through habit and study? Not only was *ingenium* thought to be the faculty by which one alighted on the middle term in a logical syllogism, but physicians thought of it as a distinctive faculty, while thinkers like Juan Luis Vives made it stand in for the whole apparatus of the soul's various faculties.[24] Thus, in different places and times, ingenuity could refer sweepingly to the mind in general, or quite precisely to certain cognitive processes. While early Renaissance physicians and philosophers simply thought of *ingenium* as the capacity for invention (logical, rhetorical, or otherwise), Goodey has observed that over the sixteenth century *ingenium* became increasingly a part of a concrete faculty psychology, often associated with the imagination and with "quick-wittedness"—fast forms of cognition.[25]

Along a third axis, ingenuity was a matter of presentation. In displaying itself, ingenuity played on the knife's edge between hiding and revealing, judgment and insouciance. As Sallust's often-repeated maxim put it, "no one ever deployed their *ingenium* without a body."[26] The hidden dimensions of wit must out in works, words, and deeds. The display of ingenuity required on the one hand a certain liberality, the sign of an ebullient fertility—hovering in and out of sight. One found something "ingenious" when the effect had an unexpected source; when unlikely parts made a surprising whole; when a contrivance elegantly worked by hidden means. Likewise, it took ingenuity to recognize such displays. One of ingenuity's great theorists, Emanuele Tesauro, explained that "it is a secret and an innate delight of the human intellect to find that it has been sportively deceived; because the transition from illusion to disillusion is a kind of learning by an unexpected way; and therefore most pleasing."[27] On the other hand, unlike Romantic genius, ingenuity had to be presented with careful judgment. Courtiers and salon-goers jousted wits in deadly play, but always within the bounds of decorum. The paradox was that a powerfully inventive mind was recognized by its play within a prudent distance of the measured rule. Thus, what distinguishes Romantic genius from pre-Romantic *ingenium*—the Nietzschean *Übermensch* from the exemplary court artist—is that the former gives sovereignty to itself. In pre-modern Europe, the sovereignty of ingenuity is given by another ruler: the prince, who gives the artist licence and is himself granted sovereignty by God.[28] This carefully choreographed dance of ingenuity introduced, however, a significant tension. Gifted individuals—the *virtuosi* and *ingegniosi* of the early modern court—had to push beyond the norms of decorum in order for their sovereign *ingenia* to be recognized. They had to push just far enough, without breaking the sovereign's trust (as Galileo infamously did, with disastrous personal consequences).

The fourth, final axis around which we see ingenuity aggregate has to do with these limits: the "artificial" or "artful." To play with rules implies mastery over them. In Latin, Italian, and Spanish, ingenuity was often paired with industry not only in contrast but sometimes in pleonasm; ingenuity displays both talent and mastery of a craft.[29] The ingenious courtier, courtesan, or clockmaker matched natural cunning with technique. In particular, the mechanical, artisanal overtones of the lexis of ingenuity stretched under divergent pressures. The artisan's ingenuity had long been ambivalent, praised for its capacity even as it was blamed for the meanness of mechanics. Neither meaning disappeared. In fact, they spread, and the notion of "artificiality" was transferred from people to objects as well. Although early humanists tended to ignore the fact that in Latin antiquity *ingenium* was also a legal ruse or a siege engine, these objective meanings multiplied in some of the early modern vernaculars to fill early modern material culture. Thus paintings, telescopes, and Richelieu's new bureaucratic machinery were all ingenious as displays of their makers' minds and hands. This went beyond artisans. One of the most powerful analogies to structure premodern knowledge, of nature as craftsman, ensured that natural objects too could be ingenious, melding art and nature into a single category. Thus not only could a gallant *salonnière* possess *esprit*, but Nature herself might display "spirited" matter.

Lexicographical Ingenuity and the Genius of Language

The practice of lexicography itself is a good example of the "artfulness" of ingenuity. The very work of lexicography exercises one's wit, Ambrogio Calepino observed in the preface to his 1502 *Dictionarium*: "Non enim tam instruendorum aliorum quam mei exercendi ingenii gratia id efficere aggressus sum" (For I set out to do this not so much to teach others as for the sake of exercising my *ingenium*).[30] In reflecting on their own practices, humanist lexicographers developed a discourse on ingenuity as the critical textual ability to elucidate meaning through gloss, commentary, emendation, the right grasp of authorial intentions, and styles. The roots of this discourse are to be found in classical rhetoric, in which *ingenium* is identified as one's natural ability to generate *copia*, that is, to find or invent the good subject-matter or *res* for oratory.[31] In this tradition, *ingenium* is the source of invention. In the early modern period, however, *ingenium* developed further into the site of sound, critical judgment.³² Artful selection and excerpting generated lexicographical *copia*, captured in the alphabetically arranged commonplaces and glosses of dictionaries. Yet this enterprise was a slippery one; fictitious etymologies and overclever wordplay were the province of the *logodaedalus*: a dangerously cunning wordsmith.³³

One of the earliest instances relating ingenuity to various humanistic practices of textual criticism is in Polydor Vergil's preface to his 1496 edition of Niccolò Perotti's posthumous *Cornu Copiae* (1489).³⁴ The first Renaissance dictionary of classical Latin, the *Cornu Copiae* was in fact a word-by-word commentary on Martial's epigrams. That the beginnings of neo-Latin lexicography should consist of the copious unravelling of Martial's witty compressions is no coincidence.³⁵ As laconic genres on the rise throughout the period, epigrams, adages, and aphorisms offer a particularly apt perspective on early modern ingenuity. Their conceits and concision displayed the quick-wittedness of their authors and demanded of their readers an equally swift mind.³⁶ Lexicographically, they are connected to the development of genius.

The Oxford English Dictionary (OED) provides three definitions of *aphorism*:

1. A "definition" or concise statement of a principle in any science.

2. Any principle or precept expressed in few words; a short pithy sentence containing a truth of general import; a maxim.

3. *abstractly*, The essence or pith. *Obs. Rare.*

Neither these nor the supporting quotes that accompany them invoke *genius* in any sense. Nevertheless, these words *genius* and *aphorism* rubbed shoulders through their shared relationship with *wit*. This is evident from the amplest lexicographical work of our period in English: Samuel Johnson's celebrated *Dictionary* (1755). Johnson's fourth sense of *genius* is "Disposition of nature by which any one is qualified for some peculiar employment." Among the entry's quotes is a famous couplet from Alexander Pope's *Essay on Criticism* (1711), which connects *genius* to *wit*: "One science only will one *genius* fit; / So vast is art, so narrow human wit." Indeed, by Johnson's time these two words were more or less synonymous, the fifth sense of his entry for *wit* being (in its metonymic sense) "A man of genius." We may understand this connection through an earlier definition of wit. In his *English Dictionarie* (1623), Henry Cockeram defined *witty* as "ingenious, atticke, pregnant," offering for "witty sayings short and pithy" none other than *aphorisms*.[37] We have arrived, with *pithy*, at the OED's senses (2.) and (3.), such that we may now state with confidence that an early modern English aphorist and his inventions could have been described using the flexible language of "genius."[38]

This kind of quick wit also lay behind the intuitive grasp of a text that Polydor Vergil praised when thanking his friend, the physician and philosopher Thomas of Venice. Unlike Vergil, Thomas was particularly good at spotting and amending scribal errors, discrepancies in style, or awkward grammatical constructions. These critical practices demanded ingenuity from readers and authors alike. For instance, in the 1536 edition of his *Thesaurus linguae latinae*, Robert Estienne left the elucidation of meaning to his reader's learned guesses: "Eam itaque lectoris iudicio maluimus divinandam relinquere, quam temere ex nostro Marte atque ingenio apponere" (And so I leave to the judgment of the reader to conjecture what he can supply from our art and *ingenium*).[39] Equally, lexicographers were painfully aware of the toil their craft required. Drawing on the classical coupling of *ars et ingenium*, or *industria et ingenium*, Estienne emphasized the industry of lexicography in the first 1531 edition of his *Thesaurus*. The painstaking compilation of his lexica from the works of Plautus and Terence was an exhaustive and exhausting job; Estienne confesses that he neglected his household and his health for two years because of it.[40] Yet this industry is a specific form of ingenuity, in the light of Estienne's own definition of the term in his *Thesaurus*: "Industria est ingenium seu ars qua aliquid struere sumus idonei" (Industry is the *ingenium* or art by which we are made suitable for composing something).[41] When applied to the problem of bilingual translation, these critical practices reveal not only the ingenuity of the lexicographer, but also the ingenuity of his mother tongue. In the preface of the Latin-French *Dictionarium Latinogallicum* (1538), Estienne insists that the purpose of lexicographical industry is to define what would later become the "genius of the French language," by making French a worthy competitor of Latin.

At first glance, the respective word histories of *ingenium* and *genius* as they unfolded in early modern dictionaries not only confirm but reinforce the humanist myth of a European culture of ingenuity resting on the linguistic foundation of a shared Latinity. As a learned language, Latin broached vernacular differences; in fact, it was the very foundation of learned vernaculars themselves.[42] Yet Latin has no significant precedence other than chronological in the European word history of ingenuity. Lexicographers themselves made this point. In his *Tesoro de la lengua castellana, o española* (1611), Sebastián de Covarrubias often uses simple words or whole quotations from Latin, either as etymologies or to further elucidate his Spanish headwords. Yet his preface argues that Latin does not hold precedence over any of the many languages from which Spanish emerged. He begins with a version of the myth of Babel that shows him keenly aware of how hybrid the vernacular is: Spanish is more a messy mixture of foreign tongues than Latin's rightful heir. Covarrubias playfully exhausts the metaphoric possibilities offered by the label of "treasure": a new epic hero, the lexicographer will meet and defeat the various "monsters" of foreign tongues before he can reach the "treasure" of the Spanish language. He fears that as soon as he brings this treasure "out of the cave" and publishes it, the critics will, like Moorish witches, turn its gems into coal; but he also hopes that well-meaning and learned readers will use their supple *ingenium* to transform those coals (back) into sparkling carbuncles and beautiful rubies.[43] For Covarrubias, the hybrid genius of the Spanish language lies ultimately in the poetic and literary uses that inspired wits make of it—its relationship to a Latin linguistic ideal is irrelevant. In polyglot Latin dictionaries, the impression that equivalence between vernacular and Latin terms is straightforward proves to be just that: an impression. The compressed definitions and translations in eleven vernaculars of Latin headwords in Calepino's *Dictionarium* (1590 edition) and the "list" layout of Hieronymus Megiser's *Thesaurus polyglottus* (1603) amount to illusions of transparency. These terms were translatable, but not so straightforwardly as some entries imply.

Ingegno and *genio*, *esprit* and *génie*, *Sinnlichkeit* and *geest*, *wit* and *genius* all featured as translations of *ingenium* and *genius* into vernaculars before they became entries in their own right in vernacular dictionaries. Their linguistic variety mirrors the diversity of the Latinate culture of ingenuity when translated into different national and vernacular contexts. Yet while Latin was often construed as the touchstone of linguistic excellence against which to pit vernacular languages, bilingual Latin > vernacular dictionaries such as Petrus Dasypodius's *Dictionarium latinogermanicum* (1535) and Estienne's Latin-French *Dictionarium latinogallicum* (1538) also highlighted the distance between Latin and vernacular languages and cultures.

Lexicographers were acutely aware of this issue. It features in the preface to one of the earliest bilingual Latin/vernacular dictionaries, Antonio de Nebrija's *Dictionarium* (1492). Nebrija explains the transience of linguistic meaning by glossing an epigram from Martial about a chestnut tree planted by Julius Caesar in Córdoba. These trees, he says, were common in Spain under Roman occupation. To the Romans, chestnut trees represented their *genii* and were of utmost importance, yet Nebrija's Spanish contemporaries barely know them and they can no longer be seen in Spain.[44] Such trees—absent from the Spanish landscape, their meaning lost to the Spanish people, yet fossilized in a Latin epigram—emblematize the difficulty of retrieving the genius of language.

Ultimately, lexicographers fashioned this genius in vernacular-only dictionaries. At first, they did so by retrieving a lost linguistic and cultural golden age. The first, 1612 edition of the *Vocabolario* of the Accademia della Crusca exemplifies this tendency. The words that made it into the *Vocabolario* had been harvested from the works of Dante, Petrarch, and Boccaccio. Dante and Boccaccio marked the boundaries of the *buon secolo*, which witnessed the blossoming of the Italian vernacular (specifically Tuscan) into perfection; according to some, that language alone was worthy to become the national standard. More than a century later, Johnson's *Dictionary* located the genius of the English language among the great writers prior to the "decadence" of the Restoration: Bacon's natural philosophical writings, John Hooker's Bible,

Raleigh's political and military oratory, Spenser and Sidney's poetry and fiction, and finally Shakespeare for the diction of common life:

> I have studiously endeavoured to collect examples and authorities from the writers before the restoration, whose works I regard as *the wells of English undefiled*, as the pure sources of genuine diction. Our language, for almost a century, has ... been gradually departing from its original *Teutonick* character, and deviating towards a *Gallick* structure and phraseology, from which it ought to be our endeavour to recal it, by making our ancient volumes the ground-work of stile, admitting among the additions of later times, only such as may supply real deficiencies, such as are readily adopted by the genius of our tongue, and incorporate easily with our native idioms.[45]

For Johnson, in these sources words (*verba*) rather than things (*res*) were at stake, that is, good diction and style. Their texts revealed the deep "Teutonic" and "Saxon" roots of English, which constituted the genius of the language and ought to be preserved from the lexicographically decadent doublets of foreign—and especially French—imports through translation.[46] Thus, Johnson's lexicographical enterprise was a critical as well as antiquarian one.

The incipient nationalism this kind of search implies could be taken to ridiculous extremes. Thus, the ingenious etymologies of Johannes Goropius Becanus made a case for the Adamic status of the Dutch language: for Goropius, Paradise spoke Dutch.[47] Goropius exemplified the excesses of lexicographical ingenuity as overly fanciful and erudite guesswork and became the laughing stock of fellow lexicographers. In his polyglot *Lexicon tetraglotton* of 1660, James Howell related the mocking reply of the Italian Accademia della Crusca to Goropius's Edenic tongue: Dutch "being a rough and cartalaginous or boany Speech in regard of the collision of so many consonants" was better suited to sentencing a man to be stoned to death.[48]

By way of contrast, the French lexicographers of the seventeenth century emphasized criticism above "mere" antiquarian truffle hunting. The delineation of the *génie de la langue*, itself an expression of the "esprit français," suffuses the prefaces of Pierre Richelet's *Dictionnaire des mots et des choses* (1680), Antoine Furetière's *Dictionaire universel* (1690), and of the *Dictionnaire de l'Académie* (first edition 1694).[49] These works stress two features: good denotation—the semantic accuracy of the French vocabulary—and good grammatical operators: its syntactical and logical precision, supposedly replicating clear thought processes.[50] The French *génie de la langue* was therefore potentially universal because it was the linguistic expression of a fundamental human trait, namely rationality. Thus, the preface of the *Dictionnaire de l'Académie* lists among the perfections of the French language "this direct construction, which, without straying from the natural order of thought, chances upon all the delicacies that art can bring to it," upholding distinctness as a lexicographical ideal in the definitions of simple words.[51]

Structure

Despite its labyrinthine theme, this book is not a maze. It consists of six chapters, each devoted to one particular language or set of languages: Latin first, followed by Italian, Spanish, French, German and Dutch, and English. As each chapter is self-contained, readers interested in one or a few specific word families may focus on the relevant sections without having to read the rest of the book. Those interested in a broader and more nuanced pan-European account will find it worth reading the chapters sequentially, so as to experience the full breadth of the book's argument.

Each chapter is structured as follows: first, a brief account of its keywords, to explain concisely the rationale behind our selection, in relation to the historiography of ingenuity in that particular language; then a section devoted to the lexicographical landscape of that chapter's language, tackling issues of chronology and periodization and, most importantly, sources. The rest of the chapter is then focused in turn on each of the ingenuity keywords, arranged in successive subsections.

Besides offering a sense of cohesion amid all its diversity, this unified structure ensures that attention is focused on the most important, but also the most elusive, elements in this story: fluctuations within and across languages. In consonance with the four semantic axes described above, the range of variations that these chapters are intended to highlight and discuss include:

- the semantic developments experienced by terms within a particular word family
- the indications of moral or object-subject reversibility exhibited by certain keywords
- the recording but also the occlusion, by period lexica, of meanings and usages recorded by other sources
- the linguistic and methodological crossovers from language to language and from national context to national context
- other culturally relevant questions, such as the emergence of national identities through language, or the association of the language of ingenuity with notions of style

The choice of Latin as the first language of ingenuity to be discussed is pivotal to the book's structure. We have already explained why, but it is worth repeating here. First, there is the centrality of Latin as Europe's learned language throughout the period under consideration. The early phases in our chronology are particularly relevant: Latin stood out as one of the most fertile working grounds for the humanist project in a moment also marked by the consolidation of European vernaculars. Second, and more importantly, the Latin case exposes the humanist concern with philological systematicity, which in lexicographical terms saw this very language, Latin, becoming a model of supposed univocality and translatability.

Given the dependence on Latin of the Italian and Spanish ingenuity word families, these languages are discussed next. Although both languages share a number of keywords (e.g. *ingegno* and *ingenio*), their lexicographical traditions differ sufficiently so as to produce very different word histories. Italian was especially concerned with the *questione della lingua*, the inflections of its keywords heavily dependent upon the construction of a literary-*cum*-linguistic canon rooted in the works of Dante, Boccaccio, and Petrarch. More than in other languages, the Italian dictionaries reflect the vigorous (and early) discussion of ingenuity as a pillar of artistic theory. In the Spanish case, the lexicographical sources offer an interesting point of contrast to the lively culture of ingenuity that marked this period, which includes the so-called Spanish Golden Age. Ingenuity did inform the work of lexicographers. However, the treatment that *ingenio* and *agudeza* receive in these sources offers but a partial view of the complexity and sophistication with which ingenuity was exercized and theorized. The French word history of ingenuity suggests that the French *âge classique* was not so much the Age of Reason but the age of *esprit*: dictionaries reflect the use of this term for the purpose of social and cultural disciplining, as well as its predominance in the debates from which our modern notion of literature and criticism emerged. Next, German and Dutch offer a counterpart to the Latinate family of ingenuity, which testifies to the pan-European nature of the culture this book describes. German lexicographers especially were acutely conscious of their Teutonic roots, and went to great lengths to eject the Latinate families of ingenuity from their lexica. Yet despite their best efforts, over the period German and Dutch developed vocabularies that bore suspicious resemblance to the word families of ingenuity elsewhere. The English chapter is situated last, since its most ambitious lexicographical project came late: Johnson's *Dictionary* (1755). The trajectories of the English keywords reflect that language's development from both Anglo-Saxon and Latin, their semantics a result of complex etymological intertwinings and the impact of loan words. As in the French case, the English language of ingenuity was notably affected by late seventeenth- and early eighteenth-century attempts to define a literary canon and criteria for good taste. English is distinctive, however, in showing the effect of local debates about the workings of the mind—which marked *wit* especially—and the inflation of *genius* in criticism and aesthetics.

1 Latin

When Erasmus's *Encomium moriae* first circulated across Europe in Latin in 1511, the humanist Republic of Letters laughed. The pope himself found it a funny read. Here was yet another brilliant display of Erasmian wit, and learned Europe got the jokes. The fanciful, allegorical tale was an immediate hit, its success suggesting at least one, humanist form of a European culture of ingenuity, in which learned puns were transparent to its Latinate audience. In rhetoric, wordplay was a well-identified expression of the *ingenium*'s inventive powers.[1] Classical Latin, a tongue that crossed the linguistic boundaries of national vernaculars, underpinned such a European culture of ingenuity.

Yet as we noted in the Introduction, the precedence of Latin over the vernaculars was purely functional and chronological. Early modern Latin dictionaries might well suggest that the European, Latin culture of ingenuity was simply another neat humanist retrieval from the classical tradition—but this would be an illusion. In fact, the Renaissance redefined *ingenium* into a capacious umbrella term that unified and accommodated an array of (originally unrelated) Greek terms and meanings. Thus *ingenium* already denoted a mental ability in the 1508 edition of Galen, where Leoniceno consistently used it to render *agchinoia* (quick-wittedness).² By 1600, however, the term had taken on a much greater semantic extension in the medical textbook of André du Laurens, royal physician to Henri IV. There *ingenium* translates a whole range of Greek terms: one's good nature, one's quick-wittedness, one's ability to learn well.³ Seventeenth-century philosophical lexica finally took stock of this lexical construction by granting *ingenium* the status of a well-defined, stable philosophical concept.⁴ In short, the Latin vocabulary of ingenuity was a neo-Latin construction that built on the recovery of a classical heritage.

Keywords

Our first keyword, *genius*, derives from the verb *gigno* (to produce), and chiefly denotes the tutelary spirit that accompanied people from birth or the deity that governed a place (*genius loci*). Ancient Romans used it to translate the Greek *daemon*, and rarely used it metaphorically to characterize one's wit.

As we shall see, while medieval dictionaries repeated the ancient association with deities responsible for fertility and generation—the rare adjective is *genialis*—fifteenth-century humanists introduced the notion that a *genius* might refer to gods that pagans had invented to explain natural powers. Sixteenth-century lexica also noted that such tutelary spirits foreshadowed the good and bad angels of Christianity. In response to the understanding of ancient genius as natural powers made gods, the seventeenth- and eighteenth-century history of *genius* highlights the possibly impious consequences of this naturalist interpretation and adopts an antiquarian mode of investigation of ancient Greek and Roman religions on the matter.

• • •

Our second and most important keyword, *ingenium*, also derives from *gigno*, and spans a wide semantic range throughout the period. Writers such as Cicero and Quintilian, approximating the Greek *phusis* (φύσις) and *euphueia* (εὐφυεία), first define *ingenium* as those inborn qualities that manifest themselves in temperament and moral traits and ground a social performance of identity; close neighbors included *natura* and *indoles*. Cicero had already asserted that one could recognize someone's *ingenium* from such performances, especially oratorical ones.[5] As we shall see, *ingenium* as the character underlying display became ever more important in the vocabulary of moral identity, temperament, and social assessment; changes in those assessments also reveal important turns in early modern cultures of ingenuity.[6] Thus the semantic trajectory of the adjective *ingenuus* (from *ingenitus*, nobly born) hints at the complex history of the social evaluation of the early modern intellectual. Shifts in the history of another adjective, *ingeniosus* (ingenious), are also particularly telling in conjuring up the ambivalent doublets of display and concealment, effortlessness and conceit. First qualifying a praiseworthy nature, it ends up condemning a socially and politically suspicious and conceited person: *ingeniosus* could mean being too clever by half. These ambivalences also permeated a culture of courtly witticisms and literary conceits.

Indeed, another important semantic node progressively emerges from the qualification of diverse *ingenia* as inborn natures: this node is a cognitive one and defines *ingenium* as wit. Categorizing the diversity of wits was a major pedagogical concern, echoed in the flurry of qualifiers of one's intellectual abilities listed in early modern dictionaries: one could have a slow or quick *ingenium*, a sharp or a dull one.[7] Teachers found *ingenium* a useful way to denote cognitive capacities: neighboring terms pointing to practical wisdom and cunning include *astutia, sollertia,* and *subtilitas*. Chief among these was *ingenium* as inborn talent and inventive power, highlighted by Cicero and Quintilian. Their manuals for oratory were read in every medieval and early modern Latin school. Cicero made *ingenium* an inborn power of oratorical invention to be set alongside *ars*, either as its opposite or as its complement, in a formulation influentially repeated by Quintilian.[8] *Ingenium* took center stage in debates over erudite and literary practices in the Republic of Letters, from Petrarch's defense of invention to Morhof's conservative apology for polymathy (1688).[9]

In progressively constructing the *ingenium* as a marker of identity—of a given temperament and social nature, and increasingly of specific cognitive abilities—humanist lexica ignored two dominant medieval meanings: *ingenium* as machine; and *ingenium* as machination, legal subterfuge. Both assume that it takes a fertile mind to engineer a machine and a trick. Historians of craftsmanship and engineering have long found *ingenium* a key word for describing and promoting the maker's mind and mechanical invention.[10] This legal and mechanical context inflects the semantic ambivalence of the adjective *ingeniosus* we noted above: *ingeniosus* thus becomes an insult for *mere* mechanical insight, or fraud.[11]

Lexicographical Landscape

The tale of ingenuity in Latin turned on the vagaries of the dictionary-making enterprise, led first by humanists buried deep in ancient texts, and then by seventeenth-century antiquarians who stuffed lemmata with examples from medieval manuscripts. Latin held a special position as a sort of metalinguistic trading zone—by no means neutral—even in the growing genre of polyglot dictionaries where diverse vernaculars took new shape.

Early humanist lexicographers also borrowed liberally from medieval dictionaries. The other grand medieval achievement alongside Isidore of Seville's vast *Etymologiae* was the *Catholicon* of Balbus, from the thirteenth century.[12] In fact, the permanence of medieval lexica well into the early modern period should not be overlooked: more than twenty-five editions of the *Catholicon* were published between 1460 and 1521. Some humanists such as Johannes Reuchlin (1455–1522) largely created their own works by copying Balbus lemma after lemma, silently adding and subtracting at will.[13] Others reworked the *Catholicon*, pulling its learning apart to reintegrate it around a new structure. In this vein, Niccolò Perotti (1429–1480) organized his *Cornu Copiae* as a commentary on the twelve books of Martial's *Epigrams*—at such prodigious length, however, that scholars immediately seized on it as an encyclopedia to Martial's antique world and Latinity.[14] From the first edition of 1489, it was outfitted with a finding apparatus including 100 pages of alphabetical index, transforming it into a dictionary.[15] Perotti took on the paradigmatic role of the philologist glossing a text, and in so doing created a reservoir of culture. In turn, Perotti supplied material for Fra Ambrogio Calepino (ca. 1440–1510), whose *Dictionarium* evolved in over 210 editions between 1502 and 1779, so that schoolboys made his very name synonymous with pedantic lexica.[16] The centrality of Perotti and Calepino to Renaissance Latinity can be measured by Erasmus, who pointed out to his friend Helius Eobanus in the 1520s that the most useful authors of the age, and therefore the most used, were not the poets but the humbler authors of reference works: Perotti, Rhodiginus, Calepino, Budé, and Volterrano.[17] Later authors of lexica rarely started from scratch. They cheerfully plagiarized each other, noisily pointing out the flaws of predecessors, but quietly borrowing what was useful. The Latinate lexical architecture of ingenuity, therefore, was that of an aging manor, constantly under piecemeal renovation.

If lexicography supplied the materials for building the language of ingenuity, Robert Estienne, the great Paris scholar-printer, was one of its chief architects.[18] His generation of humanists found the earlier resources lacking. In the 1520s Erasmus encouraged the French Hellenist Guillaume Budé to write a "really full lexicon" that included Latin idioms and tropes.[19] Budé began the work, but never published his manuscript, and instead it was the *Thesaurus latinae linguae* (first published 1531) of their friend Estienne that lexicographers and academy projects throughout Europe built on for the next two centuries. Estienne claimed he had spent three years trying to correct Calepino, but in the end was forced to produce his own dictionary by drawing up word lists from a fresh reading of ancient texts.[20] Estienne himself produced three major editions of the book, in 1531, 1536, and 1543, and developed a range of smaller dictionaries out of the *Thesaurus*, including a *Dictionarium latinogallicum* (1538) and a *Dictionnaire francois latin* (1539–1540).[21] These in turn contributed both style and substance to later dictionary projects, such as the German-Latin dictionaries associated with Conrad Gessner, André Madoets's *Thesaurus theutonicae linguae* (1573), and the famous *Thresor de la langue françoyse* of Jean Nicot (1606).[22] Estienne's *Thesaurus* was hailed already in his own day as a triumph: it offered clear, spacious typography; it gave French translations (in the 1531 edition); it eliminated non-classical elements; and, above all, it included long lists of exemplary phrases from classical Latin authors. Estienne's working methods in the *Thesaurus* are telling. For both *genius* and *ingenium*, he scooped the entirety of his definitions—and several classical examples—from Calepino.[23] Then he added examples from the works of two authors he admired and for which he had recently compiled indices: Plautus and Terence.[24] The result of this practice was a nigh-surgical excision of the medieval senses of *ingenium* (relating to machines and legal trickery). As we show below, Perotti, Calepino, and Estienne thus limited ingenuity to an individual's moral character and behavior.

In the later sixteenth century, most lexicographers put their energy into adapting Estienne and especially Calepino into polyglot reference works, rather than into new kinds of Latin dictionaries. The result was that the tidy gardens of quotations Estienne and his colleagues had collected, weeded, and nourished were invaded from every direction. By 1545 some editions of Calepino included not only Greek and Latin, but also German, Flemish, and French.[25] After 1590, editions from Basel included eleven languages.[26] The growing interest in polyglot dictionaries—often still with definitions or headwords in Latin—encouraged the accrual of more and more examples, reflection on derivations, and in some cases the return of the medieval usages that the earlier humanists had tried so hard to eradicate. Scholars who apprenticed in Christophe Plantin's cosmopolitan print shop might be thought the least susceptible to national pride; André Madoets and Cornelis Kiel (Kiliaan) wrote large Latin-based dictionaries that brought together Greek, French, German, and Dutch.[27] Nevertheless, sixteenth-century polyglots, especially bilingual dictionaries, show strains of pride in vernacular languages well before the *Vocabolario degli Accademici della Crusca* (1612) and the *Dictionnaire de l'Académie françoise* (1694). Landmarks of lexicography set vernaculars alongside Latin, for example in Antonio de Nebrija's *Dictionarium latino-hispanicum* (1492) and Estienne's *Dictionarium latinogallicum* (1538). Latin lexicography thus served local interests. As we noted in the Introduction, Plantin's friend the physician Johannes Goropius wrote a long *Origines Antwerpianae*

(1569), which expended a thousand folio pages on arguing that Dutch preserved the original language of Adam. He defended Dutch stereotypes of frugality and good measure by criticizing Terence's line—made an adage by Calepino and Estienne—that "what a wretch has saved bit by bit out of his allotment, he steals from his *genius*."[28] That is, the frugal man robs his own natural bent by not indulging it. Goropius reasoned that he who saves carefully "by no means cheats his *genius*, but rather promotes a happier sort of life, which consists in frugality rather than display." In fact, he continued, moderating one's food and drink helped rather than hurt one's *genius*.[29] Goropius here engaged a theme that was elaborated in early modern period emblems, in which the positive and negative effects of consumption were instantiated in the mottos *vinum acuit ingenium* and *crapula ingenium offuscat* (Figures 1.1 and 1.2).[30] Thus polyglot lexicography could puncture stereotypes of peoples and languages—or sharpen caricatures.

The seventeenth century witnessed two important turns in the Latin lexicography of ingenuity, in particular with regard to the word history of *ingenium*. The first turn was the antiquarian retrieval of a medieval lexicographical heritage, which brought the mechanical and legal meanings of *ingenium* as machine and trickery back into prominence. Philologists began to pore over medieval manuscripts with new appreciation for medieval Latin and its legal precedent. In principle, lawyers always had to attend to medieval precedent; this is evidenced in the growing number of legal lexica such as those by Jakob Spiegel (1554) and Johann Kahl (1600). After 1600, newer polyglot dictionaries such as John Minsheu's *Ductor in linguas* (1617) grew fat with antiquarian erudition. Charles du Fresne, seigneur du Cange, wrote a *Glossarium mediae et infimae latinitatis* (1678), which remains to this day a reference for post-classical Latin usage among medievalists.

The second significant turn in the Latin lexicography of ingenuity was of a philosophical nature. Throughout the seventeenth century, the expanding genre of philosophical lexica bore testimony to the emergence of *ingenium* as a stable philosophical concept. There were indeed significant overlaps between the definitions of *ingenium* in the philosophical lexica of Rodolph Glocenius (1613), Johann Alsted (1626),

Fig. 1.1. The emblem *Crapula ingenium offuscat* (drunkenness clouds wit), illustrated by an engraving in copper by Otto van Veen (1607), offers a warning to those too eager to believe that *vinum acuit ingenium* (wine sharpens wit).

EMBLEMA XXXIIII.

Vinum ingenij fomes.

Vuiferum Bromium, volucrem sed præpete penna,
Quid tacitæ posuistis Amyclæ?
Tollit humo ingenium Bacchus, mentem erigit altam,
Pegaseaq́; velut vehit ala.

Quæ?

Fig. 1.2. The emblem *Vinum ingeniis fomes* (wine is kindling for wits), a variant on the proverb *Vinum acuit ingenium*, represented by Bacchus prepared to soar on wings, fortified by the grape (Junius 1565).

Johann Micraelius (1653), Henning Volckmar (1675), and Stéphane Chauvin (1692). In this very specific corpus, word history does amount to tracing the formation of a concept.

Genius

The story disclosed by the word history of *genius* in early modern Latin and polyglot dictionaries is colored by obsolescence and antiquarianism. The term primarily emerges as part of antiquarian interest in the religious practices of ancient Rome.[31]

In its most prominent and stable semantic cluster, *genius* is identified as a god who imparts generative power, or as the tutelary spirit who accompanies and protects one from birth, and determines one's natural tendencies or bent. Lengthier antiquarian dictionaries then historicized this first definition of *genius,* and naturalized or Christianized the ancient god or spirit. Naturalized *genius* often became a principle of human identity, as one's generative power, moral discernment, hedonist bent, or one's soul. Sometimes in reaction against this naturalization, *genius* could be Christianized as one's good or bad guardian angel. In its second semantic cluster, *genius* is what singles out persons and things alike; this definition remains very stable and concise in all lexica. The following semantic map of *genius* in the corpus outlines the main episodes of this story:

Genius, n.
1. God or spirit.
   a. God of nature.
   Reuchlin 1478 >.
      a.1. God of marriage.
      Reuchlin > Robored 1621.
      a.2. God imparting generative power.
      Perotti 1525 >.
      a.3. God of hospitality and pleasure (eating and drinking).
      Calepino 1502 >.
      a.4. The elements as seeds of things.
      Calepino 1502 > Holyoake 1676
      a.5. The Zodiac and its influences.

Calepino 1502 > Holyoake 1676.
b. Tutelary spirit.
b.1. Pagan:
Calepino 1502 (from Censorinus): equated with Latin *lares* and *penates* >.
b.2. Christian: good and bad angel.
Nebrija 1492 >.
b.3. Tutelary spirit of a place.
Dasypodius 1536 >.
2. Individuating principle.
a. One's nature.
Nebrija 1492: "La naturaleza" > Kahl 1600: quotes Budé who assimilates it with "*ingenium*," and criticizes contemporary bishops for this assimilation > Robored 1621: ("natural de cada hum") > Pereira 1653: ("O natural, ou indole").
b. understanding.
Perotti 1525: Minerva > Calepino 1502: One's rational soul >
c. Defining quality (people and things).
Reuchlin 1478: ("honor, dignitas vigor") > Calepino 1590: ("The grace and pleasantness of a thing") > Robored 1621: ("tomase pola energia, e graça") > Holyoake 1676: ("the grace and pleasantness of a thing").

From the perspective of a pre-history of Romantic genius, the history of the Latin *genius* appears remarkably stable, without the connotations that Romanticism brings to the word. As a god, *genius* either stands for natural powers of generation or growth—hence its association with the elements as seeds of things (1.a.4) and astral influences (1.a.5)—or is a tutelary, protective spirit. As an individuating principle, *genius* comes loosely to denote the cognitive, moral, and temperamental senses that are mapped, more precisely, by *ingenium*.

These main meanings remain the same throughout the period for both semantic clusters. In fact, most dictionaries offer the same examples. Whether they are classicizing lexica (Perotti 1525; Calepino 1502; Estienne 1531 or 1543), their later polyglot instantiations (Calepino 1590, 1654, and 1718; Estienne 1538 and 1571), or more antiquarian and encylopaedic dictionaries (Kiel 1562; Hadrianus 1567 and 1620; Kahl 1600; and Hofmann 1677)—all provide the same etymologically justified definition of *genius* as a tutelary spirit, lifted from Censorinus's *De die natali*.[32] Calepino quotes Censorinus on the matter as follows: "Censorinus scribit Genios dictos quia, ut gignamur, curent, sive quod una gignantur nobiscum, vel etiam quia nos genitos suscipiant ac tueantur. Certe a gignendo genius appellatur. Eumdem esse Genium et larem multi veteres memoriae prodiderunt" (Censorinus writes that geniuses are so called because, as we are born, they look over us, either because they are born with us or because they protect and care for us once born. It is surely derived from *gigno* (to be born). Many of the ancients call to mind that a *genius* and a *lar* are one and the same thing).[33]

This definition is invariably illustrated by the same Roman oath invoking one's *genius*, first lifted from Terence's *Andria: The Fair Andrian* in Estienne 1531.[34] In cases that naturalize *genius* from being the festive god of nature and pleasure into the meaning of one's natural bent and moral discernment, dictionaries also systematically cite, in the wake of Calepino's explanatory paraphrase and Estienne's own excerptions, the contrasting pair *indulgere genio* and *defraudere genium*: to follow or to deprive one's temperament, or to "eat, drink and make merry" and "to pinch one's belly," as Holyoake 1676 translates them.[35]

The word history of *genius* also differs from the usual story of Romantic genius in its emphasis. In our lexica, the word history of *genius* disregards outstanding, singular creative ability.[36] In fact, the prehistorical traces of that story are mere asides to the main narrative of naturalization of the pagan god into a natural-philosophical, psychological, or moral power or principle of life. This narrative, and the reaction to it, frames all the meanings of the first semantic cluster of *genius* as a god or spirit. The evolution of *genius* here can be read as one particular instance of humanist naturalism in the sixteenth century, of its potentially libertine uses in the seventeenth century, and antiquarian reactions to it in the eighteenth century.[37]

This narrative and its chronology unfold neatly in the various edi-

knee.] Lampridius in Heliogabalo: Ita ut subitò vestes ad pedes defluerent, nudusque una manu ad mammam altera pudendis adhibita, ingenicularet.

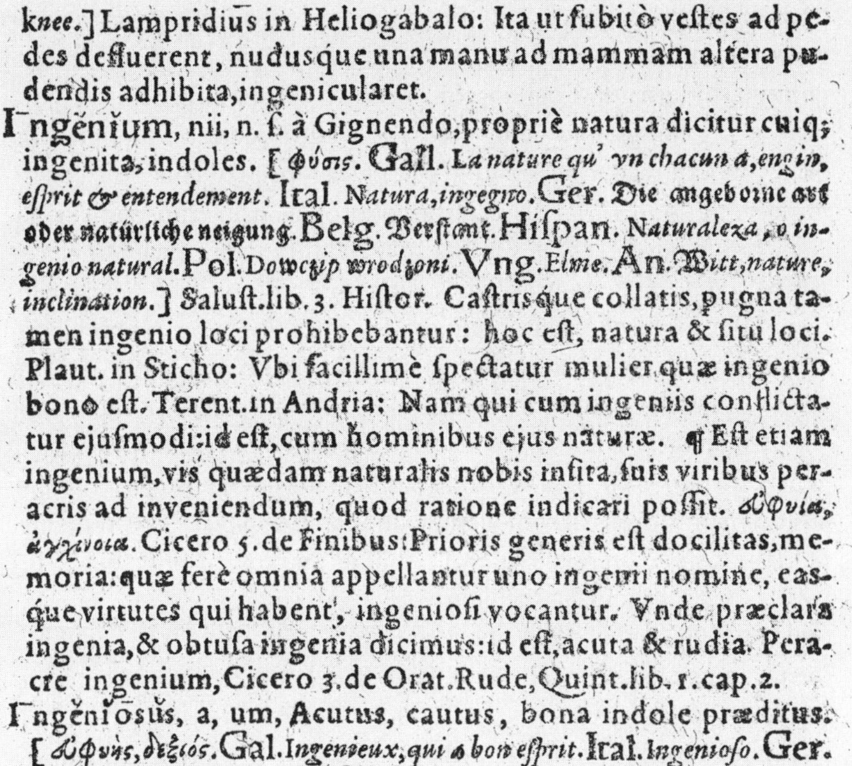

Ingĕnĭum, nii, n. f. à Gignendo, propriè natura dicitur cuiq; ingenita, indoles. [φύσις. Gall. *La nature qu' vn chacun a, engin, esprit & entendement.* Ital. *Natura, ingegno.* Ger. Die angeborne art oder naturliche neigung. Belg. Verstant. Hispan. *Naturaleza, o ingenio natural.* Pol. Dowcip wrodzoni. Vng. *Elme.* An. Witt, nature, inclination.] Salust. lib. 3. Histor. Castrisque collatis, pugna tamen ingenio loci prohibebantur: hoc est, natura & situ loci. Plaut. in Sticho: Vbi facillimè spectatur mulier quæ ingenio bono est. Terent. in Andria: Nam qui cum ingeniis conflictatur ejusmodi: id est, cum hominibus ejus naturæ. ¶ Est etiam ingenium, vis quædam naturalis nobis insita, suis viribus peracris ad inveniendum, quod ratione indicari possit. ἀφυία, ἀγχίνοια. Cicero 5. de Finibus: Prioris generis est docilitas, memoria: quæ ferè omnia appellantur uno ingenii nomine, easque virtutes qui habent, ingeniosi vocantur. Vnde præclara ingenia, & obtusa ingenia dicimus: id est, acuta & rudia. Peracre ingenium, Cicero 3. de Orat. Rude, Quint. lib. 1. cap. 2.

Ingĕnĭŏsus, a, um, Acutus, cautus, bona indole præditus. [ἀφυής, δεξιός. Gal. *Ingenieux, qui a bon esprit.* Ital. *Ingenioso.* Ger. Sinnreich/ verstendig. Hisp. *Ingenioso, y agudo de ingenio.* Polon. Dowcipni. Vng. *El més.* An. Wittie.] Cicero pro Cluentio: Homo in primis ingeniosus, & in dicendo exercitatus. Idem pro Roscio Comœdo: Quo quisque est ingeniosior, & solertior, hoc docet iracundius & laboriosius.

Ingĕnĭōsē, adverbium, Subtiliter & acutè. [ἀφυῶς. Gall. *Ingenieusement, subtilement.* Ital. *Ingeniosamente, sottilmente.* Germ. Sinnreichlich/ kunstlich. Hispan. *Sotil y ingeniosamente.* Polon. Dowcipnie. Vn. *El méssen.* An. Wittely.] Cicer. 1. de Inventione: Nam satis in ea videtur ex antiquis artibus ingeniosè, & diligenter electas res collocasse.

Ingēns, tis, om. t. Magnus & immensus. [Trab. ὑπερφυής, ὰμεγέθης. Gall. *Mout grand, fort grand.* Ital. *Molto grande.* Germ. Sehr groß. Belg. Groot. Hisp. *Cosa muy grande en su genero.* Pol.

Fig. 1.3. The multilingual Calepino reached its greatest number of languages in an eleven-language edition first published in 1590 at Basel.

tions of Calepino. Indeed, all the meanings listed under "god" end up instantiating nature or *phusis* in its various guises. The first meaning is primarily a natural-philosophical one: the natural generative power to bring things into being (see 1.a, p. 28) could only remind the early modern reader of Aristotle's definition of nature as the "genesis of growing things."[38] This natural-philosophical interpretation of the "generative power" locates it first in the ability to beget children: Perotti provides the first instance of a quote from Festus that recurs in the rest of the corpus: "genius inquit est deorum filius et parens hominum, ex quo homines gignuntur, propterea genius meus nominatur, quia me genuit" (Genius is the son of the gods, and the father of men, from whom men are born. For this reason it is called *my* genius, because it generated me).[39] The god of nature has become nature itself as (pro)creative dynamism. In early modern natural philosophy, this reproductive faculty was one prerogative of the soul as an ordered principle of life whose degrees included the ability to grow (vegetative soul), to move (animal soul), and to reason (rational soul). Thus, the naturalization of the meanings listed under 1 into those listed under 2 interprets for example the festive god of hospitality and pleasure (see 1.a.1 and 1.a.2, p. 28) and the tutelary spirit (see 1.b, p. 29) within the moral and temperamental semantic axes of ingenuity, internalizing what had previously been understood as an external entity.[40] This growing sense of *genius* as natural disposition features more frequently in seventeenth-century lexica.[41] The religious and natural meanings of genius as both festive god and natural bent toward pleasure are concomitant in the net of paraphrased references and glossed quotations of Calepino 1502:

> Ab aliquibus genius dicitur vis, et naturae deus hospitalis, et ipsa voluptas. Scintra [sic] de antiquitate verborum libro tertio. Scis geniales homines ab antiquis appellatos qui ad invitandum et largius apparandum cibum promptiores essent. Iuxta quod etiam Servius interpretatur pro voluptuoso, convivali. Virgilius libro primo Gaeorgicorum. Invitat genialis hyems. Nam quotiens voluptati operam damus Genio operam dare dicimur, sicut econtrario apud Terentium. Suum defraudans genium cum parce vivimus et a voluptate cessamus.

(Genius is said by some to be a power, and a god of a hospitable nature, and pleasure itself. Santra, *De antiquitate verborum,* book III. You know that men were called "genial" by the ancients who were quick to host and supply large meals. Hence Servius understands it to mean "voluptuous," "convivial." Virgil, in the first book of the *Georgics*: 'Convivial winter invites [them].' So that each time we attend to our pleasure, we are said to attend to our Genius, and just as we find the converse in Terence: we cheat our genius when we live parsimoniously, and abstain from pleasure).

This account, drawing from two grammatical sources (Santra and Servius), poetry (Virgil), and comedy (Terence), identifies *genius* with the festive pleasures of eating and drinking.[42] Under the influence of Estienne's reordering of his systematic excerpts from Plautus, Terence, and Persius, these examples heavily shift the balance toward the comedy and satire we find in the classicizing lexicography of both Estienne (especially 1543) and later editions of Calepino (in particular 1590). This accretion of satirical and comedic examples expanded the meaning of *genius* to include the sense of a natural drive and singularizing marker of identity—aligning it, we shall see, with the definition of *ingenium* as one's natural inclinations.[43]

Kahl 1600 provides the most striking, reflexive record of this naturalizing assimilation of *genius* and *ingenium*, envisaged now from the moral perspective rather than that of temperament. According to Kahl, Budé was the first to note that the term *genius* "was usurped by the masters of learning in our time to denote the natural dynamism and the *ingenium*."[44] Yet Budé, Kahl adds, was also well aware of the syncretistic continuities between the ancient *genii* and their Christianized versions in the discourse of old theologians, who posited the existence of good or protecting angels and of bad angels. Kahl inherits and develops Estienne's insistence on this continuity. He refers to Plutarch's account of Brutus's evil *daemon* and Plato's praise of the virtuous Socratic one; they serve him neatly as ancient anticipations of the Christian good and bad angels.[45] The argument Kahl summarizes thus amounts to a Christian reaction to the naturalization and internalization of the pagan tutelary genius into one's natural and moral disposition: the pagan genius was *not* an allegory of one's *ingenium* as moral discernment, but a dim foreshadowing of the Christian good and bad angels. This Christian definition is, glaringly, the most stable and repetitive of the word history of *genius*: dictionaries recorded the orthodox rebuttal to the possible impious consequences opened up by the natural-philosophical and anthropological naturalization of the pagan genius. *Genius sive natura* was decidedly not on the agenda; nature was not to be stripped of transcendent agency, nor was man to be made god, like Caligula the tyrant or like the *lares* and *penates*, the spirits of a family's deceased elders.[46] The dogmas that lay at the heart of Christianity were at stake, concerning both Creation and the individual, separate soul.[47] The full-blown polemic raised by the editor of Calepino 1718 in the gloss on Censorinus's definition highlights the antiquarian rebuttal of these impious views:

> Genios dictos, quia, ut gignamur, curent, sive quod una gignantur nobiscum, quod impium, et haereticum est. . . . Denique nonnulli Genium animum uniuscuiusque esse tradunt, quem Plato, Cicero, et alii Philosophi tanquam Deum quendam coli voluerunt. Nam fit a Deo per creationem, sed, cum introducitur in corpus, quodammodo, ut Vossius ait, cum homine gignitur.
>
> (Geniuses are called thus, because they care for us from the moment of our birth, either because they are born with us—this a heretical an impious view . . . Finally some say that the genius is the soul of each of us, which Plato, Cicero, and other philosophers wanted to revere like a god of some kind. In fact the soul is made by divine creation but, when it is introduced in the body, in some respect, as Vossius says, it is born with us.)

The interpretation of *curent* is what worried antiquarian lexicographers: it could be impiously construed either to mean that the pagan genius is responsible for our birth and therefore creates us, or that our genius is born with us, in which case it bears dangerous similarities with the rational soul as a God-given principle of individuation. In Calepino 1718, the lexicographer avoids these dangers by appealing to Isaac Vossius's interpretation: *genius* as one's soul is not divine in itself; however, its introduction in the body at birth is one form of God's creative power, and the true meaning of the phrase "born with us."

Individuation rather than outstanding cognitive ability is the main concern in entries about *genius* as one's rational soul; the important qualifier here is "one's" rather than "rational." The naturalization of genius into one's rational soul features in the classicizing dictionaries of Calepino and Estienne up to the seventeenth century.[48] Seventeenth-century antiquarian dictionaries, however, counter this naturalizing tendency. This definition disappears from Kahl 1600, and Hofmann 1677 inscribes it within an anti-pagan polemical tradition by attributing it to Varro via Augustine, and qualifying it as "profane," a stance also adopted by Calepino 1718 as we have already pointed out.[49] The cult of the Roman emperor's *genius*—a man made god—worried the early moderns as one of the worst forms of idolatry, likely to lead to impious free-thinking. Similarly, the possibility that the soul might be individuated by purely natural means, possibly the body, opened a whole theological can of worms regarding grace and resurrection.[50] The word history of *genius* provides us with faint echoes of some architectonic early modern debates on the theological or ontological status of subjective singularity and on the nature of the soul. The latter problem is the explicit frame of reference, as we will see, in some philosophical definitions of *ingenium* at the same period.

The semantic concern with singularity as a presentational feature displayed by things and places *as well as* people accounts for the final sense of *genius* in the corpus. Not only was *genius* an early social and physical marker of singular identity in Reuchlin 1478 (*vigor, honos, dignitas*), it is also what makes a place unique (the *genius loci*); in the case of people as much as artefacts, this uniqueness is regularly described as pleasant, and formulated as an aesthetic *je ne sais quoi*: a form of "energy," and that most elusive if omnipresent category of Renaissance aesthetics, grace.[51] That this quality should be singled out as what would make a book endure according to one of Martial's epigrams frequently quoted by our dictionaries—"Victurus genium debet habere liber" (To endure, a book must have *genius*)—is one such prehistorical trace of the modern notion of genius in the early modern word history of *genius*. Such traces appear only rarely.⁵²

Ingenium

The main strands of early modern use of *ingenium* are, with one major omission, knotted together in Perotti's *Cornu Copiae*, first published in 1489: *ingenium* as inborn nature and wit features there, but not as machine and legal trick. The *Cornu Copiae* was primarily a commentary on Martial's epigrams: the genre of commentary required Perotti to set *ingenium* within a larger word family, organically growing out of an extended discussion of its root *gens*. Perotti used the word to explore the complete vocabulary of how people are related by birth: both the notions of peoples (*gentes*) and kinds (*genera*) originate in the verb for giving birth (*gigno*). We must attend to Perotti with some care, because later dictionaries emulated his structure: first a definition, then examples from ancient literature. In the Latin printed lexica, Perotti's definitions become the building blocks of the lexicography of ingenuity:

> Item a gignendo ingenium, quod proprie significat naturam, cuique ingenitam. . . . ponitur etiam ingenium pro vi intelligendi naturali, quia saepe invenimus, quae ab aliis non didicimus, unde praeclara ingenia, et obtusa ingenia dicimus. hoc est acuta, et rudia, et homines ingeniosos, qui naturalem vim ad aliquid agendum accommodatam habent.⁵³

> (So *ingenium* is from *gignendo*, because it properly means the nature inborn in each person. . . . Ingenium is also given for the natural power of understanding, since we often invent what we did not learn from others—thus we speak of "brilliant *ingenia*" and "stupid *ingenia*," that is, of sharp and rough ones . . . and we call ingenious those men who have a natural power suited for doing something.)

Perotti also gives a definition for *ingenuus*: "Huius diminitivum est ingeniosum, item ingenuus, qui libera matre natus, nec postea servus factus est, cuius contrarium est libertinus, qui iusta servitute manumissus est. Unde interrogare hoc modo solemus, ingenuus ne an libertus est?"

(The diminutive [of *ingenium*] is *ingeniosum* or *ingenuus*, one born of a free mother, and not afterward made a slave; its opposite is the freedman, who has been emancipated from rightful slavery. Because of this we are used to asking in this way, "is he ingenuous or a freedman?")[54] *Ingenuus* entails not only a legal status, but also a social identity and its related behavior. For Perotti those various meanings formed a connected whole, in part justified by their common root in the verb *gigno* and the chief sense of *ingenium* as inborn nature. *Ingenium*-as-wit was derived from *ingenium*-as-nature, because one's inborn abilities include the cognitive power to discover something (*vis inveniendi*), that is, invention. This cognitive meaning invites further qualification; *ingenia* can be sharp or blunt, quick or slow, classifications that form the basis of the burgeoning Renaissance literature of pedagogy. *Ingenuus* at first glance seems to digress by referring to a legal status conferred at birth—an external, social gift. Is *ingenuitas* an *in*born nature, or simply a social status? Yet its relation to "inner nature" is reinforced by Perotti's explanation that free birth is revealed in one's actions; one's social behavior reflects an innate moral quality.[55]

In this section we will follow the two main threads that Perotti tied to *ingenium* in early modern Europe. The first thread was used in Latin and polyglot lexica to weave *ingenium* into increasingly sophisticated accounts of the identity of persons, either on temperamental or on cognitive grounds. A glance at a synoptic survey of these meanings will show the expansion of this moral and cognitive axis.

> *Ingenium*, n.
> 1. Temperament.
> a. Inborn nature.
> Reuchlin 1478: ("quasi intus genitum, scilicet a natura") >.
> b. Natural aptitude.
> Santaella 1499 > Perotti [1501].
> c. One's nature (metonymy for a person).
> Nebrija 1492.
> d. The nature of a land.
> Estienne 1543, "Ingenium soli, sive terrae" >.

e. Inborn nature of inanimate objects.

Estienne 1538 >.

2. Cognitive: inner intellectual power.

Balbus (13th c): (as "ingenium esse intrinsecam vim anime et naturalem") >.

a. Of invention or discovery.

Reuchlin 1478: what cannot be learned from others: ("interior vis animi quo sepe invenimus que ab aliis non didicimus") > Calepino 1502: what can be adjudicated by reason: ("vis quaedam naturalis nobis insita suis viribus praevalens ad inveniendum quod ratione iudicari possit") >.

b. Ability to learn and memorize.

Calepino 1502 >.

c. Ability to choose and judge (philosophical dictionaries only).

Glocenius 1613 > Chauvin 1692.

d. Metonymy: products of one's wit.

Holyoake 1676.

3. Mechanical.

Holyoake 1676 >.

4. Legal.

a. Cunning, fraud.

Holyoake 1676: ("artificium, techna, dolus malus") >; Du Cange 1678: ("fraus").

b. Interpretation of motives.

Du Cange 1678.

The second thread we will pursue makes up the warp of hidden medieval meanings of *ingenium* that lexica elided, then rediscovered. The early humanist recovery of ancient vocabulary entailed an erasure, too, of rich medieval senses. The list above shows that Perotti simply omitted definitions of *ingenium* as a mechanical or legal contrivance. Reference works such as Perotti, Calepino, and Estienne's *Thesaurus* were intended to replace medieval repertories of Latinity such as Isidore of Seville's *Etymologiae* (seventh century) and the *Catholicon* of Balbus (thirteenth century). As a result, common medieval meanings—

ingenium as a siege engine or malicious ruse—were ignored by most sixteenth-century humanist lexica, only to be reinstated in later seventeenth-century antiquarian and legal dictionaries.

The word history of *ingenium* in humanist lexica is characterized by the careful and complex construction of *ingenium* as singular, inborn nature, made apparent in a variety of cognitive, moral, and social markers of identity. *Ingenium* as one's individual nature was by far the dominant meaning in this corpus; we will consider the ways in which lexicographers subsumed under it the second most prominent meaning of *ingenium*-as-wit, made into yet another identity marker, alongside one's moral discernment and social behavior.

In earlier, medieval sources, *ingenium* is not an individual's nature in any grand sense, as it would become in the early modern period. Isidore of Seville's *Etymologiae*, the seventh-century encyclopaedia of word origins, picked up on the classical account of *ingenium* as the source of verbal discourse.[56] In fact, Isidore expanded on *ingenium* as the kind of natural wisdom that underlies all of the arts, citing Minerva's springing from the head of Jupiter as a metaphor for this general power of invention, "which is *ingenium*."[57] Of course this was a power *given by nature*, an "intrinsic, natural power of the soul," as the thirteenth-century work of Balbus put it.[58] Drawing on Balbus, the German Hebraist humanist Reuchlin defines *ingenium* as inventive power most clearly: "it is an inner power of the soul by which we often discover things we were not taught by others."[59] Perotti, along with his heirs the Augustinian prior Calepino and the Parisian scholar-printer Robert Estienne, generalized this inventive power to denote human nature. Because this book is concerned with such semantic transfers and the lexicographical processes they require—finding etymologies, simplifying definitions, accruing examples, choosing typographies—we should examine this specific transformation from inventive capacity to human nature in some detail.

This transformation partly happened in a new etymology, signaled at the beginning of each dictionary entry. Balbus had simply given a morphological explanation: "it is comprised of *in* and *genium*," hinting at a link to *genius*.[60] Isidore's etymology had been mythical, connecting various crafts to Jove, Mercury, and so on. The etymology shifted when Calepino and Estienne added Perotti's new derivation of the verb *gigno* and supplied the same primary definition: "properly said, it is the nature inborn to each person."[61]

The etymology was followed by a definition, which contributed to the shift to *ingenium* as nature. In fact, Perotti, Calepino, and Estienne offered a new definition. Their main definition of *ingenium* was no longer a concrete "natural power" or a kind of natural wisdom. Even the connection that Balbus made with *genius* disappeared. Instead, early modern *ingenium* became a larger term for one's inclinations and abilities, that is, for one's temperament—the whole nature one inherits at birth—hence the metonymic ability of *ingenium* to denote a person as a whole.[62]

After the etymology and definition, each of these archetypal lexicographical works then illustrated the definition with several sentences from classical authors. The new supporting quotations effected changes just as profound as the new etymology had done. The lines from ancient authors restricted the wider range of medieval meanings to just those authors whom lexicographers read, remembered, and admired. These examples did not expand *ingenium* very far beyond "innate nature." The examples Calepino gathered from Perotti were recycled in nearly every subsequent dictionary. One on the *ingenium loci* made very clear that this referred to the physical character of the land;[63] another described how men of the same sort fight.[64] Together, definition and examples both push *ingenium* toward the concept of "given nature."

Examples could even hide alternative definitions. Along the way, both Calepino and Estienne did offer a definition of *ingenium* as a power of invention, somewhat like Balbus, but they set it amidst their *exempla*. The definition was clear: "Ingenium is also a certain natural power seeded within us, and by its powers especially good at discovering what can be judged by reason."[65] The *De finibus*, a pious work of Cicero known to every medieval and early modern student, was the familiar source that backed this definition. The sentence offered a view of ingenuity especially dear to bookmen: "Together with *ingenium* one should name teachability and memory, and they call people with these qualities *ingeniosi*."⁶⁶ In the first edition of his book, Estienne picked up the same line, apparently without noticing that it actually contained a second definition; he placed it further down, submerged by the list of *exempla*. Thus even the book's typography conspired to shift the meaning of *ingenium* toward the broad definition of "nature" (Figure 1.4).

In fact, the Ciceronian citation fitted into a growing mosaic of quotations that used *ingenium* as shorthand for an individual's temperament. For the first edition of the *Thesaurus*, Estienne drew extensively on Calepino, but also enriched these *exempla* with quotations from the indices of Plautus and Terence that he had recently compiled. The comedic elements of their plays, whether the surprise of sly deceptions and witty plots or the depiction of characters as human types, assume the *ingenium* as an engine of dramatic action and the foundation of true identity hidden under display. In these examples, *ingenium* is revealed as one's true nature: "I didn't understand his nature at all,"⁶⁷ "For one who has himself loved bitterly inspects the lover's true nature,"⁶⁸ "I know he did it; he's that sort,"⁶⁹ "so long as one believes that [the secret] will out, he keeps watch; if he thinks his secret well kept, he goes back to his natural bent."⁷⁰ In each case, *ingenium* stands in for the whole true self, its innate inclinations and abilities, hidden under the costume of circumstance.

Most of the new layers of examples in the second edition of the *Thesaurus* (1543) strengthened those moral and cognitive aspects of ingenuity that Estienne had previously grounded in usages from Terence and Plautus. Yet they also exhibited more sharply the inherent divergence and tension in the core meaning of *ingenium*—in some respects, a distinct reenactment of the old "nature versus nurture" argument. On the one hand, *ingenium* is made the object of active verbs, which implies that an *ingenium* can be changed. As the lead example, Estienne refers to Quintilian, *Ars oratoria* I.4, where it is possible "to sharpen an *ingenium*" (*acuere ingenium*), suggesting that one's *ingenium*, far from being fixed at birth, can be nurtured.⁷¹ On the other hand, if *ingenium* is simply one's given nature, it can only be described. Adjectives qualifying *ingenium*—as *imbecillum*, *inquietum*, *acutus*, *miser*, etc.—make up the bulk of the ever-growing list of examples for

INFRVNITVS infrunita infrunitum, Insipiés ac stolidus, qui nulli vsui est. à græco deducta dictio. nam Græci ἄφρονα dementem dicunt.

INFRVCTVOSVS infructuosa infructuosum: Qui ne porte point de fruict, qui est sans fruict.
  Infructuosæ preces: Prieres faictes en vain. Plin. epist. 176. b, Quod vel maximè dolorē meū vicerat, obuersantur oculis cassi labores, & infructuosæ preces, & honor quem meruit.

INFVCO infucas, infucare, Colorare.
  Infucare, Pallier quelque tromperie, que on ne lappercoiue. Plaut. in Milite, 9. 46, Nunc pol ego metuo ne quid infucauerit: Si hic non videbit mulierem, aperitur dolus.

INFVLA infulæ, Fascia in modum diadematū, vt inquit Seruius, in decimū AEneid. à qua dependent vittæ ab vtraque parte. Festus, Infulas, sacerdotum flaminum vocabant antiqui. Alii scribunt infulas, filamenta alba esse, quibus templa ornabantur: & moris fuisse, vt nubentes, priusquam mariti domum ingrederentur, infulis postes ornarent.

INFVMO infumas, infumare, Ad fumū siccare. Plautus, Atque hilas infumatas, & sumen.

INFVNDO infundis, infudi, infusum, infundere, Intus fundere. Verser dedans, & comme entonner. Columella. lib. 2, Merúmque faucibus, si æstuauerint boues, infundet. Plaut. in Pœnulo, 6. 34, In fundas infundebā grandiculos globos: ego illos volātes iussi fundit arier.
  Infundere, Supra fundere. Virg. lib. 1. georg. Certatim largos humeris infundere rores.
  Infundibulum infundibuli, Instrumentum quo in vasa infundimus liquores, Vng entonnoir. Cato, In cella olearia hæc opus sunt, dolia olearia & opercula, sestarium olearium vnum, labellum vnum, infundibula duo.

INFVSCO infuscas, infuscare, Noircir, barbouiller de noir, comme sont les Mores.
  Infuscare vinum merum: Mettre de leaue au vin. Plaut. in Cistell. 2. 15, Raro nimium dabat quod biberem: atque id merum infuscabat.

INGEMINO ingeminas, ingeminare, Doubler, comme coup sur coup. Virg. lib. 1. georg. Tum liquidas cerui presso ter gutture voces Aut quater ingeminant.
  Ingeminant, pro ingeminantur. Virg. Ingeminant austri, & densissimus imber.

INGEMISCO ingemiscis, ingemui, ingemitum, ingemiscere, Gemir, ou plourer. Plin. in Paneg. 88, Quo cōstātius p. c. & dolores nostros, & gaudia proferamus: ingemiscamus illis quæ patiebamur.

INGENERO ingeneras, ingenerare, Ensemble, ou auec engendrer. Cic. lib. 1. de off. Eadémque natura vi rationis hominem conciliat homini, & ad orationis & vitæ societatē: ingenerátque in primis præcipuum quendam amorem in eos qui procreati sunt. Idem libro 3. Acad. In tanta animaliū varietate homini vt soli cupiditas ingeneraretur cognitionis & scientiæ. Plin. in Paneg. 32, Affectata aliis castitas, tibi ingenita.

INGENIVM ingenii, n. g. à Gignendo, Propriè natura dicitur cuique ingenita. Plaut. in Bacchid. 5. 42, Immò ingenium auidi haud pernoram hospitis. Ie ne cōgnoissoie point sa nature. Idem in Milite, 10. 44, Nam qui ipse amauit, amātis ægrè ingeniū inspicit. Idem in Stich. 3. 59, Vbi facillimè spectatur mulier quæ ingenio est bono? Idem in Asin. 17. 3, Nec quisquam est tam ingenio duro, nec tam firmo pectore, quin vbi quicquam occasionis sit, sibi faciat bene. Idem in Bacch. 21. 11, Scio fecisse: eo ingenio est. Il est de telle nature. Idem in Sticho, 10. 46, Ita ingeniū meū est, quicum vis depugno facilius, quàm cum fame. Idem in Adelphis, 1. 1. 46, Dum id rescitum iri credit, tantisper cauet: si sperat fore clàm, rursum ad ingenium redit. Idem, Nam qui cum ingeniis conflictatur eiusmodi. Idem in Heaut. 1. 1. 99, Ingenio te esse in liberos leni puto.
  Ingeniū habes patris quòd sapis: Tu resemble a ton pere, de ce que tu es sage. Plaut. in Pœnulo, 18. 26, Ingeniū patris habet, quòd sapit.
  Ingenium pudicum. Terent. in Hecyra, 1. 2. 77.
  Ingenium, vis quædā naturalis nobis insita, suis viribus præuales ad inueniendū quod ratione indicari possit. Cic. lib. 5. de finib. Vna ingenii appellatur docilitas & memoria: eásque virtutes qui habent, ingeniosi vocantur.
  Ingenium inhumanum, hebes, fractum, contusum, tardum. Terent. in Eunucho, 5. 2. 41, Non adeo inhumano ingenio sum Chærea, neque tam imperita, vt quid amor valeat, nesciam. Plin. epist. 167. d, Quibus ingenia nostra in posterū quoque hebetata, fracta, contusa sunt. Cic. de lege Agraria cōtra Rul. Sed quem vestrū tam tardo ingenio putauit, cui post eos consules Syllam Dictatorem fuisse in mentem venire non posset?

Fig. 1.4. By simplifying the arrangement of etymology, definition, and disciplined of *ingenium* as one's temperament or nature.

Ingenium meum rectum est ad te. Plaut. in Capt. 7.8, Vtroque vorsum rectum est ingenium meum, ad te, atque ad illum: pro rota me vti licet: vel ego huc, vel illuc vortar quo imperabitis. Iay lesperit a faire ce que vous vouldres tous deux.

Ingenia ita mihi sunt. Plaut. in Cistell. 5.7, Ita mihi ingenia sunt, quod lubet, non lubet iam id continuò. Iay lesperit si legier.

Ingeniũ subactum. Cic. de Orat. Subacto mihi ingenio opus est: vt agro non semel arato, sed nouato & iterato. Subactio autem, inquit ipse, est vsus, lectio: auditio, literæ.

INGENS ingentis, o.g. Dicitur (vt inquit Festus) augẽdi consuetudine, vt inclamare, inuocare. Quia enim gens, populi est magnitudo, ingentem per compositionem dicimus valde magnum.

Ingens animus, Vng grand courage. Plin. in Paneg. 103, Si vnius tertium cõsulatum eundem in annum, in quem tuum contulisses, ingentis animi specimen haberetur.

Ingens confidentia. Terent. in And. 5.3.5, O ingentem confidentiam!

Ingens facinus. Terent. in Adelphis, 4.7.3, Fero alia flagitia ad te ingentia boni illius adolescentis.

Ingentes gratiæ. Terẽt. in Eunucho, 3.1.2, Magnas verò agere gratias Thais mihi ingẽtes.

Ingens ignis. Plaut. in Capt. 13.74, Quid iubeam? ER. ignem ingentem fieri. AE. ignem ingentem? ER. ita dico: magnus vt sit.

INGENVVS ingenua ingenuũ, Qui in libertate natus est. Frãc. Et ab ingenuorũ antiquitate, vt Boetius in Topicis inquit, gẽtilitas ducitur. Plaut. in Pœnulo, 14.72, Quia Adelphasium, quam herus deamat tuus, ingenua est. Et paulò post, Qui eas vendebat, dixit se furtiuas vendere, ingenuas Carthagine aiebat esse. Terẽt. in Phor. 1.3.16, Vt ne addam, quòd sine sumptu, ingenuam, liberalem nactus est.

Ingenuus, relatiuum est libertini. Est enim libertinus, qui iusta seruitute manumissus est. Plaut. in Milite, 10.188, Ingenuam, an libertinam?

Ingenua facies. Plaut. in Pœnulo, 16.17, Facies quidem pol ingenua est.

Ingenua facta, Nobles faicts. Plaut. in Milite, 13.17, Vah, egone vt ad te ab libertina esse auderem internuntius, qui ingenuis factis responsaret.

Ingenuatus ingenuata ingenuatum, Ingenuè procreatus, Né en liberté. vel ingenuè vitam ducens, Noblement viuant. Plaut. in Milite, 10.135, Itidem diuos dispertisse vitam humanam æquum fuit: qui lepidè ingenuatus esset, vitam longinquam darent.

Ingenuè, aduerbium, Liberè, ac simpliciter. Vt ingenuè loqui, id est, vt liberũ hominem decet, nihil timidum, nihil habens seruile. Franchement.

INGERO ingeris, ingessi, ingestum, ingerere, Porter, ou ietter dedant. Plaut. in Pseud. 4. 24, Tu qui vrnam habes, aquam ingere: face plenum ahænum sit cito.

Ingerere dicta in dolium pertusum: Perdre ses parolles. Plaut. in Pseud. 5.134, In pertusum ingerimus dicta dolium, operam ludimus.

Ingerere dicta in aliquem. Plaut. in Asin. 16.77, Nunc vxorem me esse meministi tuam? modo cum dicta in me ingerebas, odium, non vxor eram. Quant tu mesdisois de moy.

Ingerere mala in aliquem. Plau. in Bacchid. 19.34, Atque vt tibi mala multa ingerã. Idem in Pseud. 5.124, Licet vtar manu? c. ingere mala multa. Frapperay ie? Fais luy force mal. Idem in Menæh. 16.17, Idem faciebat Hecoba, quod tu nũc facis: omnia mala ingerebat. Il faisoit du pis quil pouoit. Terent. in And. 4.1.16, Ingeram mala multa. atque aliquis dicat, Nihil promoueris. multum: molestus certè ei fuero, atque animo morem gessero.

Ingerere osculum. Suet. de claris grãmat. 23, Qui cum in turba osculũ sibi ingerentẽ, quanquam refugiens deuitare non posset, Vis tu (inquit) magister quoties festinantem aliquem vides abligurire?

Ingerere pugnos in aliquem. Terent. in Phor. 5.8.95, Nisi sequitur, pugnos in ventrem ingere, vel oculum exculpe.

INGITAS ingitatis, Quasi indigitas, paupertas, inquit Perottus. Plaut. Profectò ad ingitatem lenonem rediget, si eas adduxerit.

INGLORIVS ingloria inglorium, Qui nulla laude aut gloria dignus est, & veluti ignobilis. Virg. lib. 4. georg. Ille horridus alter Desidia, latámque trahens inglorius aluum.

Inglorius militiæ. Tacitus, Nec ipse inglorius militiæ. Syllius, lib. XI, Inglorius ausi. Id est, nequiens efficere quod tentabat.

fff. iiii.

examples, the typography of Estienne's *Thesaurus* (1531) focuses on a definition *ingenium* that filled several columns of tiny print in folio pages in early modern editions of Estienne up to the eighteenth century.[72]

These lists of adjectives fed a culture of classing people, in which *ingenium* denoted an inner nature only inasmuch as it ended up being consciously "outed," constructed, disciplined, and displayed. This presentational dimension of ingenuity and its ambivalences pervade our lexica. Cicero measured orators by their *ingenia*; early modern authors measured themselves and their rivals in the same way. In his *Invectives* Petrarch presented his literary persona in comparison to the "excellent" *ingenia* of the past (Plato, Aristotle, Homer, Virgil, Cicero . . .), and defended his *ingenium* as a gift from God, to be contrasted with the stupidity of his enemies.[73] In an important letter on literary invention, Petrarch revealed that he saw *ingenium* precisely as the faculty of composition.[74] Less famous writers emulated without end. One would be hard pressed to find a letter circulated in the Republic of Letters that did not praise the recipient's *ingenium*, demur the writer's modest *ingenium*, or offer fervid commentary on the *ingenia* of friends and foes. Authors of such letters could turn to later editions of Calepino and Estienne to find a bottomless well of adjectives.

Ingenium-as-nature also accrued social significance: in the lexica, one's status and behavior became social manifestations of one's inborn nature. Perotti was explicit on the relationship between *ingenium* and *ingenuatus*. Yet these terms also revealed a tension that emerges from putting so much weight on *ingenium* as "inborn nature"—a *hereditary* quality of unique individuals.[75] How do you inherit something that you alone possess? On *ingenuatus*, we see Estienne wrestling to match his definitions to his copious examples. At first he simply defines the adjective based on the etymology, as one who is "né en liberté" (freeborn). Yet examples from Plautus pull the definition of *ingenuatus* away from the idea of birth toward the axis of presentation, in conduct. In the light of these, Estienne therefore nuances his definition in the 1543 *Dictionarium latinogallicum*: "in particular, it means to lead life ingenuously, *to live nobly* in French."[76] He uses one specific line from Plautus to support this definition: "and so it was that the gods apportioned out human life, and to those who were ingenuous they gave long life."[77] Estienne seems to have reasoned that the gods awarded long life only for good behavior, and so Plautus must have meant conduct rather than birth when he used *ingenuatus* (ingenuous).

Polyglot dictionaries straightforwardly assumed the focus on *ingenium* as an individual nature. Cornelis Kiel's (Kiliaan) important *Tettraglotton*, published in 1562 by Christophe Plantin, simply quoted the first bit of Latin and French from Estienne's *Dictionarium latinogallicum* (1538), with direct translations in Dutch: "Ingenium, ij, n.g. φύσις, εὐφυεία. *La nature qu'un chacun a. Esprit et entendement qu'on a de nature.* De nature die een iegelick heeft. Gheest ende verstant dat men van naturen heeft" (Ingenium, nature, good nature [Greek]. The nature that each person has, the spirit and understanding one has by nature [French, repeated in Dutch]).[78] Likewise, the entries for *ingenue*, *ingenuitas*, and *ingenuus* feature people as their main explicit topic: their moral status set by birth and comportment: "*l'honnesteté et noblesse* belonging to *personnes franches.*"

In the seventeenth century, legal lexica became an important source here, not least because ingenuity turned out to be a useful way to class individuals. Court cases turned, after all, on whether one was noble, or well-intended.[79] This partly contributed to the use of *ingenium* as shorthand for one's cognitive abilities. In his massive compilation of earlier dictionaries, the *Lexicon iuridicum* (1600), the Heidelberg law professor Johannes Kahl used the humanist lexicographical arguments to draw out these legal implications. Whereas *ingenui* is usually derived from *gens* and denotes a child born from free parents, and by extension, of noble birth, Kahl relegated this "ancient" definition to a first entry on *ingenui* and then presents a second entry with an alternative etymology: he derived *ingenui* from *ingenium*, denoting those who can make a living out of their own wit and judgment: "INGENUI: ab ingenio dicuntur, quod Latinis naturam significat. Is ingenuus est, qui suo arbitratu, et ingenio potest vivere: aut qui naturae bono fruitur, nulla vi afflictae et mutatae" (Said to be from ingenium, which means nature in Latin. A man is ingenuous who can live by his will and ingenuity, or who enjoys the fruit of nature without being struck down or changed by any power).[80] Thus Kahl reinvented the etymology of *ingenuus* as from *ingenium* rather than *gens*. We might wonder if this redefinition obliquely promotes a new socioeconomic category, namely the scholar, who subsists by putting his mind to work; the scholar monetizes wit. In this respect, Kahl's legal lexicon offers an ideologically loaded example of how humanists remodeled the lexis of ingenuity to suit their own interests.

Legal lexica also moved quite far beyond ingenuity as an individual's cognitive powers. Law concerns the status of animals, lands, and above all the intentions of people. Kahl parsed *ingenium* in four definitions.[81] The first reprises the notions of inborn nature and power of invention belonging to the core meanings identified in the great humanist lexica, though already here his classical examples hint at a technical, legal sense of *ingenium* as a malign intention.[82] In the same way, the second definition highlights increasingly wide uses of *ingenium* as a marker of biological and even ecological identity: it applies to animals (a mule can have an *ingenium* that makes it ill-suited to the yoke), but also to inanimate objects, such as the *ingenium* of a region or place, illustrated by references to Virgil's *Georgics*, Sallust, and Tacitus in which *ingenium* comes close to the better known *genius loci*. The final two definitions widen the semantic extension in another direction. Both refer the reader to key legal texts on *ingenium*, namely the *Digest* and the *Code* of Justinian: in the phrase *magno ingenio* (by great ingenuity) in relation to the fabled legal acumen of the ancient lawyer Scaevola;[83] and the phrase *subtili ingenio* (by a subtle stratagem),[84] which illustrates the technical, legal sense of *ingenium* as evil intention, namely the cunning and trickery one should use to obtain gold from barbarians.[85]

• • •

The legal meaning of fraudulent intention and trickery in Kahl's *Lexicon* is of medieval origin.[86] It was not among the stable, core meanings of *ingenium* in the humanist lexica listed above, but became prominent during the second half of the sixteenth and in the seventeenth century among those who no longer despised the Middle Ages. When the Catholic-leaning Cambridge antiquary Henry Spelman collected his *Archaelogus* in 1626, he confined his attention to underrepresented medieval meanings, aligning *ingenium* with the French *engin*, or various devices (*techna, dolus*, etc.), as well as mechanical objects (*machina* and *instrumentum*). From an example of *ingenium* describing a city's various ornaments, Spelman deduced that it might also mean a spectacle.[87]

This alternative lexis of ingenuity became stable in the first early modern dictionary of medieval Latin, the *Glossarium mediae et infimae latinitatis* (1678) of Charles Du Fresne, seigneur du Cange. His two lengthy entries on *ingenium*, replete with examples from medieval charters, summarize with special clarity. The first entry of du Cange defines *ingenium* as the (covert) intention behind an evil deed, which a court would assess to determine guilt. In this legal definition, *ingenium* is tied to fraud (*fraus*), cunning, and a scheming mind (*machinatio, malum ingenium*). The second entry defines *ingenium* as a specific technology: a war machine.[88] Du Cange mentioned Tertullian's *De pallio*, which became a standard Patristic source for the mechanical meaning of *ingenium* in early modern lexicography. Du Cange therefore recognized forms of ingenuity that purist lexicographers of Latin had ignored, but which were likely an enduring feature in early modern culture. As we shall see in later chapters, most vernaculars gave cunning and machines a prominent place within their vocabularies of ingenuity.

Though he made medieval senses systematically available, in a roundabout way du Cange came to share some conclusions with the humanist lexicographers. In particular, he equated the unusual form *genium* with *ingenium* in one of his entries. He reasoned that *genium* was equivalent to *ratio*, here seen as the reason, sense, or spirit of a legal text. This brought *ingenium* very close to *genius*—an unusual move in the late antiquarian tradition, as we found in the word history of *genius*.

In increasingly bulky humanist and antiquarian lexica, we found that the trajectory of the Latin *ingenium* over the sixteenth century is one of semantic expansion and dissemination: the cognitive meaning of *ingenium* is swallowed up by others, concerning individual nature and identity. A very different story is told in the philosophical lexica of the seventeenth century. There the word history of *ingenium* is characterized by increasing analytical focus on the term's cognitive meaning. These lexica responded to and informed growing debates in learned circles about the nature of ingenuity, whether in poetics, artistic theory, natural philosophy, or medicine. By their very nature, these lexica are concerned with setting conceptual boundaries. Henning Volckmar's *Dictionarium Philosophicum, hoc est enodatio terminorum ac distinctionum celebriorum in philosophia occurentium* (1675) reminds us of this analytical purpose by its very title; its definitions are meant to elucidate (*enodatio*) conceptual distinctions (*distinctiones*), that is, to tease concepts apart. The corpus of classical authorities these lexica adduce is therefore markedly different from those discussed above: medicine, not rhetoric; Aristotelian epistemology and natural philosophy, not *res literaria* or law, are set as the bases of pedagogical theory.

Not only did they rely on Aristotelian faculty psychology to specify the cognitive operations of the *ingenium* (invention, judgment, memory, and reminiscence), these lexica also drew on Galenic and Aristotelian humoural accounts of temperaments in order to provide a philosophical content to the Sallustan commonplace according to which "there is no exercise of the *ingenium* without the body." All these lexica agree on a basic definition of the *ingenium* as an inborn power of the mind performing its operations in accordance with a specific disposition of the body. In all these lexica, the *ingenium* is located at the joint between body and soul. Thus Glocenius 1613 states that "the variety of *ingenia* depends in some part on the temperament of the body, in other part on the variable disposition of the soul and the constitution of organs, and on the rules of faculties, like the imaginative one," while Chauvin 1692 asserts that "the powers of the *ingenium* stem from the body."[89]

Early instances of these lexica echo the copious lists of qualifiers of one's *ingenium*—sharp or blunt, quick or slow—in the Calepino and Estienne dictionaries, yet overall they provide increasingly precise definitions of *ingenium*.[90] They adopt three strategies to add precision.

First, they superimpose the specialized philosophical vocabulary of scholastic philosophy onto the terminology of their humanist predecessors. Thus, the early definition of *ingenium* as inner power or dynamism (*vis ingenita*) is specified by the philosophical term *conatus* (the natural tendency to persist into being) in both Glocenius 1613 and

Volckmar 1675. Similarly, Chauvin 1692 appeals to *habitus* (ease of operation) in order to distinguish *ingenium* from its usual near-synonym in humanist lexica, *indoles*, which for Chauvin stands for a more fundamental or essential nature than *ingenium*.

Secondly, these lexica analyze the concept of *ingenium* in typologies and distinctions borrowed from medicine and faculty psychology. Indeed various typologies of *ingenium* structure those entries: typologies of *ingenia* as temperaments and typologies of the cognitive operations of the various faculties involving the *ingenium*. Thus Micraelius 1653 equates *ingenium*—which, for him, translates εὐφυεία, good nature— with "the faculty to know and to retain such knowledge" (facultas cognoscendi et cogita conservandi). He then provides a typology of the abilities subsumed under this faculty: the ability to learn with ease (ἐυμάθεια), the subtle power of invention of the mind (ἀγχίνοια), dexterity in adjudicating and good judgment (δύναμις κριτικὴ), and finally memory (μνήμη). This cognitive typology recurs in almost all entries. Chauvin 1692 provides chapter and verse for its Aristotelian sources: *ingenium* as judgment features in the *Nicomachean Ethics*, *ingenium* as logical invention (equated with *sollertia*) can be found in the *Posterior Analytics*.[91] Conversely, the identification of *ingenium* with the ability to learn (*docilitas*) and with memory derives from Cicero's *De finibus* 5, which we have already encountered as a commonplace in the humanist lexicography of *ingenium*.[92]

Humoral medicine had an equally significant impact on philosophical definitions. Volckmar 1675 thus superimposes onto the typology of faculty psychology another typology of temperaments: a sanguine temperament breeds a superficial *ingenium* (which is only able to dabble in invention, learning, or judgment), whereas a choleric one fosters sharp insight, a melancholic one good judgment, and a phlegmatic one a contemplative ability ill-suited to worldly business.[93] This typology bears traces of the Galenic humoral tradition as well as of the pseudo-Aristotelian Problem 30.1 on melancholy—but mostly of their reception by the Spanish physician Huarte de San Juan.[94] The medical typology of *ingenia* in the *Examen* was geared toward social usefulness, allocating individuals to their most productive place:

specific *ingenia* were more suited to specific activities. A second, pedagogical typology finds its most systematic expression in Volckmar 1675, who distinguishes between ordinary *ingenia* and extraordinary or heroic *ingenia*.[95] Ordinary *ingenia* get the most attention: some are moderate; some are fiery, which are precocious and need to be grounded; and some *ingenia* are earthly, being stable, heavy, and in need of lifting up.

The third and final strategy for making the concept of *ingenium* precise was to locate definitions within specific, ongoing philosophical debates. The importance of medical frameworks especially highlights the growing philosophical awareness of the embodied nature of the *ingenium* and the conceptual problems it raises. Thus, while embodiment is mostly implicit in Alsted 1626, it is explicit in Glocenius 1613 and Volckmar 1675, and becomes the main philosophical problem tackled in the final part of Chauvin 1692. Summing up the debate, Chauvin reminds the reader that Alexander of Aphrodisias provided an intellectualist answer to the question of the localization of *ingenium*, identified with the operation in the body of a God-given intellect.[96] Chauvin dismisses this answer and reinterprets Aphrodisias's view: the *ingenium* is a *habitus* (operative disposition) of the soul, determined by the body. This determination accounts for differences between the cognitive and moral performances of individual human souls. Chauvin's very elaborate definition of *ingenium* testifies to its importance in a major early modern philosophical debate, namely the mind-body relation.

Conclusion

In the humanist Latin and polyglot lexica there emerged a sophisticated account of *ingenium* as "inborn nature" made manifest through a variety of identity markers: one's wit, moral discernment, behavior, and social status. This elaboration helped to disseminate *ingenium* as an increasingly varied and copious term, a testimony to its rising importance and omnipresence in European cultures of ingenuity.

By comparison, philosophical lexica highlight the importance of the cognitive meaning of *ingenium*. They do so with increasingly precise, analytical definitions of *ingenium* that mark the emergence of *ingenium* as a stable concept within the philosophical debates of the period. As a result, the notion that one's *ingenium* amounts to one's *genius* as a tutelary spirit (*inexistens intelligentia*) features only as an antiquarian oddity (Chauvin 1692). Even so, such lexicographical antiquarianism helped to retrieve and promote the medieval mechanical meaning of *ingenium*, tying it to intellectual ability and craftsmanship: early modern lexicography reveals unexpected continuities between arcane bookish knowledge and the rise of the new sciences.

By contrast, the early modern Latin word history of *genius* was almost entirely an antiquarian matter, sometimes instrumentalized in contemporary theological arguments about the nature of the soul. This history bears very little trace of the term's pre-Romantic fortunes: that story took place in vernacular dictionaries.[97]

2 Italian

In the earliest Italian treatise devoted exclusively to *ingegno*—*Trattato dell'ingegno dell'huomo* (1576)—the natural philosopher Antonio Persio offers a curt etymology of the term denoting his subject: "Questo nome d'ingegno, alla guisa di tanti altri vocaboli della nostra lingua volgare, di corpo è tutto Latino, & volgare solamente d'accidenti... & però lo diriveremo come se fosse Latino in tutto" (This word *ingegno*, like many other words in our common tongue, is entirely Latin in form, and vernacular entirely by chance... and therefore I shall derive it as if it were wholly Latin).[1] Accordingly, he begins by citing the "most noble authors of antiquity," who take *ingenium* to mean the quality given normally to men by nature. Yet as Persio's definition unfurls, the linguistic and semantic impurity of his topic starts to become apparent. First, he seeks to narrow the meaning of the word by appealing to the acme of good Latin, Cicero, misquoting *De finibus bonorum et malorum* by borrowing from the entry for *ingenium* in Calepino's Latin dictionary:

> da gli stessi auttori particolarmente anchora è preso per quello che communemente è significato da' volgari, come il prese Cicerone, quando e'

disse, *prioris ingenii est docilitas, memoria: quae fere omnia appellantur uno ingenii nomine: easque virtutes qui habent, ingeniosi vocantur.* Perchioché noi per ingegno intendiamo propiamente quella parte dello spirito, per la quale siamo atti a comprender le cose, di cui è questo mio ragionamento.[2]

from these same authors [of antiquity] in particular is taken also that meaning which is equally signified by the vernacular ones, such as Cicero takes it, when he says "To the former *ingenii* belong receptiveness and memory; and practically all such excellences are included under one name of *ingenii* and their possessors are spoken of as 'talented.'" Therefore by *ingegno* we understand properly that part of the spirit through which we come to understand things, which is my argument.

Then, having set out his stall, Persio asserts that *ingegno* should never be defined simply as "nature" despite the fact that "the best in our language" have done so, quoting Boccaccio's *Decameron* as an example of misuse.[3] Indeed, part of the problem with *ingegno*, as he sees it, is the way ancient and modern writers have muddled its meaning according to their whims:

& perche in ogni lingua gli approvati scrittori, secondo o necessità, o vaghezza, od altro accidente gli conforta, un vocabulo dal suo propio significato sogliono traportare in uno alquanto straniero, è avenuto, che cosi gli scrittori Latini, come i volgari questo nome d'ingegno preso per tutto l'huomo alcuna volta, si come quando disse, *Qui cum ingeniis conflictatur eiusmodi, & paenè stulta est inhaerentium oculis ingeniorum enumeratio, inter quae maximè nostri aevi eminent princeps carminum Virgilius* &c che disse Paterculus; &, come disse il Petrarcha,
 Et quale ingegno ha sì parole terse?

and because in all languages the best writers, according to need or beauty or some other supporting reason, are wont to change the proper meaning of a word into one which is quite strange, it has happened that the Latin writers, just like the vernacular ones, have sometimes taken this word *ingegno* to mean the whole of man, such as when it is said "When one is involved with characters like this,"

and "it is almost folly to enumerate men of talent who are almost beneath our eyes, among whom the most important in our age are Virgil, *etc*." as Paterculus says; and as Petrarch says also:
"And what man has words ready enough?"[4]

In his treatise, Persio aimed to clear up what he considered to be simple and widespread misconceptions about *ingegno* by offering a thorough, intellectually rigorous, and extended account of his subject. Doubtless he would have bridled at the entry for *ingenium* in Francesco Priscianese's contemporaneous *Dictionarium Ciceronianum* (1579), in which—despite its title—the word is glossed simply as "ingegno, naturale."

The differences in genre, scope, and ambition between Persio's and Priscianese's works highlight the gulf that existed between dictionaries and other kinds of writing—most notably natural philosophy, but also artistic theory and poetics—in which *ingegno* was treated in Italy throughout the sixteenth and seventeenth centuries. Nevertheless, Persio's remarks draw attention to commonalities between the *trattatisti* and the lexicographers, rooted in the most significant language debate of the era: the *questione della lingua*. Like most dictionary-makers of the period, Persio acknowledged "the best" vernacular authors, quoting two of the *Tre Corone*—Boccaccio and Petrarch (the third is Dante)—whose works provided the material for constructing a normative vocabulary of Tuscan. Yet in the best-known instance of the *Tre Corone*'s authoritative lexical status—the Accademia della Crusca's famous *Vocabolario* (1612)—we find a very different definition of *ingegno* to Persio's. Citing as examples of use different passages from Boccaccio and Petrarch (along with Dante's *Divina Commedia* and commentaries upon it), the compilers rejected both Persio's technical definition as the "power of understanding" and Pricianese's simple translation as "nature," opting instead for a primary sense of *ingegno* that is much closer to the notion of artistic creativity: "Ingegno. Acutezza d'inventare e ghiribizzare, che che sia, senza maestro o avvertitore. Latino Ingenium" (*Ingegno*. Sharpness in inventing and creating anything whatsoever, without a teacher or prompter. Latin *Ingenium*.).

By grounding their definition upon exemplary literary sources, the compilers of the first, authoritative monolingual Italian dictionary cemented a specialized and poetic sense of *ingegno*, derived from the meaning of *ingenium* as inborn generative power in the classical tradition. Their somewhat rarefied definition persisted throughout the early modern and into the modern era, pushing *ingegno* close to those aspects of heightened creativity that would eventually be reified as Romantic genius. Thus, the modern standard dictionary of the Italian language — the *Grande Dizionario della Lingua Italiana* (GDLI) — defines *ingegno* as: "Potenza creatrice dello spirito umano che costituisce la massima espressione del talento e dell'intelligenza; capacità creativa, ispirazione artistica o poetica; genio" (Creative power of the human spirit, which constitutes the greatest expression of talent and intelligence; creative capacity, artistic or poetic inspiration; genius). Although modern, this definition captures a strong thread in the earlier fortunes of *ingegno*, which was connected (and very occasionally synonymous) with *genio*, as well as a host of neighbouring terms such as *spirito*, *intelletto*, and *natura*.

Keywords

The definition of *ingegno* in the GDLI indicates that by the modern age, *ingegno* had come to be equated with *genio*, via Romantic genius. In early modernity, these two words—like their Latin roots—stood apart from each other and were only occasionally connected by virtue of their relationship to the human spirit and its inclinations.

The word *spirito* is, properly speaking, a neighbor of ingenuity in Italian rather than one of its keywords. Its meanings sprawled across territory that—as with the French *esprit*—lies adjacent to the domain of ingenuity: the Holy Ghost and angels, demonic possession, "medical" spirits, and other kinds of subtle bodies.[5] While in some contexts these could be connected to the culture of ingenuity—demonic possession in relation to poetic fury; the spirited matter of an artisan's medium; the "divine" *ingegno* of supremely gifted artists; the elusive "air" (*aria*) of individual style—these relationships are not readily apparent in our corpus.[6] Furthermore, although *spirito* was often defined as or in relation to the soul, its connection to ingenuity was limited to specific functions of the human mind—namely "intellect" or "understanding"—or to a person's natural inclination. Only very occasionally was *spirito* considered synonymous with *ingegno* or *genio*, the latter defined consistently as a deity or spiritual being.[7]

Although *genio* occasionally registered as a synonym of *ingegno*, its association with superior mental or artistic ability comes late, and then largely via French and English influence. As the Italian translation of Chambers's *Cyclopedia* (1747–9) put it: "un *Genio*... cioè una mente superiore &c.... E frase più tosto Francese e Inglese, che Italiana" (a *Genius*... which is a superior mind, etc.... is more a French or English phrase than an Italian one). Indeed, *genio* did not even merit an entry as a headword in the first edition of the Accademia della Crusca's *Vocabolario* (1612), eventually making a belated appearance in the third edition (1691), but even then only as "inclination." The reason for this is twofold. First, the proximity of Italian to Latin meant that Latinate senses of the ingenuity family of words tended to be transferred directly into the vernacular—despite Persio's complaint about vulgar impurities. Thus, as we will see also in the case of Spanish, when *genio* did appear in our corpus it tended to be in the conventional sense of the Latin *genius*, as defined in humanist dictionaries.[8] Second, the compilers of Crusca 1612 evidently considered *genio* to be a minor term in the writings of the *Tre Corone*, unlike *ingegno*, which already in the sixteenth century had attracted the attention of commentators on Dante and Boccaccio. As such, the "genial melancholy" of artistic inspiration—evident in Italian natural philosophical, medical, and some art theoretical writings—is entirely absent from the dictionary definitions of *genio*, although implied in those of *ingegno*.[9]

Where the "ingenuity" senses of *genio* are limited and fairly tidy, those of *ingegno* are diverse and rangy. This reflects not only the transferral of Latin *ingenium*'s semantic scope but also the particular linguistic and lexicographical contexts of Italy in our period. The valences of the word *ingegno* in the writings of the *Tre Corone* encouraged lexicographers to dwell on its definition and to connect it especially to poetic senses of invention and inspiration. However, the drive to establish normative meanings rooted in the literary authority of a small canon occluded the vigorous expansion and transformation of *ingegno* in non-lexicographical domains. In Italy more than in other regions, the gap between dictionary definitions and more widespread word use is vast. For instance, the important role of *ingegno* in emerging notions of artistic freedom and changes in the status of artisans barely registers in the dictionaries.[10] Likewise, the elaborate entanglement of *ingegno* with the development, during the fifteenth and sixteenth centuries, of aesthetic notions such as *acutezza, difficultà, disegno, grazia, maniera, meraviglia, sprezzatura, terribilità*, etc., took place in other genres: in treatises on painting and architecture, guides to courtliness, poetry and criticism, and artists' biographies.[11] Because the authors writing in these genres who used *ingegno* in especially novel or prominent ways (e.g. Alberti, Castiglione, Michelangelo, and Vasari) were only acknowledged—if at all—by the Accademia della Crusca very late in our period, their influence on our corpus was minimal.[12] Equally, the massive inflation of *ingegno* in the poetics associated with *concettismo* (especially the writings of Emanuele Tesauro) in the early-mid seventeenth century, or Giambattista Vico's revitalization of the word in his philosophical work at the end of that century, had little to no impact on dictionary definitions.[13] Notably, the dictionaries also singularly failed to acknowledge the highly gendered nature of the period debates about creativity and intelligence, in which *ingegno* was to the fore.[14]

This does not mean that the lexicography of *ingegno* was stable or straightforward. The preservation of Latin senses of *ingenium* in the vernacular, the diversity with which the *Tre Corone* used the word, and the subtle pressure exerted by the aforementioned aspects of the wider culture of ingenuity ensured that its fortunes were lively—at least up to the publication of Crusca 1612. Moreover, ancient and modern senses of the word competed and collided with medieval ones, such as those connected to deceit, cunning, and the arts of war. The latter proved especially salient. Italian lexicographers' emphatic definition of *ingegno* as a machine, device, instrument, weapon, and stratagem reflects not only the heritage of romance usage but also the preeminence of Italian military engineers in the period, who innovated in the design and construction of fortifications and war machines, while copiously writing treatises on their art.¹⁵ Indeed, although the inventiveness ascribed to *ingegno* was, for most lexicographers, pronouncedly poetic, the *ingegnere* was nevertheless considered a paradigmatic possessor of ingenuity, as the tautological definition in Crusca 1612 shows: "Ingegnere. Ingegnoso ritrovator d'ingegni, 'e di macchine. Latino machinator, architectus" (Engineer. Ingenious inventor of devices and machines. Latin *machinator*, *architectus*).¹⁶ The prominence of this sense of *ingegno* ensured a complex interplay between liberal and mechanical senses of creative potency in the dictionary definitions, while feeding the "deceitful" or "tricksy" aspects long ascribed to poets and artisans alike.

Lexicographical Landscape

Lexicography in early modern Italy was dominated by the *questione della lingua*: the extended dispute over what the shared, "national" language (*lingua comune*) for the Italian peninsula should be. This was no less than a debate about the *genio* of the Italian tongue and hence the nature of its people.¹⁷ Extending from the fourteenth through the seventeenth century, the *questione della lingua* preoccupied not just lexicographers but also poets, scholars of many stripes, courtiers, lawyers, and theologians. Focused initially on the competing dialects spoken in different regions of Italy—whether one, several, or all of them should be used as the foundation for a standardized language of literature—the debate quickly extended to questions of correct usage and literary style (*buon gusto*), identity and social status, oral versus written language, and matters of orthography. Indeed, the *questione* had implications for so many aspects of cultural, social, and political life that we can only offer a brief summary of its key features here.¹⁸

Dante was the first to turn his attention to the problem of a standard vernacular, proposing in *De vulgari eloquentia* (ca. 1303–5) that literary language should be based not on a single dialect (though he himself wrote in a Florentine variety of Tuscan) but should draw on the "best elements of all, in order to achieve the universal quality to which he aspired as a stylistic ideal."¹⁹ Subsequently, the *questione*—which intensified in the sixteenth century following the spread of printing—revolved around the respective merits of archaic Tuscan (favoured by Pietro Bembo), contemporary Tuscan (promoted by Claudio Tolomei), an archaic common language evident in several dialects (proposed by Girolamo Muzio), and the contemporary language of the Italian courts: the *linguaggio cortigiano* championed by Giangiorgio Trissino.[20] Bembo's argument in his *Prose della volgar lingua* (1525) that writers of Italian should model themselves on Boccaccio for prose and Petrarch for verse (a vernacular equivalent to humanist Latin modeled on Cicero and Virgil) proved decisive. The Florentine Tuscan used by those authors (and Dante) was established as the Italian monolect by the end of the sixteenth century, a position solidified by the Florence-based Accademia della Crusca, whose purist *Vocabolario* (1612) laid out a normative vocabulary based on and citing extensively the writings of the *Tre Corone*.

Even prior to Crusca 1612, most Italian dictionaries responded directly to the concerns of the *questione*. There are a few exceptions: the early Latin-Italian glossaries in manuscript, which reflect the regional variety of Italian but tend to offer only pairs of words, sometimes with simple explanations; the earliest Italian-Latin dictionaries with dialect words in headword positions, such as Tranchedini 1455–70 (Tuscan) or Scoppa 1511/55 (Neapolitan), both of which provide strings of synonyms; and the earliest monolingual dictionary, Volpi 1460–66—a list of learned Italian words and proper nouns with brief definitions.

By contrast, the impact of Bembo's intervention in the *questione* was reflected immediately in the production of printed dictionaries, such as Nicolò Liburnio's *Tre Fontane* (1526). Published just one year after Bembo's *Prose*, the "three fountains" of this book's title refer to the writings of the *Tre Corone*, from which, Liburnio suggested, the Italian language should spring. In a similar vein are Lucilio Minerbi's *Vocabolario* (of Boccaccio), published in his edition of the *Decameron* (1535) and Francesco Alunno's influential *La fabrica del mondo . . . ne quali si contengono le voci di Dante, del Petrarca, & del Boccaccio* (1539), the latter a compendious repository of the language used by the *Tre Corone*, including careful definitions of significant words. With a few exceptions (such as Luna 1536, which included Ariosto) barely any sixteenth-century dictionaries challenged the orthodoxy of the *Tre Corone*'s authority.

INGIGNIRE. Ingegnoso ritrouator d'ingegni, e di macchine. Lat. *machinator, architectus*. Gr. ἀρχιτέκτων. Stor. Aiolf. Molto si turbò l'aria, e cominciò a piouere: allora il maestro Ingegnére s'affrettò.

INGIGNEVOLE V.A. sust. Ghiribizzo, astuta inuenzione. Lat. *dolus, astus*. Guid. G. Ma quella, che regnaua con molto sagace ingegno, si studiaua, con sagaci ingegneuoli, di mantenerlo in isperanza.

INGEGNO. Acutezza d'inuentare, o d' apprendere, che che sia. Lat. *ingenium*. But. Ingegno e vna virtù interior d'animo, per la quale l'huomo da se truoua quello, che da altri non ha imparato. E altroue. Ingegno chiamano gli autori lo naturale intendimento, che l'huomo ha, e arte quella, che ammaestra l'huomo con regole, e ammaestramenti. E di sotto. Ingegno è quella virtù dell'anima, con la quale lo'ntelletto fa L'operazione, e gli atti suoi, e impropriamente si dice ingegno quel delle mani, ma deesi chiamare attitudine. Bocc. n. 19.7. Non ti sento di sì grosso ingegno, che tu non auessi, ec. Lab. n. 22. Si possono da'più sublimi ingegni comprendere. Dan. nf.c.2. O Muse, o alto ingegno or m'aiutate. Petr. Son. 18. Però lo'ngegno, che sue forze stima, Nell'operazion tutto s'agghiaccia. ¶ Per inganno, astuzia, stratagemma. Lat. *dolus, astus*. Liu. M. E, per tale ingegno, fu la legge tutto l'anno gabbata. ¶ E a'ngegno, posto anuerbialm. ingannevolmente, astutamente. M. Vill. 7.8. Gl'Inghilesi, maestri di baratti, aueuano mandati caualieri de' loro a'ngegno, che tornauano la notte per quel cammino. ¶ Per istrumento ingegnoso, e, per lo più, si dice d'i serrature, o da aprir serrature. Lat. *machina, machinamentum*. Bocc. n. 19.25. E, con certi suoi ingegni, apertala, chetamente della camera uscì. Franc. Sacch. rim. Benchè sauio non sia, e le mie chiaui Non abbian tanti ingegni. Lib. Maccab. M. Vsciron della Città de', e affocaron gl' ingegni [cioè macchine]

INGEGNOSAMENTE. Con ingegno. Lat. *ingeniose, acute*. Filoc. lib. 1.58. Noi ingegnosamente gliele sottraemmo. Vit. S. Padr. Ringraziòe Iddio, e San Maccario, che così ingegnosamente l'auea fatta limosiniera.

INGEGNOSISSIMO. Superl. d'ingegnoso. But. Ella ebbe Dedalo, il quale era ingegnosissimo.

INGEGNOSO. Dotato d'ingegno. Lat. *ingeniosus, acutus*. N. ant. 82.5. Il demonio, ch'è ingegnoso, e reo d'ordinar di fare quanto male e'puote. Tes. Br. E però fae l'huomo rosso, ingegnoso, acuto, fiero, e leggieri. Petr. Son. 123. Oue

Fig. 2.1. The entry for *ingegno* in the Crusca, *Vocabolario* (Venice, 1612).

This canonical status was effectively fixed in Crusca 1612.[21] A dictionary several decades in the making, the *Vocabolario* set a new standard in Italian lexicography. Published under the collective authorship of the Florentine Accademia della Crusca, this handsome folio asserted typographical and lexicographical clarity, with short, clear definitions backed up by carefully selected quotations, which the reader could trace in a bibliography of sources. Sublemmas distinguished between primary and secondary meanings of headwords, while the dictionary's alphabetical structure ensured ease of reference.

The *Vocabolario* was a product of the growth and strengthening of literary academies in Italy during the sixteenth century. In Florence, the Accademia Fiorentina had already established a powerful tradition of literary, philosophical, and artistic education and publication, in which the *Tre Corone* (all of whom wrote in Tuscan) featured prominently. Indeed, the Crusca *Vocabolario* was influenced significantly by Vincenzo Borghini's *Annotationi* (1579) to a censored edition of Boccaccio's *Decameron*, sponsored by Cosimo de' Medici. In the *Annotationi*, Borghini undertook careful philological work on selected words from Boccaccio's text, which he treated as part of the living language of contemporary Florence. His labors informed the next generation of Tuscan lexicographers, chief among them Lionardo Salviati—a founding father of the Accademia della Crusca and its most influential early member. Shortly after his election to the Accademia Fiorentina in 1564, Salviati delivered an oration "nella quale si dimostra esser la fiorentina favella e i fiorentini autori superiori a tutte le altre lingue, sì dell'antichità che moderni, e a tutti gli aultri autori" (in which it is shown that the Florentine speech and the Florentine authors are superior to all other languages, ancient and modern, and to all other authors).[22] From this starting point, and inspired by Borghini's example, Salviati produced his own edition of the *Decameron* in 1582, followed by a two-volume *Avvertimenti* (1584–6). In the *Avvertimenti*, Salviati stated that he had begun to compile a "vocabolario della toscana lingua," in which would be gathered all the words and phrases used by the best authors writing before 1400. This project—left unfinished on Salviati's death in 1589—probably formed the nucleus of the Accademia della Crusca's dictionary. Founded with Salviati's encouragement in 1584, the Accademia began formally to plan the *Vocabolario* in 1589 and by the early 1590s had begun to apportion ranges of words to its members. Minutes of the academy's meetings reveal not only that the dictionary was to be based on the writings of the *Tre Corone*, but that—at least in its early phase—the compilers were especially interested in "establishing the precise sense of a given word rather than illustrating it from the authorities."[23] While the academy made good on this aim, providing precise definitions rather than strings of synonyms or endless quotations, it nevertheless relied heavily on literary authorities.[24] In the end—perhaps due to differing degrees of familiarity with vernacular philology among the academicians—the *Vocabolario* emphasized archaic rather than contemporary usage. Technical terms were shunned and encyclopaedic material avoided, which meant the mechanical arts were especially underrepresented. Thus, the vocabulary published in 1612 was in essence already "classical": the record of an elite, literary language which was at best a rarefied version of the Tuscan spoken by contemporary Florentines.

This aspect of the *Vocabolario* drew some stinging responses. Prior to the dictionary's publication, John Florio had sparred with academy members, offering in *A Worlde of Wordes* (1598) and *Queen Anna's New World of Words* (1611) translations that, for all their imprecision, captured far more fully the lively variety of contemporary Italian.[25] Closer to home, the Paduan Paolo Beni published, within a year of the *Vocabolario*'s appearance, *L'Anticrusca* (1612): a polemical work lambasting the archaisms of the Crusca dictionary, its Florentine bias, and its neglect of recent authors, notably Tasso. Despite controversies such as this, however, the *Vocabolario* dominated Italian lexicography for the next several centuries. Subsequent editions were expanded with new entries, enlarged definitions, and additional authorities—for instance, the *Rime* of Michelangelo in the second edition of 1624 and the writings of Galileo in the third edition of 1691. Even though archaic words started to be labeled "V.A." (*voce antica*) more frequently by the third edition, the *Vocabolario* remained at heart a dictionary reflecting the literary canon inherited by the Florentine Renaissance.

It was left to others to attempt to fill the gaps. Yet whereas Filippo Baldinucci published a dictionary of the visual arts—the *Vocabolario toscano dell'arte del disegno* (1681), dedicated to the Accademia della Crusca—and a number of polyglot dictionaries and translations of foreign lexicographical works appeared in the seventeenth and eighteenth centuries (notably Ménage 1669, Chambers 1747–9), the dominance of the Crusca went largely unchallenged up to and beyond its fourth edition (1729–38). Very few other monolingual dictionaries were published, and those that were largely copied the Crusca. This, as we shall see, had significant consequences for the fortunes of our keywords. Their meanings as terms of art in natural philosophy and contemporary artistic theory were effectively suppressed in favor of the archaic senses in which they had been used by Dante, Petrarch, and Boccaccio. Moreover, the ways in which they were defined barely changed following the publication of the first edition of the *Vocabolario*.

Genio

Our first keyword, *genio*, did not even merit a lemma in Crusca 1612—although it appeared in the definition of a sublemma in the entry for *natura*: ("per genio, e costume") and, in Latin, in the entries for *diportare* ("genio indulgere") and *godere* ("indulgere genio").[26] When eventually included as a headword in the third edition of the *Vocabolario* (Crusca 1691), *genio* was given as the translation of the Latin *genius* and *ingenium* followed by a quotation from the minor Florentine poet Giovanmaria Cecchi's comedy *Il servigiale* (1561), which features a personification of "Genio."[27] In offering *genio*-as-deity as its primary example, Crusca 1691 simply reflected the way in which the word had been defined previously in Italian dictionaries.

Genio, n.
  1. A spirit or deity.
    a. A deity born together with man.
    Pulci ca. 1460–66 >.
    GDLI, 1: "Spirito buono o cattivo che secondo la mitologia romana assisteva ogni uomo dalla nascita alla morte ispirandone le azioni e tutelandone particolarmente la virtù generativa."

b. An angel or spirit, whether good or evil, that attends each man.
Scobar 1519 (*genius*) >.
GDLI, 1.
c. A deity of all things.
Scobar 1519 >.
Not in GDLI.
d. The god of study.
Luna 1536 >.
GDLI, 2: "2. Essere simbolico che presiede alle arti (in particolare alla musica, alla poesia, alla pittura) o al pensiero; personificazione delle arti stesse."
e. The soul of man.
Porcacchi 1584 >.
Not in GDLI.
f. God as the natural elements.
Porcacchi 1584 >.
Not in GDLI.
g. An entity between a deity and man.
Porcacchi 1584 >.
Not in GDLI.
2. Nature or inclination of an individual.
Canal 1603 >.
GDLI, 4: "Figurativo Spiccata disposizione naturale; propensione, attitudine, inclinazione; gusto . . . Istinto (di un animale)."
3. Natural aptitude, talent, or inclination toward a specified thing.
Chambers 1747–9 >.
GDLI, 7: "Capacità e abilità intellettiva limitata a una determinata disciplina o a un determinato campo dell'attività pratica; ingegno, intelligenza; acutezza d'istinto, finezza di gusto.—Anche: qualità d'ingegno superiore al comune."

From the mid-fifteenth century onward, Italian lexicographers departed little from humanist Latin definitions of *genius* as a spirit—tutelary or otherwise—that attends man. Hence, Pulci ca. 1460–66 defined *genio* as "l'iddio che nasce insieme coll'uomo" (the deity born together with man), while Scobar 1519 offers both a pagan and Christianized definition of the Latin *genius*: "Lu deo di onna cosi particolari . . . Lu angilo bono" (The god of any particular thing . . . The good angel). This meaning remains stable across the relatively few dictionaries in our corpus that register *genio* as a headword. Some specify that it is connected to nature, whether of a person (i.e. their temperament, inclinations) or, more literally, as the elements themselves. Venuti 1561 is an early example "Dio della natura di alcuno," while Porcacchi's entry in the 1584 edition of Alunno's *Fabrica* (Porcacchi 1584) is especially full:

> Genio. Latino *genius*. Scrivono gli antichi, che il Genio non è altro, che l'anima ragionevole di ciascuno: & altri vogliono, che 'l Genio sia Dio, compreso sotto nome di fuoco, aria, acqua, & terra, che sono i semi delle cose, chiamati elementi. Ma Platone nel Simposio parlando de' Genii copiosamente; dice che la natura d'essi è di mezo fra gli Dei, & gli huomini, e interpreta, & riporta le cose humane a gli Dei, & le divine a gli huomini: di quelli, i precetti, le sacre solennità, l'istituzioni, & l'ordine: & di questi le preghiere, e i sacrifici.

> *Genio*. Latin *genius*. The ancients wrote that the Genius is nothing more than the rational soul of something, and others have it, that the Genius is a god, known by the name of fire, air, water, and earth, which are the seeds of things and are called elements. But Plato in the *Symposium* talks extensively about Genii; saying that their nature is to be between the gods and men and to interpret and report the things of men to the gods, and the divine to men, among which are the laws, sacred ceremonies, institutions, and the order of things; and of these, prayers and sacrifices.

To support his definition, Porcacchi quoted from sixteenth-century texts: Ariosto's *Orlando furioso*, Annibal Caro's *Apologia*, and Jacopo Sannazaro's *Arcadia*. The Caro quotation indicates that Porcacchi considered *genio* to be close in sense to *spirito* and that the word suggested also the temperament that underlies a "congenial" nature: "M'hanno fatto conoscere, che voi siete d'un Genio conforme al mio, cioè d'uno Spirito" (You have made me understand that you are of a genius like mine, that is [we are] of one spirit). This sense of *genio* as an individual's nature caused it to be confused and conflated with *ingenium* (hence the inclusion of the latter in the entry in Crusca 1691), which may go some way to explaining Persio's frustration that *ingenium/ingegno* tended to be glossed simply as *natura*.

A further potential confusion with *ingenium*—routinely defined in the Latin lexica as the capacity for learning—is suggested by Porcacchi's use of Plato, specifically the "tutelary" aspect of the *genii* in the *Symposium*. This connection helps to explain the association of *genio* with study, as in Luna 1536: "il dio del studio," and why Crusca 1691 gave the Latin *studium* as a synonym for *genio* in its primary definition: "Per Inclinazione d'animo, Affetto. Latino *studium*, *voluntas*." Rather more curious is the second synonym—*voluntas*.[28] This association is rare in the Latin dictionaries, occurring occasionally in relation to *ingenuitas* but not in relation to *genius*.[29] There may be a trace of this in Crusca 1623, where, in the entry for *bevanda*, it is put in parallel to doing an activity: "Diciamo anche beva. E del vino, la sua beva è nel tal tempo, cioè egli è buono a ber nel tal tempo: onde metaforicamente esser nella sua beva, si dice di chi fa che che molto volontieri, e secondo il suo genio." (We also say "beva." And of wine, that it is ready to be drunk, that is, it is good to drink at a given moment; metaphorically, therefore, "to be in one's drink" is to say that one does something very readily, and according to one's genius.)[30] Here, *genio* is clearly used in the familiar sense of "following one's genius" (*andare a genio*): a phrase that appears prominently in the entry for *andare* in Crusca 1691.[31] This free following of one's nature connects *genio* directly to *geniale*, as in Ruscelli's definition of the latter as "Che è da nozze, di piacere" (Which is of weddings, of pleasure; Ruscelli 1588).[32] The pleasurable, cheerful aspects of *geniale* return us to the example of "congeniality" in Caro, quoted by Porcacchi, which toward the end of our period surfaced in the new noun *genialità*, defined in Bergantini 1745 as "Astratto di geniale, Simpatía."

Thus, the early modern history of *genio* in Italian effectively mirrored that of *genius* in the Latin lexica.[33] Toward the end of our period, *genio* was defined as a specific talent for one thing or another, as in Chambers 1747–9: "*Genio* in senso più ristretto si prende anco per un talento, o disposizione naturale ad una cosa più che ad un'altra. Nel qual senso diciamo, un *genio* per il verso, per le scienze, &c." (Genius, is also us'd in a more restrain'd Sense for a natural Talent, or Disposition to one Thing more than another. In this Sense we say, A *Genius* for Verse; for the Sciences, &c. [Chambers 1728]). This definition—notably an Italian translation of an English encyclopaedia—built on the already established notion of "following one's nature." Yet for most of our period, the sense of particular or even "special" talent belonged properly to another word: *ingegno*.

Ingegno

Ingenium and its various vernacular translations are routinely distinguished from, if not opposed to, art (*ars*).[34] That is to say, *ingenium* is an innate gift rather than an acquired skill, which needs to be coupled with other virtues (diligence, care, study, etc.) in order for its possessor to perfect his abilities and to work optimally.[35] In a characteristically pithy account of the subject, David Summers explains: "*Ingenium* (and the Italian *ingegno*) means talent, the natural gift of a person, and is opposed to what may be learned from art and experience. In a long tradition . . . the term *ingenium* was used to refer either to natural talent in general or to the results of the exercise of talent."[36] Hence the classical antithesis *ars et ingenium*, a topos in rhetoric (derived chiefly from Quintilian), which in the hands of fifteenth-century Italian humanists became a commonplace term of praise not only for writers but increasingly (and controversially) for visual artists.[37]

According to Michael Baxandall, this coupling constituted the very "system" for structuring the arts and artists in the Latin-learned culture of the Italian Renaissance, the formulaic aspects of which he distinguished from the looser, more inventive expressions of praise for painters in the vernacular—notably, Boccaccio's salute to Giotto's *ingegno*, manifested in the painter's extraordinary mimetic ability.[38] Baxandall suggested that as the fifteenth century progressed, *ars/arte* and *ingenium/ingegno* were bound together not as incommensurable opposites but as interdependent notions, creating a semantically fluid space for working out the changing relationships between rule and licence, learning and talent.

This fluidity may explain why the earliest occurrence of *ingegno* as a headword in an Italian lexicographical work elides the distinction between *ingenium* and *ars*.[39] In the Italian-Latin *Vocabolario* (1455–70)—a manuscript lexicon by Nicodemo Tranchedini, ambassador of the Sforza—we find the following: "*Inzegno*: Ingenium; studium; industria; machinacio; machinamentum; machina. *Inzegnoso, insegnero*: ingeniosus; studiosus; industrius; ingenierus; architeta; operum gnarus; architetus; perspicax; elaboratus (*vel* excultus) *vel* secundo (*vel* liberali) ingenio."[40] While *ars* is absent in these entries, it appears—along with *ingenium*—in that for *industria*: "Industria: industria; negociacio; exercitum; ingenium; ars; opera; cura; studium; perseverancia; gratia; diligentia; vigilantia; solercia; prudentia."[41] Tranchedini probably did not intend strict synonymy for his extended lists of defining terms, which seem to operate by way of allusion as much as by strict equivalence, while rattling across the cognitive, artificial, moral, and presentational axes of ingenuity. Nevertheless, his definition of *ingegno* as "study" and in connection to "art" was not idiosyncratic. A full century later Giovanni Marinelli offered for *ingegno*: "vale senno, intelletto, avedimento, valore, & virtù: & latino prudentia, intelligentia, ratio, ars," while his entry for *industria* is: "arte, diligentia, astutia, ingegno, consiglio & cautela sono d'una significatione istessa & latino industria, ars, ingenium, solertia, diligentia, sedulitas, labor" (Marinelli 1565).[42] We have seen already that *genio* was associated with studiousness. Likewise, *ingenium* was considered to be the site of teachability. It may well be that meaning of *ingenium* that drove Tranchedini to link *ingegno* to learning. His spelling of the adjective *insegnero* (via the archaic variant *inzegno*) brings the word close to *insegnare*: "to teach."[43] Equally, therefore, his definition of *ingegno* as its apparent opposite—*studium*—may reflect Cicero's association of *ingenium* with memory and learning (although Cicero distinguished it sharply from *industria*, as other lexicographers noted).[44] This connection is pronounced in Latin dictionaries, so it is not surprising that the humanist Tranchedini should have made a similar connection, nor indeed that his contemporary Matteo Palmieri could claim that the *arti dingegno* [sic] were flourishing for the first time in a thousand years because of study.[45]

Rather more intriguing, however, is Tranchedini's explicit association of *ingegno* with the mechanical arts, specifically architecture and engineering. His definition of *ingegno* as "machine" was likely informed by the medieval heritage in which this meaning was prominent and which persisted in Italian dictionaries (e.g. Calepino 1553; Fenice 1584), even though Latin lexicographers of the sixteenth century tended to suppress it.[46] This and the obvious etymological relationship doubtless prompted his inclusion of *ingenierus* as a synonym for *ingegnoso*, swiftly followed by *architetus*. Yet the fact that Tranchedini compiled his dictionary in Florence in the 1450s and 1460s (during which time he moved in Medicean circles) suggests another possible reason for these connections: the Vitruvianism of Leon Battista Alberti and his colleagues.

While Alberti's *De re aedificatoria* (1485) was published after Tranchedini composed his *Vocabolario*, the ideas expressed in that treatise had begun to circulate some decades before (Alberti wrote his book ca. 1443–52). Following Vitruvius, Alberti associated *ingenium* with the good judgment required of the architect.[47] Here and in his treatise on painting (*Della pittura*, written 1435/6) he emphasized that while a good artist must possess natural gifts (*ingenium/ingegno*), those gifts must be tempered and brought to fruition through diligence and study, drawing on Vitruvius's maxim "Neque enim ingenium sine disciplina aut disciplina sine ingenio perfectum artificem potest efficere" (Neither talent without training nor training without talent may produce the perfect craftsman).[48] Indeed, Tranchedini's association of *ingegno* with the skillfully honed abilities of the architect and the engineer—as indeed with the products of their crafty proclivities, machines (*machina*)—reflects the rapidly rising socio-intellectual status of mechanicians in mid-fifteenth-century Italy.

Filippo Brunelleschi—celebrated creator of the dome of Florence Cathedral—is a good example. In his lifetime, Brunelleschi was singled out as a man who surpassed even the ancients in his *ingegno*. Alberti, for instance, praised his "ingegno maraviglioso," while "the original plaque over his burial place in Florence Cathedral identified him as a *magni ingenii vir*, and the inscription beneath his more conspicuously honorific portrait in Florence Cathedral states that his excellence in the 'daedalic art' is documented as much by the 'marvelous dome of this most renowned church as by the many machines his divine *ingenium* invented'."[49] Around the same time, Federico da Montefeltro, Count and later Duke of Urbino, honored his court architect Luciano Laurana in a patent explicitly connecting his profession to ingenuity, by way of the mathematical components of the liberal arts: "We deem as worthy of honour and commendation men gifted with ingenuity [*ingenium*] and remarkable skills, and particularly those which have always been prized by both Ancients and Moderns, as has been the skill of architecture, founded upon the arts of arithmetic and geometry, which are the foremost of the seven liberal arts because they depend upon exact certainty. It is an art of great science and ingenuity, and much esteemed and praised by us."[50] Tranchedini echoes these sentiments in binding together *architetus, ingenierus, architeta, operum gnarus* (expert in works) and *liberali* (even if the latter is only in parentheses) in his definition of *ingegno*. Strikingly, however, this association with liberty—whether personal or professional—is entirely absent from all subsequent instances of *ingegno* in our corpus of dictionaries until it resurfaces (and even then obliquely) in the "masterless" capacity to invent of Crusca 1612. Indeed, a notable feature of the Italian lexicographical case is the lack of overlap or confusion between the *ingeni-* and *ingenu-* senses of ingenuity. In Italian dictionaries, the "freeborn" and "liberal" senses of *ingenuità* are kept firmly separate from *ingegno*, whether as headword or in definitions. This is the case even in those entries where we might expect such a connection to be made, such as in the common definition of *ingegno* as "nature" or "natural condition."

This "natural" sense of *ingegno* dominates our corpus throughout the sixteenth century. A typical entry is Alunno 1557, in which *ingegno* is "naturale, la natura" (but also *mente*—a point to which we shall return). Some extended this aspect to define *ingegno* as "natural discourse," such as Venuti 1561: "discorso naturale ... Hoc Ingenium ... Haec Industria p.c. Cicero."[51] Although Venuti cited Cicero by way of authority, it may be that this sense of *ingegno* was informed by the reception of Isidore of Seville's *Etymologiae*, which noted classical accounts of *ingenium* as the source of verbal discourse.[52] More significantly, though, the relationship of *ingegno* to discourse lent itself easily to the most powerful and important definition of the word in our period: the capacity to create, associated initially with poets and men of letters but increasingly with artists of all sorts. This, as we have seen, is the primary definition given in Crusca 1612: "Acutezza d'inventare e ghiribizzare, che che sia, senza maestro o avvertitore."

Immediately, we notice the extent to which this definition deviates from other senses we have encountered thus far, in that it qualifies *ingegno*

1. As a specific capacity to invent (*inventare*) or to create (*ghiribizzare*).
2. That this capacity is apparently boundless (*che che sia*).
3. That it is sharp or pointed (*acutezza*).
4. That it does not require training or prompting (*senza maestro o avvertitore*).

Such specificity is at odds with previous lexica, not least Giacomo Pergamino's *Memoriale* (1602), which was the most extensive monolingual Italian dictionary published prior to the *Vocabolario*. Pergamino comments on the variety of meanings of *ingegno*, observing that it "ha diversi significati: Arte, Astutia. Industria, Inteletto. Natura, Senno, Prudentia." This is a clear hangover from the Latin lexicography, in which *ingenium* was translated in myriad—often contradictory—ways. All of Pergamino's synonyms crop up with greater or lesser frequency in the sixteenth-century monolingual and bilingual Italian dictionaries, such that *ingegno* was readily associated with all aspects of the human mind (*mente*).

The Crusca 1612 definition accommodates some of these senses while departing from others. It is strikingly different to the *ars et ingenium* tradition we encountered in Tranchedini and subsequent lexicographers, in which *ingegno* was associated with study and believed to require art in order to be perfected.[53] By the turn of the sixteenth century, *ingegno* alone had become a sufficient prerequisite for any creative endeavor. This and the other aspects of *ingegno* suggested by the Crusca definition are evident in other words in the European family of ingenuity (notably "sharpness," as per the Spanish *agudeza*), but the Italian case is particular for the sources on which the Accademia's entry rests. To support their interpretation of *ingegno*, the *accademici* quoted what they considered to be representative passages from Boccaccio, Petrarch, and Dante:

> Non ti sento di sì grosso ingegno, che tu non avessi, eccetera [Boccaccio, *Decameron*, II.9]
>
> Si possono da più sublimi ingegni comprendere. [Boccaccio, *Laberinto*, 21]
>
> O Muse, o alto ingegno or m'aiutate. [Dante, *Inferno*, II.7]
>
> Però lo 'ngegno, che sue forze stima, Nell' operazion tutto s'agghiaccia. [Petrarch, *Canzoniere*, XX. 7–8][54]

On the face of it, these quotations do not immediately elucidate the definition, which may explain why they are preceded by an extract from Francesco da Buti's *Commento* (1394/5) on the line from Dante: "Ingegno è una virtù interior d'animo, per la quale l'huomo da se truova quello, che da altri non ha imparato" (*Ingegno* is an internal power of the soul through which someone may find by himself that which he has not learned from others).[55] The truncated way in which Buti is quoted in Crusca 1612 obscures the fact that he based his definition of *ingegno* on the eleventh-century Italian lexicographer Papias—an important source for many early moderns concerned with the nature of ingenuity:

> Ingegno secondo Papia è una virtù interiore d'animo, per la quale l'uomo da sé trova quello che dalli altri non à imparato; e perchè l'autore trovava cose nuove, che mai da altrui non avea imparate, però dice; o alto ingegno, or m'aiutate; cioè aiutate me Dante a componere questo poema. E per questa invocazione si de[v]e intendere essere invocata la grazia di Dio, la quale ministra e dà li nove gradi significati per le muse e per l'ingegno.
>
> *Ingegno*, following Papias, is an internal power of the soul, through which one may himself find that which he has not learned from others; and because the author found new things that he had not learned from others, that is why he says "O high *ingegno*, help me"; that is, help me—Dante—to compose this

poem. And one should understand this invocation as calling on the grace of God, which aids and gives the nine gifts signified by the muses and by *ingegno*.[56]

The first part of Buti's account offers a familiar, technical definition of *ingegno* as a power of the soul; the second implies the Neoplatonic notion of intellectual ascent. The "finding out" aspect expressed the dynamism of human *ingegno*, which in the verb form *ingegnare* was defined variously as "to search with diligence" or "to investigate or scrutinize something."[57] Glossing the same passage from Dante in his later *Comento . . . sopra la Comedia di Dante Alighieri* (1487), Cristoforo Landino defined *ingegno* similarly as a forceful part of the soul. As he explains:

> E latini chiamono ingegno quel forza dell'animo per laquale siamo capaci della doctrine. Et queste ha due parte: una quella acume col quale ciassottigliamo a investigare & imprendere. Et questo gli antichi chiamano docilita & moderni Philosophi apprensiva. La seconda quella virtu per laquale retegnamo quello che habbiamo inteso, & che sta e detta memoria. Adunque o alto ingegno cioe potentia dell'anima apta a conseguire la cognitione delle gran chose. Il perche dixe Augustino che ingegno e quella potentia dell'animo con laquale l'animo saguza & exercita conoscere quello che ancora non cognosceva. Onde e diffinito Quod ingenium sit extentio intellectus ad incognitorum cognitionem. Adunque lo ingegno investiga e la ragione giudica le chose investigate da lo 'ngegno: & la memoria le ripone chosi guidicate.[58]

> The Latins call *ingegno* that force of the soul through which we are capable of learning. And this has two parts: one the acumen with which we discriminate in investigating and comprehending. And this the ancients called "teachability" and the modern philosophers "apprehensive power." The second is that power through which we remember what we have understood, & which is called memory. Therefore, "O alto ingegno" is that capacity of the soul suited to attaining the understanding of great things. Which is why Augustine says that *ingegno* is that capacity of the soul with which the soul sharpens itself and comes to know that which it did not already know. Thus it is defined: "Let *ingenium* be the

> reaching of the intellect towards thinking unknown things." And so the *ingegno* investigates and reason assesses the things investigated by the *ingegno*, and memory retains the things assessed.

Landino's commentary was a widely known source that had been used by previous lexicographers to define *ingegno*, but it was rejected by the Accademia della Crusca in 1612 in favor of Buti's explanation. Yet in the second edition of the dictionary (Crusca 1623), in the only change to the entry for *ingegno*, Landino's "apprehension" replaced "fantasy" (*ghiribizzare*): "Ingegno. Acutezza d'inventare e apprendere, che che sia, senza maestro o avvertitore." It seems likely that in the first edition the *accademici* opted for Buti because he fused the "faculty" sense of *ingegno* with the broader notion of divine gifts. There is, of course, a demiurgic root to this aspect of *ingegno*, which in the sixteenth century led some commentators not only to celebrate the seemingly godlike creative powers given to artists through God's grace, but to call artists themselves *divino* (divine).[59] The latter, given the context, brings the quotation from Dante very close to notions of *furore poetico*, defined by Cesare Ripa in his popular guide to iconography, the *Iconologia*, as: "Una soprabbondanza di vivacità di spiriti, che arricchisce l'anima de numeri, e de concetti meravigliosi, i quali parendo impossibile che si possono havere solo per dono della natura, sono stimati doni particolari, e singolar grazia del Cielo" (A superabundance of liveliness of the spirits, which elevates the soul to the gods, and to wondrous conceits, which it is impossible to have through the gifts of nature alone, it is a particularly revered gift, given from Heaven alone).[60]

In their selection of quotations, the Crusca only implied this sense rather than making it explicit—perhaps owing to the strong association of *ingegno* with "nature" as much as through a desire to avoid the more controversial aspects of the *furor poeticus*.[61] Nevertheless, not only is there a clear sense of the "uplifting" or "elevated" aspects of *ingegno*-as-inspiration in the passage from Dante, but also of Ripa's *concetti meravigliosi* in the *ghiribizzare* of the Crusca definition.[62] This difficult-to-translate term—*ghiribizzare*—has a range of meanings in the period: to invent, to imagine, and to sketch being but a few.⁶³ Notably, *ghiribizzo* appears as a headword in Crusca 1612, under which the reader is referred to the entry for *capriccio*, defined as "pensiero, fantasía, ghíribizzo."⁶⁴ There can be no doubt, then, that in the 1612 definition of *ingegno* we are firmly in the realm of artistic creativity, specifically the artist's power to invent anything—even, presumably, those things not found in nature ("che che sia").

By including this element in their definition, the Crusca academicians drew on the well-established scholastic distinction between animal and human ingenuity. This distinction, evident in Aristotle, was magnified by Aquinas: "every swallow makes a nest in the same way and every spider a web in the same way, which would not be the case if they acted by intellect and art [intellectu et arte]. For not every builder makes a house the same way, because the artisan judges the form of the thing built and can vary it."⁶⁵ A person's capacity to deviate from the telos of invention directed by necessity and to vary his creations was elevated by certain fifteenth- and sixteenth-century theorists to become a boundless capacity to create.⁶⁶ An early example is the architect-engineer Francesco di Giorgio Martini, who contrasted the natural imperatives that drive animals to create "architectural" works with the inventions of the human intellect: "quasi infinite, infinito varia" (almost infinite, of infinite variety).⁶⁷ In the latter part of the sixteenth century and into the early seventeenth, this intellectual power—or *virtù*, an occasional synonym of *ingegno*—underpinned the aesthetics of the marvelous, which in poetry was purveyed most influentially by the *virtuoso* Giambattista Marino.⁶⁸

While the *accademici* della Crusca did not cite Marino or other contemporary poets in 1612, their definition of *ingegno*—soon inflated into the *sine qua non* of poetic invention by Marino's disciple, Emanuele Tesauro—reflects the era's fascination with the untrammeled powers of the imagination to produce things of striking novelty and variety, here given an exemplary medieval pedigree.⁶⁹ Of course, in choosing Dante as their prime example the Crusca not only participated in the consensual view that Dante was the prince of poets, but availed themselves of a notably Florentine example of *ingegno*.⁷⁰ In so doing, they situated themselves in a long tradition of patriotic hero-worship, visual as well as literary. This is perfectly illustrated by an anonymous engraving of ca. 1470 (after Domenico de Michelino's 1465 fresco), depicting Dante as poet of the *Divina Commedia* standing next to the city of Florence—resplendent with the prominent Duomo recently completed by the "new Archimedes," Brunelleschi. Inscribed beneath the image is a motto, quoting from the same passage in the *Inferno* as the Crusca entry: "DANTE ALLEGHIERI POETA FIORENTINO CON ALTO INGEGNO . . . ." Not accidentally, this image combines visually the literary *ingegno* of Florence's most famous poet with the technical *ingegno* of its most celebrated architect-engineer, and where Dante's *ingegno* was "elevated" (*alto*) sufficiently for him to imagine the wonders of heaven, Brunelleschi's was manifest in the tallest edifice of the age: his dome a model of the heavens.

What, though, of the "masterless" aspect of *ingegno* in the Crusca 1612 definition? Dante could hardly be considered a poet led simply by *ingenium* without *industria*.[71] Here, perhaps, we may catch a glimpse of a shadow that hovers over the *Vocabolario* but which is never fully discernable: that of Baldassare Castiglione's *Libro del cortegiano* (1528)

Fig. 2.2. Dante as ingenious poet with the Duomo of Florence (ca. 1470).

and the "courtly language" of the *questione della lingua*. Castiglione was neither acknowledged nor quoted in the Crusca's dictionaries, but the notion that *ingegno* is a capacity to invent "without teaching or prompting" smacks of *sprezzatura*: the seemingly effortless display of supposedly natural talent.[72] Given their obsession with Boccaccio, one wonders whether the *accademici* della Crusca knew the passage in the *Cortegiano* in which Castiglione praises Boccaccio for his "natural" *ingegno*: "anchor che 'l Boccaccio fusse di gentil ingegno, secondo quei tempi, & che in alcuna parte scrivesse con discrettione, et industria, nientedimeno assai meglio scrisse quando si lassò guidar solamente dall'ingegno, et instinto suo naturale, senz'altro studio o cura di limare i scritti suoi, che quando con diligenzia, & fatica si sforzó d'esser piú culto e castigato" (while Boccaccio was of a noble wit, who according to the times and in certain parts [of his oeuvre] wrote with discernment and industry, he never wrote better than when he let himself be guided solely by his wit and by his natural instinct, without any other work or care to polish his writings, than when with diligence and effort he tried to be more cultured and disciplined).[73] In a sense, Castiglione's praise for the "best," most "natural" parts of Boccaccio's oeuvre accords with the definition of *ingegno* as "natural discourse," which in the popular handbook *De arte rhetorica libri tres* (1562) is bound up with the relationship between *ratio* and *oratio*. Its association with the "sharp" aspects of *ingegno*—the *acutezza* of Crusca 1612—is more ambiguous. In the above quotation, Castiglione rejects polish in favor of carelessness, yet as is well known, the *sprezzatura* he advocated was itself archly refined: the province only of the innately talented. Moreover, there is more than a hint of the "sharp wit" of the courtly game—so prominent in the *Cortegiano*—in the *acutezza* of *ingegno*, an aspect that comes clearly into focus when we realize that the secondary definition of the term given in Crusca 1612 is *inganno*: "Per inganno, astuzia, stratagemma."[74]

The word *inganno* is worthy of study in its own right.[75] For our purposes, we should note first that in the Crusca definition *inganno* is not shown to relate to the Latin *ingenium*: they offer instead *dolus* and *astus* as synonyms. Thus, like the English ingenious/ingenuous, *ingegno*'s connection with *inganno* came via etymological fudging.[76] As Ménage put it: "*ingegno* non credo che vaglia altro che *inganno*. I Latini pure de' tempi bassi in tal sentimento usano la voce *ingenium*" (I do not think *ingegno* should be anything but *inganno*. The Latins even in the old times used the word *ingenium* in this sense) (Ménage 1669).[77] This twinning lies along the cognitive axis of ingenuity—the quick thinking (and sharp dealing) that is closer to cunning and subtlety than "apprehension." As early as Anon. 1435–60 and Tranchedini 1455–70, *ingegno* had been connected to subtlety (*subtilitas/solertia*), while the "sharp" aspects of *ingegno* are evident in the entries for *aguzo* and *cima* in Scoppa 1511/15.[78] Not until Marinelli 1565, however, are the morally ambivalent aspects of these links raised. In his entry for *astutia*, Marinelli lists *inganno* and *ingegno* alongside potentially negative terms such as *laccio* (snare) and the decidedly nasty *malitia* (malice).[79] This moral reversibility is the province of "deceit, a guile, a fraude, a cosening tricke," as Florio translated *inganno* in *A Worlde of Words* (1598).[80] Here we have entered into the arena of ancient *métis*: the ruse implicit in a stratagem (military, political, or otherwise), the duplicity of the conman, the insincerity of the wooer, the guile of the diplomat, and the craftiness of the artist.[81] As Paul Barolsky reminds us, this kind of wit has a fast-paced, slippery meaning, difficult to pin down precisely in words. It is the *ingegno* of the quick conceit, the subtle allusion, the indeterminate image, and the false friend, depicted with most fiendish aplomb in Bronzino's London *Allegory* ca. 1545.[82] In the right hands, and for the right audience, such deceits can of course be pleasurable (*geniale*), even beautiful. Hence, in the third edition of the Crusca *Vocabolario* (1691) the entry for *ingegnoso* directs the reader to *bellissimo*, the entry for which is illustrated with quotations from Boccaccio, Petrarch, and the *Cento novelle antiche* about beautiful, enticing, and potentially captivating women and men.[83] These associations affirm that *ingegno* was positioned prominently on the "artificial" axis of ingenuity. Indeed, the adjective *ingegnose* was defined as *artificiose* in Alunno 1539b followed by Crusca 1623.

The "artificial" sense of *ingegno* is especially pronounced in relation to *ingannare* when the latter refers to objects that, through the skill and ingenuity of their maker, sportively deceive the senses, provoking both wonder and curiosity. This ancient, thaumaturgical topos pervaded the aesthetics of the marvelous in the early modern period, the material manifestations of which were legion: from clockwork automata that rivaled the self-moving tripods of Vulcan to paintings so startlingly true to nature that they bested even the grapes of Zeuxis.[84] Such objects were routinely described and praised using the language of ingenuity, but never appear in the dictionary entries for *ingegno*. Nevertheless, one of the most pervasive and stable definitions of *ingegno* was a material one: as machine, device, or instrument.

We find this sense first in Tranchedini, who calls *inzegno* "machinacio; machinamentum; machina," a meaning that persists up to Crusca 1612 and beyond—the fourth definition in the first edition of the *Vocabolario* is "Per instrumento ingegnoso, e, per lo più, si dice di serrature, o da aprir serrature. Latino machina, machinamentum" (As an ingenious instrument, and, mostly, it is said to be locks, or to open locks. Latin *machina, machinamentum*). The Crusca definition highlights colloquial usage in which an *ingegno* was, specifically, a key or part of the mechanism of a lock, stating that this is the most common "device" sense of the term. In support of this assertion we may cite Filippo Baldinucci, whose *Vocabolario toscano dell'arte del disegno* (1681) was a repository of the language of technology and the arts that filled in some of the gaps left by the more literary Crusca. In the entry for *ingegno* (interestingly, *genio* does not appear in the book as a headword) Baldinucci offers first a standard definition: "Una certa forza da natura in noi inserta, per ritrovar tutto ciò, che si può con la ragione giudicare" (A certain power of nature placed within us, for discovering anything, which can then be judged by reason). Where we might expect him to follow this with "artistic" examples of *ingegno*-as-artefact (paintings, sculptures, etc.), he does not.[85] Instead, he tells us that contemporary artisans use *ingegno* to refer specifically to a particular device: "E ingegno dicono i nostri Artefici quel pezzo di ferro, per lo più di forma quadra, intaccato o traforato, che appiccandosi alla chiave, e immediatamente passando per altri ferri (che sono appiccati alla toppa) che pure anche essi si dicono ingegni, fa l'uficio di aprire e serrare"

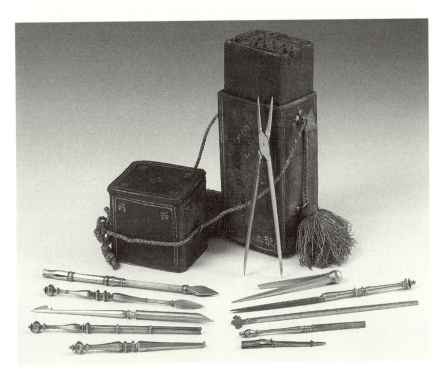

Fig. 2.3. A *stuccio* of Italian drawing instruments, early seventeenth century.

(And our artisans call an *ingegno* that piece of metal, normally in the shape of a square, notched or perforated, which, when fitted to a key, and instantly passing through other bits of metal (which are attached to the keyhole), which are also themselves called *ingegni*, works to open and lock [a mechanism]). Hence, *ingegno*-as-artefact tended to refer to small, mechanical objects or tools: other dictionaries define it simply as *ordegno* (contraption; Pergamino 1602), *attifici[o]* (artefact or artifice; Calepino 1553), and *instrumento* (instrument; Politi 1614).[86] In parity with other senses of *ingegno*, the case used by architect-engineers to carry their mathematical and drawing instruments was called a *stuccio*: an abbreviation of *astuccio*, which is close to the *astutia* or "acutezza" of ingenuity.[87] That the implements contained in these cases—compasses, styluses, and scissors—were sharp neatly reflected the acumen of their possessors: the *ingegneri*, considered to be prime purveyors of *ingegno*. These are the men who best exemplified the indeterminate position of *ingegno* between the mind and the hand: drawing on their natural talent, honed by study, they used instruments to invent (i.e. design and manufacture) engines and artefacts as useful for the stratagems of war as they were pleasing to the *otium* of peace.[88]

Ingegno, n.
    1. Industry, study, diligence.
    Tranchedini 1455–70 (*inzegno*) >.
    GDLI, 3: "Zelo, cura, premura, diligenza; applicazione intelletuale"; 4: " . . . Studio, disciplina, dottrina."
    2. Art, artfulness, skill.
    Tranchedini 1455–70 (*inzegno*) >; Ruscelli 1588 (*ingegno*) >.
    GDLI, 9: "Arte, attività; opera, lavoro."
    3. Man-made contrivance.
        a. Machine, engine.
        Tranchedini 1455–70 (*inzegno*) >.
        GDLI, 11: "Disus[are]. Strumento, arnese; macchina, ordigno; contegno, meccanismo. — In partic[olare] Antico: macchina da guerra, macchina da getto."
        b. Instrument, contraption.
        Calepino 1553 >.
        GDLI, 11.
        c. Artifice, invention.
        Calepino 1553 >.
        Not in GDLI, but connected to 11.
        d. Lock, key.
        Brocardo 1558 (*ingegnosa*) >.
        GDLI, 12: "Parte della chiave perpendicolare al fusto, munita di scontri e a profilo variabile, che viene introdotta nella toppa della serratura per metterne in azione il meccanismo e provocare anche il movimento del chiavistello."
        e. Caprice, oddity, fantastic invention.
        Ménage 1669 (*arzigogolo*).
        Not in GDLI.
    4. Nature, natural condition.

a. Nature.

Scoppa 1511/15 >.

GDLI, 6: "Disposizione naturale; propenzione, attitudine, inclinazione"; 7: "L'insieme delle qualità innate, delle caratteristiche psichiche, delle inclinazioni, delle tendenze spirituali di una persona; natura, indole, temperamento."

5. Argument.

Scobar 1519 (*ingegnu*) >.

Not in GDLI.

6. The mind, intellect.

Alunno 1539b >.

GDLI, 4: "Mente, intelletto, intendimento, senno. — Anche: animo, pensiero, spirito, cuore."

a. Natural discourse.

Venuti, 1561 >.

Not in GDLI.

b. Sharpness, acuity.

Crusca 1612 >.

GDLI, 1: "L'insieme delle qualità intellettuali e delle facoltà naturali della mente (considerate per lo più dal punto di vista della loro forza e acutezza); intelligenza — In partic.: facoltà dello spirito di intuire, apprendere, penetrare, giudicare le cose con prontezza e perspicacia"; 2: "Potenza creatrice dello spirito umano che costituisce la massima espressione del talento e dell'intelligenza; capacità creativa, ispirazione artistica o poetica; genio."

c. A force of the soul or spirit that enables learning and invention.

Crusca 1612 >.

GDLI, 1.

d. Judgment.

Franciosini 1620 >.

GDLI, 1.

7. Practical intelligence.

a. Astuteness.

Calepino 1553 >.

GDLI, 10: "Inganno, frode, astuzia; stratagemma; intrigo, trama; raggiro, imbroglio."

b. Guile, deceit.

Crusca 1612 >.

GDLI, 10.

c. Stratagem.

Crusca 1612 >.

GDLI, 10.

d. Prudence.

Marinelli 1565 >.

Not in GDLI.

8. Sense.

Marinelli 1565 >.

GDLI, 4.

9. Power, force (*virtù*).

Marinelli 1565.

GDLI, 2: ". . . Valore, virtù, capacità innata."

Conclusion

Although it was widely acknowledged throughout the early modern period that engineers—practical men, well versed in the mathematical arts—exemplified *ingegno*, this connection was suppressed in many of the dictionaries of our corpus. As we have seen, in the mid-fifteenth century Tranchedini reflected humanists' interest in rehabilitating the mathematical arts and elevating mechanical aspects of *ingegno* to liberal status. Yet while this process accelerated throughout the sixteenth century and continued into the seventeenth, the *questione della lingua* pushed *ingegno*—in the hands of lexicographers—away from the wider culture of ingenuity in theory and practice. In particular, the erudite approach to lexicography exemplified by the Accademia della Crusca enshrined a poetic sense of *ingegno*, the examples of which dwelt on literary rather than mechanical (or even natural philosophical) aspects of invention. Thus, just at the moment when mechanical artists of all kinds were officially being granted liberal status—joining and founding academies, gaining courtly freedom, writing and publishing on their arts—the *ingegno* they possessed and professed was being defined according to late-medieval norms. It was left to the polyglot dictionary-makers to capture properly the semantic range of *ingegno* and the importance of mechanical artists in its early modern rise to prominence, as the entries in Torriano's adaptation of Florio's dictionaries show:

> *Ingégno*, The nature or inclination of any man, but chiefly used for wit, engine, art, skill, cunning, knowledge, discretion, understanding or sagacity. Also any kind of engine, machine, frame or work-house, where sugar is made, a sugar-house, or any other water-works, also any tool or implement.
>
> *Ingegniére*, an Engineer, a deviser or maker of engines, machins, or witty structures. A man skillful in fortifications or buildings, also a maker of tools or instruments.[89]

3 Spanish

In the first chapter of Miguel de Cervantes's novel *Don Quijote* a memorable passage sees the main character, Alonso Quijano, getting ready for his chivalric adventures by inventing a new name for himself and his horse, and deciding who his love interest would be. But the first thing the soon-to-be knight-errant is concerned about is his weapons:

> And first of all he caused certaine olde rusty armes to be scoured, that belonged to his great Grand-father, and lay many ages neglected, and forgotten in a by-corner of his house; he trimmed them and dressed them the best he mought, and then perceived a great defect they had; for they wanted an helmet, and had only a plain morrion: but he by his industry [*industria*] supplied that want, and framed with certaine papers pasted together, a Beaver for his Morrion. True it is, that to make tryall whether his pasted Beaver was strong enough, and might abide the adventure of a blow, he out with his sword, and gave it a blow or two, and with the very first did quite undoe his whole weeks labour: the facility wherewithall it was dissolved liked him nothing; wherefore to assure himselfe better the next time from the like danger, he made it

anew, placing certaine iron barres within it, in so artificiall manner, as he rested at once satisfied, both with his invention, and also the sollidity of the worke; and without making a second tryall, he deputed and held it in estimation of a most excellent Beaver.[1]

As scholarship on *Don Quijote* has shown, an important aspect of the comic element in this passage is the fact that Quijano uses "certaine papers pasted together"—the equivalent of what we would today describe as *papier-mâché*—to create a new beaver for his helmet. It has been argued that besides its carnivalesque resonances this passage is interesting because it reveals an often overlooked feature of its protagonist: his manual dexterity, here expressed in terms of *industria*.[2] Associated with the kind of practical ability and workmanship possessed by artists, artisans, and engineers, *industria* was an important term in the ingenuity word family in Spanish at the time of Cervantes's writing. This association would not have gone unnoticed to the novel's early modern readership. Notably, in his rather free English translation of 1687, John Philips adds the word *ingenuity* to amplify this sense where it is absent in the original: "However his Industry and *Ingenuity* supplied that defect, by pasting together several pieces of Brown Paper; of which he made himself a most complete Vizor to defend his Nose and Eyes. This Invention pleased him wonderfully . . . which he did so artificially, as if he had been *Tubal-Cain* himself, that now fully satisfied with the strength and sufficiency of his Workmanship, he resolved to confide in his *Ingenuity*, without any farther Trial."[3] In conjunction with other, better-known references to more salient aspects of Quijano's personality—e.g. his exalted imagination and his temperament, well studied from the perspective of early modern medicine and faculty psychology—this seemingly casual but certainly explicit allusion to his dexterity bears testimony to Cervantes's cunning ability to assemble a wide range of ingenuity-related attributes under one single character: the *ingenioso* Don Quijote. Never in the novel is this epithet explained or justified, and rightly so. A source of fascination for readers of all periods, and a challenge to translators and interpreters, past and present, this most distinctive feature remains one of the most brilliant articulations of the early modern culture of ingenuity in the Spanish language.[4]

EL INGENIOSO
HIDALGO DON QVI-
xote de la Mancha.

Compuesto por Miguel de Ceruantes Saauedra.

DIRIGIDO AL DVQVE DE
Bejar, Marques de Gibraleon, Conde de Benalcaçar, y
Bañares, Vizconde de la Puebla de Alcozer, Señor
de las villas de Capilla, Curiel,
y Burguillos.

...resso con licencia, en Valencia, en casa de
Pedro Patricio Mey, 1 6 0 5.

a costa de Iusepe Ferrer mercader de libros,
delante la Diputacion.

Fig. 3.1. Title page of Cervantes's *El ingenioso hidalgo Don Quijote de la Mancha* (Valencia, 1605).

The Hispanic context is rich in other captivating cases. A quick survey of period sources reveals a profusion of ingenuity-related terms and expressions used in all sorts of settings: from the literary products of Spain's and the New World's most celebrated *ingenios* to countless prefaces, sermons, proverbs, nicknames, and jokes. The presence of such language has long attracted the attention of scholars of the Hispanic world, not least because two of the most influential contributors to the early modern debate on ingenuity were Spanish and wrote in Spanish. On the one hand, the physician Juan Huarte de San Juan, author of the widely disseminated *Examen de ingenios para las ciencias* (1575) — a key treatise to understand Don Quijote's ingenuity.[5] And, on the other hand, the Jesuit scholar Baltasar Gracián, the most sophisticated Spanish theorist of *ingenio*.[6] Moreover, the ingenious language of *agudeza* was said to be the specialty of the Spanish, according to several early modern authors, including Gracián.[7] This view was reinforced by the fact that Martial, the famous Roman epigrammist, had been born in the Iberian Peninsula. In dialogue with this cultural backdrop, the ingenuity word family in early modern Spanish is rich and nuanced. But when it comes to determining its keywords, two terms and their cognates stand above the others: *ingenio* and *agudeza*.

Keywords

Ingenio and *agudeza* have received extensive scholarly attention, particularly in the context of the study of major literary figures of the so-called Spanish Golden Age such as Cervantes, Lope de Vega, Góngora, Quevedo, Sor Juana, and Gracián.[8] Of the two terms, *ingenio* (including variants such as *engeño*, *engenio*, or *yngenio*) is the most significant and versatile.[9] Derived from the Latin *ingenium*, the word is associated early on with a series of attributes and abilities with which a person is endowed, including the capacity for invention, understanding, dexterity, or skill in contriving. Depending on the lexicographical sources, these attributes are described differently. Some dictionary entries are minimal, to the point of being almost uninformative. In the more prolix dictionaries, however, the definitions can be rather expansive — capturing aspects of the word's range of meaning normally discussed in more technical literature, such as the philosophical and medical treatises in the tradition of Juan Luis Vives and the above-mentioned Huarte de San Juan.[10] Also, in a manner similar to the original Latin form, *ingenio* is said to equate to an individual's nature in the sense of natural disposition and inclination—a register that resonates with the use of the Spanish term *natural* as an equivalent of *ingenio* in, for example, treatises on poetics, where it is discussed in relation to *ars/arte*.[11] In general these are positive, or at least neutral, qualities, although certain cognates and neighboring terms may sometimes have negative connotations: when *ingenio* as dexterity is understood in terms of deceptive craftiness, for example. Additionally, *ingenio* is shown to denote the products (material or immaterial) of the aforesaid capacities, as in the case of the machine-like devices used in warfare, one of the earliest uses of the word on record, amply studied by experts in military history as well as historians of technology and engineering.[12] Finally, the widespread metonymic use of *ingenio* to denote an individual is also registered in the sources, although, rather surprisingly, only as late as the early eighteenth century.

It is important to note that, throughout our timeline, this set of interconnected meanings remains largely stable, although certain semantic modulations and displacements can be detected. One worth mentioning is the gradual loss of currency of *ingenio* as natural inclination, a sense that, in turn, becomes increasingly associated with the term *genio*. Such displacement is recorded early in the lexicographical sources, although much more emphasis is given to the antiquarian meaning of *genio* as a deity or spirit, or, in a Christianized sense, as guardian angel. This shift, and the fact that none of the Spanish lexica explicitly record the sense of special and extraordinary talent associated with the Romantic notion of genius, seems like a good justification to regard *genio* as an important neighboring term rather than a keyword in the ingenuity word family in Spanish.[13] Interestingly, it is often via the entries devoted to neighboring terms, rather than in the definitions of *ingenio* and its cognates, that suggestive variations are made apparent. Another good example is offered by the term *maña*, which together with *habilidad*, *industria*, *artificio*, or *engaño*, serves to highlight the positive and negative aspects of *ingenio* as practical ability. But *ingenio* always stands as the guiding semantic thread connecting all these varying senses. In this regard, the main characteristics of *ingenio* as a keyword are two: its capacity to accommodate a number of interconnected meanings and its relative stability.

Before turning to *agudeza*, it is interesting to note the easy *convivencia* in the lexica between *ingenio* and its cognates and the set of words associated with the Latin *ingenuitas* and *ingenuus*, which often appear in close proximity in the sources.[14] In fact, such Spanish terms as *ingenuidad*—which might have led to a confusing overlap of senses—only appear as headwords in the eighteenth century, despite having enjoyed prominence, several decades earlier, in the debates on the status of painting as a liberal art, for instance.[15] Instead, the most frequently used Spanish equivalents of *ingenuitas* are *libertad* (freedom), *nobleza* (nobility), and, interestingly, *hidalguía*, which immediately brings to mind the pairing *ingenioso-hidalgo* famously associated with Don Quijote.[16] As this last reference to Cervantes's novel reminds us, it is worth emphasizing the important social and moral dimensions linked to *ingenio* and its attributes in relation to *ingenuitas*, which may be traced back to the range of meanings of *ingenium* and *iudicium* associated with an individual's moral character and behavior.[17] These considerations are critical to our understanding of early modern Spanish culture and society. First, at a "local" level, e.g. Gracián's thought as articulated in other publications in addition to those specifically devoted to the art of *ingenio*, like *El Héroe* (1637), *El Político* (1640), *El Discreto* (1646), *Oráculo manual y arte de prudencia* (1647), and *El Criticón* (1651–1657).[18] Second, at an "imperial" level, as in the case of the significance of *ingenio* in the context of the debates on the anthropological status of New World inhabitants and their cultural products.[19]

Our second keyword is *agudeza*. Centered on the familiar notion that the piercing qualities of a sharp object can be equated to a range of attributes in a person, such as perspicacity and shrewdness, the term has many elements in common with *ingenio*. In fact, its most prevalent occurrence in the Spanish dictionaries is as a qualifier, as in the expression *agudeza de ingenio* (sharpness of wit). Both keywords also share associations with quickness and conciseness, best captured by the sense of *agudeza* as a witty saying (*una agudeza*). *Agudeza*'s semantic range, however, is much narrower, according to the lexicographical sources. This makes it a much more precise term, and hence less subject to inflections. In the dictionaries this often results in rather succinct entries, which hardly see variations across our timeline. Such pithiness contrasts with the much more sophisticated use of *agudeza* in other contexts—the work of Gracián being, again, a case in point.[20] However, as in the case of *ingenio*, *agudeza*'s semantic stability in the lexica is occasionally punctuated by the definitions of neighboring terms, which add these missing textures to its meaning. A late but interesting example, as we will see, is the neighboring term *concepto*, here discarded as a keyword given its scarce occurrence (in its ingenuity-related sense) in the lexicographical sources, but whose relevance for the study of the European language of ingenuity Mercedes Blanco has superbly shown.[21]

Lexicographical Landscape

The development of lexicography in the early modern Spanish context tends to be outlined on the basis of three major landmarks. First, the publication of Elio Antonio de Nebrija's bilingual dictionaries (Latin-Spanish, 1492, and Spanish-Latin, ca. 1495), usually regarded as the starting point of Hispanic lexicography with Spanish as the first language. Second, the publication of the first monolingual dictionary in Spanish: Sebastián de Covarrubias's *Tesoro de la lengua castellana, o española* (1611). And, third, the publication of the multi-volume dictionary of the Real Academia Española—*Diccionario de la lengua castellana*, also known as *Diccionario de autoridades*—between 1726 and 1739.[22] While following this narrative closely, our survey considers a wider range of lexicographical materials, as the aim is to give preference to those sources where the language of ingenuity features most suggestively.[23]

Of the three earliest dictionaries in our survey, all published in the last decade of the fifteenth century, Alfonso Fernández de Palencia's *Universal vocabulario en latín y en romance* (1490) stands out as the most profuse and interesting.[24] Largely based on the *Elementarium doctrinae rudimentum* by the eleventh-century Italian lexicographer Papias, this Latin-Spanish dictionary sits somehow ambivalently between the medieval lexicographical tradition and the renewed interest in language and classical sources promoted by the early humanist culture. Crucially, an important selection of ingenuity-related entries is recorded there—a fact that has not gone unnoticed by scholars working on Spanish *ingenio*—so this work deservedly constitutes our point of departure, chronologically and also thematically.

The richness in Fernández de Palencia's *Universal vocabulario* contrasts with the concision that characterizes Nebrija's Latin-Spanish and Spanish-Latin dictionaries, published just a few years later.[25] Appreciated for its innovative format—concise entries, absence of quotations, uniform structure—the first of these works, conventionally referred to as the *Diccionario latino-español* in modern scholarship, enjoyed considerable editorial success and was soon adopted as an effective lexicographical model. Followed by the so-called *Vocabulario español-latino* of around 1495—the first dictionary to feature Spanish as the starting language—this and other works by Nebrija, including his ground-breaking treatise on Spanish grammar (1492), turned him into one of the most prominent contributors to the blend of political reconfiguration and philological activism that marked the early stages of European humanism.[26]

Designed to fulfill a didactic aim, both Nebrija's *Diccionario* and his *Vocabulario* were particularly successful in establishing precise lexical associations between Latin terms and their equivalents in Spanish. This is, for instance, the case with regard to the ingenuity word family, for which these works provide a sort of lexical backbone. However, the model set by Nebrija—brief entries based on word-to-word equivalences rather than explanatory definitions—raises a number of challenges, including the problem of determining the semantic reach of ingenuity-related terms whose transition from Latin to vernacular left them almost unchanged, e.g. *ingenium* > *ingenio*.[27] It is important not to be misled by the almost perfect overlap, in terms of usage and semantic range, that this kind of isomorphism seems to imply. Actually, the precise semantic specificities of such terms are difficult to ascertain based on the dictionary entries alone, given their somehow tautological character. It is an irony, then, having to acknowledge the quasi-labyrinthic circularity of the work of this Daedalus of Spanish lexicography—who, incidentally, also cared to include the entry for *logodaedalus* in his work.[28] This, in fact, can be said to be one of the key features—not exclusive to the Spanish context—of the lexicographical landscape in this period. Furthermore, many bilingual and polyglot dictionaries published throughout the sixteenth and seventeenth centuries pose a similar challenge.[29] Nevertheless, a wide selection of such lexica has been included in our survey, as a means of tracing the trajectory of the Spanish ingenuity family across our timeline.

The case of Covarrubias's *Tesoro de la lengua castellana, o española* (1611) is very different, given its monolingual nature. As Manuel Alvar Ezquerra has put it, the journey leading to this lexicographical landmark was a long one.[30] On the one hand, we should consider the strong program of systematization and promotion of the European vernacular languages that was developed throughout the sixteenth century, which in the Spanish context owed its impetus largely to Nebrija's philological legacy. But, on the other hand, we should also take into account the interest in etymology and the debates on the origins of languages that sprang out of this pan-European lexicographical culture. Indeed, Covarrubias's *Tesoro* must be considered as part of a decades-long investigation into the genesis and constitution of the Spanish language, which saw the publication of, among other works, Bernardo José Aldrete's *Del origen y principio de la lengua castellana o romance que hoy se usa en España* (1606), where the Latinate origins of Spanish are defended. More specifically, it should be noted that the *Tesoro* was primarily conceived as an etymological dictionary. In fact, it is well known that the book's original title was *Etimologías de la lengua castellana*, in reference, perhaps, to Isidore of Seville's celebrated work. As Covarrubias himself suggests in his address to the reader, he adopted the term *tesoro* to emulate the titles of some of the most important dictionaries at the time.[31] In any case, his became the first extensive etymological dictionary to be published in Spanish, considering that its most notable antecedent, Francisco del Rosal's *Origen y etimología de todos los vocablos originales de la lengua castellana*, was ready for printing in 1601 but was left unpublished.

Written and published in a moment of extraordinary cultural activity, Covarrubias's *Tesoro* constitutes a privileged record of the currency and usage of many ingenuity-related expressions in the language of the period. Much of its value as a resource has also to do with its etymological drive—etymology being at the time thought by some to be powered mainly by ingenuity.[32] In this regard, its status as one of the most important sources for our study is unquestionable. Its chronology, and its editorial fortunes, however, create an important void in our survey: first published in 1611, the *Tesoro* did not see a second edition until 1674, when an expanded version—with no added entries relevant to the ingenuity word family—appeared.[33] In the following years, no further editions of Covarrubias's work would be published. In fact, no other monolingual dictionary in Spanish would appear until the publication of the *Diccionario de autoridades* by the Real Academia Española in the early decades of the eighteenth century. Our study has tried to fill this gap by considering other lexica, mostly bilingual dictionaries of the second half of the seventeenth century and the first half of the eighteenth. This sense of vacuum that marks the course of Spanish lexicography in this period can be cited as the second key feature of the lexicographical context in Spanish.

In fact, historians of lexicography often refer to the sense of patriotic shame that motivated the members of the Real Academia Española to write the *Diccionario de autoridades*.[34] In 1726, when the first of its four volumes appeared, Covarrubias's *Tesoro* was more than a hundred years old. Despite its limitations, this work had deservedly won praise both in Spain and abroad—so the academicians state in a brief *History of the Real Academia* published at the beginning of their dictionary. However, "because ingenuity finds it easy to improve what others have invented already," they add, "the French, Italians, English and Portuguese have enriched their nations and languages with superb dictionaries, whereas we have lived with the glory of being the first, and the embarrassment of not being the best."[35]

Inspired by the models of the Italian Accademia della Crusca (founded in 1582) and, more importantly, the French Académie française (1635), the Real Academia Española (founded in 1713, granted royal support in 1714) was established first and foremost to produce a "copious and exact dictionary," a work intended to capture the "greatness and power of the [Spanish] language" and "the beauty and fecundity of its voices."[36] While the use of a large cast of *autoridades* is a feature shared with the Accademia della Crusca's *Vocabolario della lingua italiana* (1612), the Academia's dictionary was primarily modeled after late seventeenth- and early eighteenth-century French lexicography: the *Dictionnaire* of the Académie, in particular, but also the dictionaries of Richelet, Furetière, and Trévoux.[37] Interestingly, the methodological guidelines produced by the Academia stipulated that when it came to determining the *autoridad* of an entry, the use of dictionaries, except for a few cases, should not be allowed. This was because using vocabularies as sources might bring to memory words that had long been forgotten.[38] Thus, the *Diccionario de autoridades* was mostly concerned with capturing uses of Spanish when the language was at its prime.[39] This ethos is best expressed by the Academia's emblem: a crucible over the fire, with the motto *Limpia, fija, y da esplendor* (Cleans, sets, and casts splendour): a reference to the purifying effects of solid lexicographical work.[40] Exhibiting the systematicity and comprehensiveness of a large academic dictionary, the *Diccionario de autoridades* constitutes a powerful tool for the analysis of the ingenuity word family in Spanish. But it is important not to forget the circumstances of its creation—an Enlightened context marked by Neoclassical tastes and a nationalistic drive—and the effect that these might have had on the way the language of *ingenio* (and its *autoridades*) ended up featuring in it.[41]

Ingenio

As indicated above, the main characteristics of *ingenio* as an ingenuity keyword in Spanish are its semantic capaciousness and its stability in terms of usage and currency. Both traits feature in the following list, which summarizes the semantic reach of *ingenio* across our timeline, as recorded in our survey of lexicographical sources.

Ingenio, n.
> 1. Attribute or capacity possessed by an individual.
>> a. Capacity associated with intellectual prowess and invention.
>> Fernández de Palencia 1490 >.
>> b. Capacity to contrive or make something; ability, dexterity, sometimes associated with craftiness and trickery.
>> Fernández de Palencia 1490 >.
>> c. Natural wisdom, prudence.
>> Fernández de Palencia 1490 ("natural sabidoria"); Fernández de Santaella 1499 ("prudencia," headword *ingenium*).
>> d. Subtlety, mostly in a positive sense.
>> Fernández de Santaella 1499 (headword *ingenium*) >.
>> e. Pleasant, delightful.
>> Fernández de Palencia, 1490 (headword *genialis*).
> 2. Product or outcome.
>> a. A machine designed for a specific purpose, e.g. warfare or construction.
>> Fernández de Palencia 1490 (headword *mechanica*) >.
>> b. A device or contrivance that facilitates the execution of a challenging task.
>> Covarrubias 1611 >.
>> c. A scheme, a plan.
>> Covarrubias 1611 >.
> 3. Nature, natural condition, natural inclination.
> Fernández de Palencia 1490 (headword *natura*) >.
> 4. An ingenious individual.
> *Diccionario de autoridades* 1726–1739.

The accommodating nature of *ingenio* is recorded early in the lexicographical tradition. Differing in format and rationale, all the earliest dictionaries in our survey—Fernández de Palencia 1490, Nebrija 1492 and ca. 1495, and Fernández de Santaella 1499—provide multiple entries on this keyword or at least register its semantic reach. Of the four sources, Fernández de Palencia's *Universal vocabulario* stands out as the most profuse and suggestive.

Worth mentioning first is the double definition of *ingenioso* and *ingenio* (headword *ingenioso*) in terms of "capacity to produce an artifice" and "inner force linked to invention," respectively: "Ingenioso. se dize por que tenga dentro fuerça o vigor de engendrar algun artificio. Ca ingenio es fuerça interior del animo con que muchas vezes inventamos lo que de otri no aprendimos" (Ingenioso. It is said of whoever possesses the inner force or power to produce an artifice. Ingenio is a power within the soul with which we often invent things not learned from others). Complementing this important, and often-quoted, record of the pairings *ingenio*/artifice and *ingenio*/invention—which Fernández de Palencia copies from Papias—the entry registers the association of *ingenio* with nature or the natural, e.g. in the sub-definition of *ingenio* as natural wisdom, and by referring to its inner/en-gendered qualities (*dentro engendrado*). These allusions are supplemented by two quotations from Sallust, and a brief note on the use of *ingenium* by the ancients (again, as nature). Further references in other entries accentuate or expand these points, e.g. Vulcan's works are said to surpass human *ingenio* (headword *Vulcanus*), and Daedalus is characterized as *artificioso* (headword *Daedalus*). There is even a hint to specialness in the entry for *geniolus*, where the verb *resplandecer* (to shine) is used to refer to a person that excels in *ingenio*.[42]

The sense of semantic range recorded in Nebrija's Latin-Spanish and Spanish-Latin dictionaries is rather different, largely due to the succinct format of their entries. In the 1492 *Diccionario latino-español*, *ingenio natural* appears as one of the definitions of the Latin *ingenium*, together with the terms *naturaleza* (nature) and *condición natural* (natural condition)—each definition featuring as a separate entry. Additionally, the variant *engeño* is used several times in the *ingenio*-as-machine sense, for example in the entry for *machina* (cf. also *machinator*, defined as *engeñero*). The *Vocabulario español-latino* of ca. 1495 provides further evidence of the coexistence of these two variants, *ingenio* and *engeño*, by having different—but largely overlapping—entries for each:

Engeño para combatir: machina.
Engeño para edificar: machina.
Engeño naturaleza: ingenium.
Engeñoso deste ingenio: ingeniosus.
. . .
Ingenio fuerça natural: ingenium.
Ingenioso cosa de ingenio: ingeniosus.

Finally, our last example of how *ingenio* can be associated with a wide range of meanings is offered by the *Vocabularium ecclesiasticum* of Rodrigo Fernández de Santaella, published in Seville in 1499.[43] According to this popular Latin-Spanish dictionary, originally intended as a pedagogical manual for the clergy, *ingenium*, defined as *ingenio*, is sometimes understood as *entendimiento* (understanding), *sotileza* (subtlety), *prudencia* (prudence), *habilidad* (ability), or *inclinacion* (inclination). Despite its emphasis on the polysemous *ingenium* rather than *ingenio*, this entry successfully acknowledges the primacy of the latter as the former's most immediate—and hence most versatile—equivalent.

Mainly through the influence of Nebrija's dictionaries, the trajectory of *ingenio* as the dominating keyword continued throughout the sixteenth century.[44] Indeed, when Jean Steelsius published his 1545 Antwerp edition of Nebrija's *Diccionario latino-español*, its selection of ingenuity-related words and their associations remained practically unchanged.[45] A survey of other Spanish-Latin lexica published throughout the second half of the sixteenth century—Barrientos 1570, Sánchez de la Ballesta 1587, or Bravo 1599—shows no significant variations either, although it is worth noting the consolidation of *abilidad* [sic] (ability) as one of the Spanish equivalents of *ingenium*—in an unspecific but practice-oriented sense.[46] This is corroborated, for instance, by the occurrence of *habilidad* in the subtitle of Huarte de San Juan's *Examen de ingenios para las ciencias*, where the term is used as a general capacity linked to the natural talent of an individual:

> *Examen de ingenios, para las sciencias. Donde se muestra la differencia de habilidades que hay en los hombres, y el genero de letras que a cada*

uno responde en particular. Es obra donde el que leyere con attencion hallara la manera de su ingenio, y sabra escoger la sciencia en que mas a de aprouechar: y si por ventura la hubiere ya professado, entendera si atino a la que pedía su habilidad natural.[47]

(Examination of wits, according to professions. In which the range of human capacities are described, together with their corresponding areas of learning. Whoever reads this work attentively will comprehend the nature of their wit, and so will be able to choose the most appropriate profession; and if, by any chance, they practiced a given profession already, they will figure out whether it suited their natural talent.)

• • •

The primacy of *ingenio* within the word family of ingenuity in Spanish is less apparent if we examine the terms that appear in other lexicographical compilations of the period. Let us consider, first, the multilingual vocabularies that were included in the very popular guides to learning foreign languages. The ingenuity word family, not just in Spanish but in any language, is variously present in these compilations, whose contents range from body parts to household items. For example, the 1533 edition of the *Quinque linguarum*, one of the first of such multilingual guidebooks to feature Spanish since it was first published by Franciscus Garonus in 1526, does not offer much besides a reference to *entendimiento* (headword *intellectus*) and *inclination* [sic] (headword *inclinatio*).[48] Similarly, the 1556 edition of *Dictionarium quatuor linguarum* published by the printer Bartholomeus Gravius—an extended version of Noël de Berlaimont's successful Flemish-French *Vocabulaire*, which would feature up to eight languages by 1585—records terms such as *abilidad, astucia*, or *agudo*, but does not provide any context or analysis.[49] Published a few years later, though, another version of Berlaimont's *Vocabulaire*, the *Dictionaire, colloques, ou dialogues en quatre langues* of 1565—featuring Flemish, French, Spanish, and Italian—does register *ingenioso* (headword *constich*) as well as *sotil* (headword *scalck*); but, again, the precise semantic specificities of these terms are difficult to ascertain.[50]

Scarcity and pithiness also seem to be the norm when we look at the large multilingual dictionaries of the period. The 1590 Basel edition of Ambrogio Calepino's *Dictionarium*, featuring eleven languages, includes entries on *ingenium* (*naturaleza, o ingenio natural*, in Spanish) and *ingeniosus* (*ingenioso, y agudo de ingenio*); but apart from acknowledging the continuing stability of these lexical equivalences we can hardly extract any more information from them.[51] At best there are instances of semantic variation, as, for example, in the *Nomenclator octilinguis* (Paris, 1606)—an early seventeenth-century edition of Hadrianus Junius' *Nomenclator* (first edition 1567)—where *ingenio* features in the entry for *mechanicus* (*artesano ingenioso*), denoting skillfulness or craftsmanship, and in the entry for *trochlea* (*carillo o polea engenio*), to designate a mechanical contraption.[52]

The vernacular-oriented bilingual or multilingual dictionaries of the period offer a similar, albeit slightly richer, account. In Italian *ingegno* stands for *ingenio*, as in Cristóbal de las Casas's *Vocabulario de las dos lenguas toscana y castellana* (1570), the first Spanish-Italian dictionary published in Spain. In English, *ingenio* is shown to be equivalent to *witte* or *ingenium*, whereas *engeño* is used for *engine* or *machina*, according to Richard Percivale's *Bibliotheca hispanica* (1591), which includes the first extensive Spanish-English (and Latin) dictionary of the early modern period. And with respect to French, *ingenio* stands for *engin* and *l'esprit*, or *engin* in the sense of machine, according to the influential *Recueil de dictionaires francoys, espaignolz et latins* (1599) published by the Flemish scholar Henricus Hornkens—a dual definition that would be preserved and amplified by later Spanish-French dictionaries like Palet's *Diccionario muy copioso de la lengua española y francesa* (1604) and Cesar Oudin's *Tesoro de las dos lenguas francesa y española* (1607).[53] Interestingly, some of these late sixteenth-century lexica record a number of adjectival variations, as in Hornkens's entry for *esprit*: "*esprit fort aigu; ingenio que transciende; ingenium penetrans*," "*gentils esprits; lindos ingenios; macti ingenio*," "*esprit triste; ingenio mustio; triste ingenium*," "*grans esprits; ingenios refinados y sublimados; ingenia summa*," and "*lourd esprit; tosco, torpe ingenio; retusum, hebes ingenium*." Similarly, John Minsheu's 1599 expanded version of Percivale's dictionary provides some longer, nuanced entries, as is the case of *engeño*: "*Engeño*, or *Engenio*, an engine or cunning instrument, a craftie or politike invention, a scaffold in building, a framing or workemanship, a place over the scaffold whence some Angell appeared."⁵⁴

In sum, throughout the sixteenth century *ingenio* and its cognates constitute a moderately stable catalogue of terms. However, except for a few cases, the format adopted by the majority of the sources in our survey leaves many uncertainties about the range of application of these terms. This sense of partialness is accentuated when we consider the way *ingenio* is defined in other, non-lexicographical sources of the same period. A good, and frequently invoked, example is the definition of *ingenio* featured in the annotated edition of Garcilaso de la Vega's poetry that the Spanish poet Fernando de Herrera published in 1580.⁵⁵ Like many works of this kind, Herrera's book is rich in lexicographical information: the annotations are structured around particular words, for which a short explanation is provided. Herrera's commentary on *ingenio* consists of three somewhat overlapping definitions.⁵⁶ First, *ingenio* is defined both as a "natural force and power" and "a native ability to understand with ease" that enables us to undertake unusual or rare tasks ("operaciones peregrinas") and grasp the subtleties of elevated matters ("la noticia sutil de las cosas altas"). Herrera specifies that this form of *ingenio* originates from the right temperament of soul and body, a likely borrowing from Huarte de San Juan's *Examen*, published a few years earlier. Second, Herrera states that *ingenio* properly signifies an "inborn power of the soul," a "natural ability," which "cannot be acquired through art and industry"—an interesting inflection with respect to the usual view that ingenuity as ability requires training in order to be perfected. Finally, Herrera indicates that the ancients refer to *ingenio* as the "nature of any thing." Again, it is worth noting the emphasis, in the first two definitions, on the innate, and unteachable, character of this ability. Also worth underscoring is how specific the range of applications of this natural force is: *ingenio* is associated with out-of-the-ordinary activities; a sign, perhaps, of its uniqueness and exclusivity. In sum, Herrera's definition offers an interesting contrast to the concision and narrowness of the lexicographical accounts of the period.⁵⁷

The publication in 1611 of Covarrubias's *Tesoro de la lengua cas-*

tellana, o española did much to compensate for the brevity described above. In a period when *ingenio* and its cognates are omnipresent in the textual, oral, and visual cultures of the time, Covarrubias succeeds in capturing the multi-layered semantic range of the term. More importantly, he also manages to add more texture to its characterization.

As with other terms in his dictionary, Covarrubias's first step is to determine the etymology of the word *ingenio*. He associates the term with the Latin *ingenium* and the idea of birth and generation implied by the gerundive *gignendo*, as well as with the notion of inborn nature or *indoles*.[58] These associations are supported by other entries in the dictionary, e.g. the important neighboring term *natural*, which is defined as *ingenio* or inclination; or *condicion natural*, which stands for *ingenium*. Next, Covarrubias provides the first definition of *ingenio*: "Vulgarmente llamamos ingenio una fuerça natural de entendimiento, investigadora de lo que por razon y discurso se puede alcançar en todo genero de ciencias, diciplinas, artes liberales, y mecanicas, sutilezas, invenciones y engaños" (Commonly by ingenio we mean a natural force of the understanding, which investigates what through reason and discourse can be achieved in all areas of learning, disciplines, liberal and mechanical arts, as well as subtleties, fabrications, and trickeries).

It is a powerful definition, in line with the one advanced by Fernández de Palencia in his *Universal Vocabulario*. It maps the usual semantic reach associated with *ingenio* as primarily a mental faculty—which had become the dominant sense by this time—but notably also extends its range of application, from intellectual endeavors to trickeries.[59]

The second definition of *ingenio* refers to war machines and, by extension, to any contraption that has been skillfully invented ("inventadas con primor"). As we saw already, this is an old meaning of the word in Spanish, not much of a novelty in lexicographical terms. However, Covarrubias enriches the entry by providing a few key examples. The first one would have been identifiable by the majority of his readers. The so-called *Artificio de Juanelo* was the much-celebrated water-lifting machine that supplied the *alcázar* in Toledo. It was named after its inventor, the Italian engineer Janello Torriani, also known as Juanelo Turriano—a "second Archimedes," according to

Covarrubias.⁶⁰ Likewise, the *ingenio del açucar*, his second example, would have also been known to his readers, particularly those familiar with the commercial activities in the New World, where this sugarcane-processing machine was implemented (cf. also the entry for *trapiche*).⁶¹

Lastly, Covarrubias provides a third definition of *ingenio*. Although kept deliberately vague, it is a suggestive one: "Finalmente qualquiera cosa que se fabrica con entendimiento, y facilita el executar lo que con fuerças era dificultoso y costoso, se llama ingenio" (Finally, *ingenio* stands for anything that is produced by the understanding, and facilitates the execution of tasks which would be difficult and costly to complete by force). This sense of accomplishment and overcoming difficulty may be associated with other terms in the *Tesoro*, such as the adjective *capaz*, which according to Covarrubias applies to anyone who is able to understand and execute a task, and hence possesses the necessary ingenuity, understanding, and ability (*ingenio*, *entendimiento*, and *habilidad*) for it. An interesting feature of this type of ability is the moral ambivalence associated with its use. As we saw in the first definition of *ingenio*, this capacity may lead to achievements in various disciplines, but it can also be used to trick, to deceive. Covarrubias alludes to this association in his entry for *engaño*, where—following the work of Charles de Bovelles—he registers an etymological link between *engaño* and *ingenio*, since the person who deceives must be "ingenioso y astuto."⁶²

Another interesting word associated with *ingenio* and the possession of certain skills and abilities, which would carry a similar sense of ambivalence, is *maña*. Covarrubias defines it as dexterity (*destreza*). Etymologically, he derives it from the Latin *manus*, since the hand is "the instrument of instruments." In positive terms, the adjective *mañoso* stands for whoever does something with "dexterity and liberality."⁶³ Hence the important neighboring term *industria*, which Covarrubias defines as the "maña, diligencia y solercia" with which some accomplish tasks with less effort than others. Don Quijote, in this regard, would be *mañoso* as well as *ingenioso*.⁶⁴ In its other (negative) sense, *maña* stands for ruse, trick, or deception ("ardid, astucia, y engaño"). Hence the word *artimaña*, which according to the *Tesoro* stands for a concealed and cunning deception ("engaño hecho con dissimulacion y cautela"). In similar ways, *artificio* in the *Tesoro* is defined as anything that is done with art ("hecho con arte"), but also stands for appearance, simulation, and deception.

These associations linking ingenuity, workmanship, and deception were certainly not new in early seventeenth-century lexicography. In fact, they had been recorded by many dictionaries before, from those by Fernández de Palencia and Nebrija onward.[65] But the fact that Covarrubias registers them in the *Tesoro*—as well as in his other major publication, the treatise on emblems *Emblemas morales* (1610)—is more than an indication of their continuing currency at the beginning of the seventeenth century.[66] In a way, it also corroborates an expansion in the semantic range of *ingenio* in Spanish at the turn of the century. On the one hand, the emphasis on *artificio* and *habilidad* seems to be linked to the widespread culture of appreciation of artistic and literary products, which during the Spanish Golden Age were reaching unprecedented levels of sophistication. On the other hand, the interpretation of *ingenio* in terms of *maña* and *engaño* captures the connotations of craftiness associated, for example, with rogue culture, as expressed in the highly popular picaresque novel tradition.

In fact, in the following decades most dictionaries featuring Spanish as one of the leading languages would record and preserve these semantic associations and the moral reversibility of these terms. For example, the definition of *maña* in the second edition of Cesar Oudin's *Tesoro* (1616) includes such terms as *addresse, dexterité, engin, fraude* and *mal-engin*. Similarly, Minsheu's *Vocabularium Hispanicolatinum et Anglicum copiosissimum*—included in his *Ductor in linguas* of 1617—defines *maña* as "wit, deceit, cunning."[67] In Lorenzo Franciosini's *Vocabolario español e italiano* (1620), *maña* is defined as *lestezza, agilità, grazia, garbo* and also as *inganno, malizia, furfanteria*. Finally, the 1628 edition of Bartolomé Bravo's Spanish-Latin *Thesaurus* connects *ingenio* to *maña* and *habilidad*, and defines them in terms of s*olertia, industria*, and *ingenium*.

Such semantic richness informed the most complete account of *ingenio* as a keyword in the ingenuity word family in Spanish: the entry

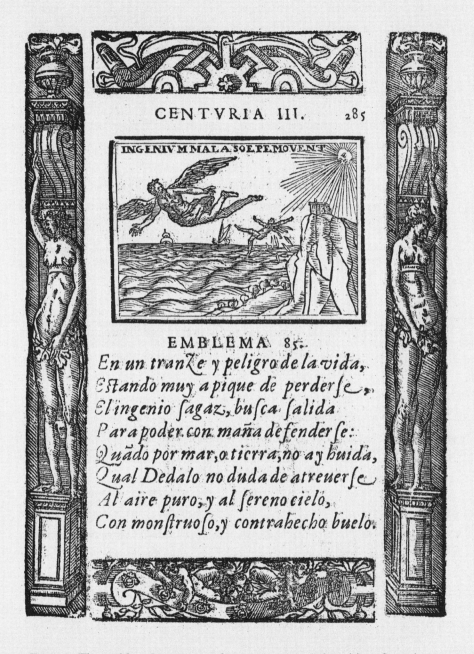

Fig. 3.2. The emblem *Ingenium mala saepe movent* (adversities often stir one's wit) (Covarrubias 1610).

nar al Maeſtro mayór, aparejadór y oficiales de cantería, albañilería y carpintería lo que han de hacer. ULLOA, Poeſ. pl. 203.

O tu ingeniéra ſagáz
de las máchinas de Marte,
hambre ſagrada del oro,
qué rieſgo no perſuades?

INGENIO. ſ. m. Facultad ò poténcia en el hombre, con que ſutilmente diſcurre ò inventa trazas, modos, máchinas y artificios, ò razónes y argumentos, ò percibe y aprehende fácilmente las ciencias. Viene del Latino *Ingenium*. MARIAN. Hiſt. Eſp. lib. 7. cap. 15. Hombre docto y de *ingénio* agúdo. SAAV. Empr. 84. Segúra es la guerra que ſe hace con el *ingénio* y peligroſa y incierta la que ſe hace con el brazo.

INGENIO. Se toma muchas veces por el ſugéto miſmo ingenioſo: y aſsi ſe ſuele decir de las comedias de un ingénio, de dos ò tres ingénios. Lat. *Ingenioſus homo*. QUEV. Entremeſ. El Poéta de los pícaros ſe fué à reveſtirſe en el cuerpo de los Poétas mechánicos, *Ingénios* cantonéros, y Muſas de alquilér como mulas.

INGENIO. Se toma tambien por las miſmas trazas, mañas ù artes de que ſe uſa para conſeguir alguna coſa. Lat. *Machina. Artificium.*

INGENIO. Se toma por las miſmas máchinas è inſtrumentos artificioſos inventados por los Ingeniéros. Lat. *Machina*.

INGENIO. Llaman los Libréros cierto inſtrumento que conſta de un tornillo de madéra, que paſſa por otros dos maderillos ò tablas, llamados Meſas, y en la primera de ellas eſtá figurada la hembra del tornillo, y en uno de los cantos ſe afirma una lengüeta de acéro, de ſuerte que al movimiento del tornillo vá la una meſa acercandoſe à la otra: y la lengüeta corta los cantos del libro que ſe ha de enquadernar. Lat. *Machina libraria*. LOP. Dorot. f. 153. No habeis viſto aquel inſtrumento con que los Libréros cortan los libros que enquadernan? pues aquel ſe llama *ingénio*.

INGENIO DE AZUCAR. Es una máchina compueſta de tres ruedas grandes de madéra, con diverſidád de dientes, en que ſe incluyen unas vigas grandes atraveſadas, que llaman puentes, ò vírgenes de la molienda, con que ſe muele ò aprieta la caña, cayendo el zumo ò liquór en unas caldéras grandes, en que deſpues le cuecen para depurar el azúcar. Lat. *Machina ſacchari extractoria*. ACOST. Hiſt. Ind. lib. 3. cap. 24. La grangería deſtas Islas es *ingénios de azúcar* y corambre. TORR. Hiſt. de los Xerif. cap. 3. Plantaron por la comarca muchas cañas *de azúcar*, è hicieron un *ingénio*, que fué el fundamento de perſeverar en aquella poblacion.

INGENIOS DE PÓLVORA. Se llaman las diverſas eſpécies de cohétes, y artificios de pólvora que han inventado, y de que uſan los artifices de la pólvora feſtiva. Lat. *Inventiones pyrotechnica*.

INGENIOSAMENTE. adv. de modo. Sutíl y mañoſamente, con ingénio, primór y habilidád. Lat. *Ingenioſè*. CRUZAD. Cort. Sant. tom. 3. Paſſ. de deſeo, ſeſſ. 1. Eſto es lo que *ingenioſamente* decía el panegyrico à Conſtantino el mozo.

INGENIOSIDAD. ſ. f. Propriedad ò calidad del ſugéto, que le hace diſcurrir ù obrar con ingénio ò ſutileza. Lat. *Ingenii vis, vel acumen*. CORN. Chron. tom 4. lib. 2. cap. 41. Como ſi la *ingenioſidád* pudieſſe formar doctos, ſin darſe la mano con la aplicacion..... Se celebran como flores de la *ingenioſidád*; y ſon eſpinas que laſtiman los corazónes de los que las oyen con inteligencia y deſengaño. ALCAZ. Vid. de S. Julian, lib. 1. cap. 1. Enlazaronſe en él, con union concorde, las codiciadas prendas de *ingenioſidád*, literatúra, magiſtério y urbanidad.

INGENIOSISSIMO, MA. adj. ſuperl. Mui ingenioſo. Lat. *Valdè ingenioſus*. LOP. Philom. f. 59. Fueron en eſto los Eſpañoles *ingenioſiſsimos*. ESPIN. Eſcud. Relac. 3. Deſc. 4. La invención cierto era *ingenioſiſsima*, y mui conforme à la philoſophia natural.

INGENIOSO, SA. adj. Habil, ſutíl, y que tiene ingénio. Lat. *Ingenioſus*. SAAV. Empr. 43. El mas *ingenióſo* en las ſoſpechas es el que mas lejos dá de la verdád. QUEV. Muſ. 8. Sylv. 25.

Y ſiendo yá deſprécio de las Parcas
En nuevo parto de ingenióſa *vida,*
Su poſtrer Padre fueron los pincéles.

INGÉNITO, TA. adj. No engendrado. Es atribúto que ſe dá preciſamente à la primera y tercera Perſona de la Santíſsima Trinidad; à diſtincion de la ſegunda que es el Unigénito. Es del Latino *Ingenitus*, que ſignifica eſto miſmo. CALD. Aut. A Dios por razon de eſtado.

.... *Es del alma*
parte no engendrada, ſiendo
el ingénito *de adonde*
el nombre toma...............

INGENITO. Significa tambien lo que es connatural, y como nacido con el miſmo ſugeto, ò en él. Lat. *Ingenitus*. HORTENS. Mar. f. 52. Acabemos, con que la paciencia ſe ha hecho en Dios naturaleza; y excelencia de una *ingénita* propriedad. SART. P. Suar. lib. 4. cap. 10. No le baſtó à eſte Exímio Maeſtro aquella ſu *ingénita* ſerenidád para librarle de hombres borraſcóſos.

INGENTE. adj. de una term. Exceſsivamente grande. Es voz Latina y de poco uſo. Lat. *Ingens*. MEN. Coron. Prolog. Del prudentiſsimo, magnánimo è *ingente* Caballero Iñigo Lopez de Mendoza.

INGENUAMENTE. adv. de modo. Libre, ſinceramente, con ingenuidád, y ſin dobléz ò engaño. Lat. *Ingenuè*. SAAV. Empr. 48. Premie el Príncipe, con demoſtraciones públicas, à los que *ingenuamente* le dixeren verdádes. MANER. Apolog. cap. 43. Confeſſaré *ingenuamente* quienes ſon los que ſe querellan de los Chriſtianos.

INGENUIDAD. ſ. f. Sinceridad, realidad en lo que ſe dice ò hace. Viene del Latino *In*-

Fig. 3.3. Entry for *ingenio* in the Real Academia Española's *Diccionario de la lengua castellana* (Madrid, 1734).

included in the fourth volume of the *Diccionario de autoridades,* first published in 1734. Most semantic strands (but not all, as we shall see) are registered here, and both the entry and some adjacent headwords incorporate some interesting additions. The first sub-entry, which is specifically associated with the Latin *ingenium*, defines *ingenio* as a faculty or power: "Ingenio: Facultad o potencia en el hombre, con que sutilmente discurre o inventa trazas, modos, máchinas y artificios, o razones y argumentos, o percibe y aprehende facilmente las ciencias. Viene del Latino Ingenium" (Ingenio: Faculty or power in an individual, with which he subtly thinks or invents schemes, plans, machines, and artifices, or reasons and arguments, or through which he perceives or comprehends with ease. It derives from the Latin Ingenium).

As in Covarrubias's *Tesoro*, it is interesting to note the expansiveness of the definition: *ingenio* is associated with the capacity to generate a wide range of outputs, from plans and schemes to devices and arguments. Additionally, *ingenio* is described as the capacity to comprehend with ease, a quality associated with sharpness and quickness. The *Diccionario de autoridades* then records the metonymical use of *ingenio*, as applied to a person: "Ingenio: Se toma muchas veces por el sugeto mismo ingenioso: y assi se suele decir de las comedias de un ingenio, de dos o tres ingenios. Latín. Ingeniosus homo" (Ingenio: It is often said of the ingenious individual himself; hence the expression "comedias of one, two, or three ingenios." Latin. Ingeniosus homo). This use of the word had had a long history by then, but, surprisingly, no formal acknowledgment as a headword in the Spanish lexicographical context.[68]

Having defined *ingenio* as the faculty to invent plans and schemes, these plans and schemes themselves are identified with the word *ingenio*. "Ingenio. Se toma tambien por las mismas trazas, mañas u artes de que se usa para conseguir alguna cosa. Latín. Machina. Artificium" (Ingenio. It is said of the schemes, contrivances, and arts used to achieve something. Latin. Machina. Artificium). Finally, the sense of *ingenio*-as-machine is recorded: "Ingenio. Se toma por las mismas machinas e instrumentos artificiosos inventados por los Ingenieros. Latín. Machina" (Ingenio. It is said of the machines and artful instruments invented by engineers. Latin. Machina). In a manner that echoes Covarrubias's entry in the *Tesoro*, the *Diccionario de autoridades* then provides three examples of such *ingenios*: a cutting device used by bookmakers, simply called *ingenio*; the better-known *ingenio de azucar*; and the *ingenio de pólvora*, as fireworks and other pyrotechnic devices. In addition to these key entries (and others involving cognates, which do not register significant variations), it is worth noting the inclusion of the term *ingeniosidad*, defined as the quality that allows an individual to think or act with *ingenio o sotileza*.[69] The term *ingenuidad* is also registered: first, it is defined as sincerity and linked to the Latin term *ingenuitas*; second, its legal meaning is provided: natural freedom, as opposed to acquired freedom (Latin: *Ingenuitas legalis*).[70]

Another noticeable, and important, feature of this otherwise rather rich set of definitions is the lack of explicit reference to *ingenio* as natural disposition or natural inclination. In the *Diccionario de autoridades* this meaning now appears as one of the definitions of the term *genio*. This sense of the Spanish word *genio* had been recorded by the lexica before.[71] Though primarily dedicated to the antiquarian meaning of the term, the entry in Covarrubias's *Tesoro* registers it. There, toward the end, we read that *genio* stands for "the symmetry and commensuration of the elements, which conserve the bodies of humans and any other living thing." Following this definition, *genio* is described as a "force and influence from the planets that inclines us to do this, or that;" this type of *genio* is not exclusive to human beings: plants, buildings, and—as Martial put it—books have it too.[72]

Outside the lexicographical context, this semantic shift is widely recorded. Nowhere is it better illustrated than in the works of Gracián, where a significant emphasis is placed on the balanced interplay between *ingenio* and *genio*. A well-known example is Gracián's discussion of these two terms at the very beginning of *El Discreto* (1646)— "Realce I: GENIO Y INGENIO"—where they are presented as two complementary features of the talented individual.[73] The gist of Gracián's argument is neatly summarized in the second aphorism of his *Oráculo manual y arte de prudencia* (1647), also entitled "Genio y ingenio": "Genio y ingenio. Los dos ejes del lucimiento de prendas; el uno sin el otro, felicidad a medias. No basta lo entendido, deséase lo genial. Infelicidad de necio errar la vocación en el estado, empleo, región, familiaridad" (Genio and ingenio. The two axes around which talents shine: one without the other, only half the happiness. Intelligence is not enough; you need also the right disposition. The misfortune of fools: to make unsuitable choices regarding their position in society, occupation, dwelling-place, and friendships).[74] In the early eighteenth century, this sense of *genio* appears to have consolidated in the lexicographical sources, as in Stevens 1706: "*Genio*, a man's angel or spirit, or rather his natural disposition: his genius." The entry in the *Diccionario de autoridades* is slightly more nuanced, as it defines *genio* as a natural inclination and disposition toward a particular activity: "Genio: Natural inclinación, gusto, disposición y proporción interior para alguna cosa: como de ciencia, arte, o manifactura. Latín. Genium" (Genio: Natural inclination, taste, inner disposition for a given thing: be it related to science, art, or craft. Latin. Genium).[75]

It is worth noting that the first of the two *autoridades* for this entry is Fernando de Herrera, the above-mentioned poet and commentator of Garcilaso de la Vega's poetry. Interestingly, the sentence cited in this entry is taken from Herrera's lengthy annotation on the term *genio*.[76] As in the case of his definition of *ingenio*, Herrera's description is rich in semantic modulations. First, *genio* is said to designate a power or property that is specific to a particular individual ("virtud especifica, o propriedad particular de cada uno que vive"). This is the part quoted in the entry of the *Diccionario de autoridades*. Second, Herrera provides the meaning that best explains Garcilaso de la Vega's use of this term: *genio* should be understood in the Platonic sense as a sort of frenzy—although Herrera suggests that it would not be erroneous to identify Plato's genius with Aristotle's "agent intellect."[77] More specifically, Herrera writes, *genio* "is offered to divine *ingenios* and gets inside them, so with its light they can discover the meanings of secret things, on which they write." Herrera indicates that it is often the case that once this "heavenly heat" (*calor celeste*) has cooled down, these authors display admiration for their writings and do not recognize them as their own. Even the meaning of the words dictated to them eludes them sometimes. Third, according to Herrera *genio* can also designate "nature itself, and the spirit that moves us to do good" as well as "angel, spirit, or intelligence." This ample definition is then followed by a quotation by Censorinus, in which *genio* is described as a tutelary being.

To return to the *Diccionario de autoridades*'s entry for *genio* as natural inclination, other definitions in this dictionary reinforce these associations. The intriguing term *pergeño*, for example, is defined as "disposition, ability, or dexterity in the execution of things," and is said to derive from the union of the preposition *per* and the noun *genium*, as in "por genio." Also, the term *espíritu*, in one of the sub-entries, is defined as the "genio, inclinación, hábito, y passión" that inclines us to certain actions above others.[78]

As these examples suggest, *genio* in the *Diccionario de autoridades* records and highlights a larger semantic range compared to occurrences of the same word in previous Spanish dictionaries.[79] However, this is a far cry from the inflation that other dictionaries in other vernaculars record. In fact, for many decades, and in subsequent editions, the entry for *genio* in the *Diccionario de autoridades* would remain practically unchanged—except for the inclusion of the term *índole* as the main descriptor. It is only in the second half of the nineteenth century, in the 1869 edition, that a significant addition appears: "Genio: Dícese hoy particularmente de los talentos de primer orden, que tienen la facultad de crear, inventar o combinar cosas extraordinarias" (Genio: Today it is said particularly of those first-class talents who are endowed with the faculty to create, invent, or combine extraordinary things).[80]

To return to *ingenio*, it is interesting to note how effectively the *Diccionario de autoridades* captures and connects the web of ingenuity-related associations displayed by neighboring terms in earlier sources. *Industria*, for instance, first defined as "dexterity and ability in any art," is also shown to be linked to ingenuity in the morally ambivalent sense of subtlety and artifice (and *maña*) that Covarrubias, among others, had recorded.[81] Similarly, the entry for *artificio* distinguishes between meanings that could be regarded as neutral or even positive, as in the case of a work that is executed with novelty, skill, and subtlety—like the *artificios* designed by Juanelo Turriano, who is explicitly mentioned—

and meanings that, like many dictionaries in the past, associate *artificio* to astuteness and dissimulation.⁸²

In other cases, terms that had featured rather scarcely in earlier sources now appear to consolidate their ingenuity-related sense. One such term is *talento*. In the Spanish lexicographical corpus, its association with *ingenio* had been explicitly recorded in the early seventeenth century, in Francisco del Rosal's *Origen y etimología de todos los vocablos originales de la lengua castellana*—a treatise that, as we indicated earlier, was left unpublished.⁸³ *Talento* is not defined in Covarrubias 1611, although it features in some entries; but it does appear as a headword in the 1674 expanded version of the *Tesoro*, where it is defined as a divine and temporal gift and also, tangentially, as "natural ability."⁸⁴ The *Diccionario de autoridades* records this tropological sense too, defining *talento* as a natural or supernatural gift bestowed to men by God. But the term is also said to stand metaphorically for such gifts of nature as ingenuity or prudence that certain individuals—the entry appears to suggest—possess above others.⁸⁵

Another interesting example is offered by the term *capricho* and its cognates. The association of this important word family with the language of ingenuity in Spanish has an illustrious point of departure: Huarte de San Juan's *Examen de ingenios para las ciencias*, where the adjective *caprichoso* is used to characterize the inventive *ingenios*, whose attitude resembles the behavior of a goat: "The Inventive Wits are termed in the *Tuscan* Tongue Capricious [*caprichosos*], for the resemblance they bear to a Goat. Who takes no pleasure in the open and easy Plains, but loves to Caper along the Hill-tops, and upon the Points of Precipices, not caring for the beaten Road, or the Company of the Herd."⁸⁶ The term *capricho* as a headword appears slightly later in the Spanish lexicographical sources, and is defined by a range of ingenuity-related equivalents in other vernaculars: Oudin 1607: "*caprice, fantasie*"; Vittori 1609: "*caprice, fantasie, capricio, fanticasteria, bizzaria*"; Minsheu 1617: "*capricio, a humour, a fansie, a toy in ones head.*"⁸⁷ The entries in the *Diccionario de autoridades* capture some of these inflections, as in the case of *caprichoso*, which is said to refer to an "ingenious" individual: "whoever conceives, finds or makes a thing with novelty and good taste is said to be capricious and of rare phantasy."[88] The *autoridad*, in this case, is the painter and theorist Antonio Palomino, and the quotation is extracted from his important art theoretical treatise *El museo pictórico y escala óptica*, published a few years earlier (1715–1724). Incidentally, the *capricho* word family appears often in this work, sometimes in relation to the work of Diego Velázquez, as Fernando Marías has shown. One relevant example is Palomino's characterization of Velázquez's *Las Meninas* as "*capricho nuevo*," which Marías takes as the starting point of a suggestive interpretation of this most ingenious painting.[89]

Other definitions that reinforce the interesting associations that link *ingenio* and *capricho* appear, for example, in the entries *mania* and *phantasia*. The *Diccionario de autoridades* defines *mania* as a "disease of the fantasy," but also as an "extravagance" or "caprice" of genius.[90] And in the fourth definition of *phantasia* this term is said to mean "fiction, tale, novel, or elevated and ingenious thought, hence the *phantasias* of poets and painters."[91] Although still in the realm of *ingenio*, we begin to see here the gradual convergence of certain characteristics—loftiness, whim, inventiveness—around certain social types: writers and artists. In other linguistic contexts, e.g. English, this shift would soon be articulated using the language of genius.

Agudeza

Of the many references to the association between *agudeza* and ingenuity in the dictionaries, Covarrubias's explanation in his entry for *agudo* (sharp) in the *Tesoro* is among the clearest and most straightforward. After referring to the qualities of any sharp object, he writes: "applying this to the soul, we say *agudo* of whoever possesses a subtle and penetrating *ingenio*."[92] This old and powerful idea constitutes the foundation on which all the uses of *agudeza* as an ingenuity keyword are built. These uses, as they are recorded in the Spanish lexicographical corpus, can be summarized as follows:

Agudeza, n.
  1. The quality of being ingenious, witty.
  Nebrija 1492 (headword *argutus*) >.

2. Sharpness of wit (as in the expression *"agudeza de ingenio"*).
Fernández de Palencia 1490 (headword *perspicatia*) >.
3. Astuteness, sagacity.
Fernández de Palencia 1490 (headword *sagax*) >.
4. Perspicacity, shrewdness.
Casas 1570 >.
5. Witty saying.
Sánchez de la Ballesta 1587 >.
6. Quickness, readiness of wit.
Minsheu 1599 >.
7. Subtlety in speech or action.
Oudin 1607 >.
8. Restlessness.
Covarrubias 1611; Stevens 1706.
9. Property related to satire.
Diccionario de autoridades 1726–1739.

Encapsulated by the often-quoted expression *"agudeza de ingenio,"* the notion of "sharpness of wit" remains practically the same throughout our timeline, although slight semantic inflections may be detected, depending on the sources. Most Latin-based dictionaries tend to associate *agudeza* and its cognates with a relatively close set of interconnected terms. In Fernández de Palencia 1490, for example, the expression *"agudez de claro ingenio"* appears in the definition of *perspicatia*, and *"agudes ingeniosa"* is used to define *calliditas*. Nebrija equates *agudeza* to *acumen* (Nebrija 1492, Nebrija ca. 1495), gives *"cosa aguda o ingeniosa"* as equivalent to *argutus* (Nebrija 1492), and defines *"agudeza de ingenio"* as *"perspicacitas, acritas, acies ingenii"* (Nebrija ca. 1495).[93] These slight semantic inflections can also be detected in the vernacular-based lexica, as in the case of the emphasis on *agudeza* as subtlety, which several dictionaries register, e.g. Covarrubias 1611, which equates *agudeza* to *sutilidad*; Franciosini 1620, which defines *"agudo de ingenio"* as *"acuto, e sotil, d'engegno"*; or Howell 1660, which defines *"Ingenious, or witty"* as *"Ingenieux, aigu, subtil; Suttile, ingenioso, acuto, accorto; Agudo, sutil, ingenioso."*

Another important case of semantic modulation is the association of *agudeza* with promptness and readiness, a sense closely linked to the idea of *agudeza* as a quick, witty saying.[94] An early reference appears in Minsheu 1599, where *agudeza* is defined as "quicknes of conceit" and "readines of wit." A century later, Stevens 1706, following Minsheu, refers to it again in the entries on *agudeza* ("readiness, quickness of wit") and "To be sharp of wit" ("ser agudo de ingenio, tener ingenio vivo"). Finally, Pineda 1740, which is an expansion of Stevens's dictionary, includes "A Flash of Wit" as an entry, and defines it as "agudeza, promtitud de entendimiento." This sense of dynamism resonates with an intriguing but relatively inconspicuous meaning of the adjective *agudo* recorded by Covarrubias's *Tesoro*: *agudeza* as restlessness, the feature of an active and unquiet individual.[95] Interestingly, as in the case of Huarte de San Juan's goat-like *ingenios* (*ingenios caprichosos*), here we have an association that connects action with unruliness, and, perhaps, also impulse and whim—characteristics of the ingenious person that later on will be attributed to the creative genius.[96]

Most of the above-mentioned semantic variations (perspicacity, subtlety, promptness) are neatly registered in the entries on *agudeza* and its cognates included in the *Diccionario de autoridades* of the Real Academia. *Agudo*, for instance, is said to stand, metaphorically, for "witty, prompt, perspicacious and subtle," qualifying not just individuals but also their actions.[97] And *agudeza* is defined as "the subtleness, promptness and easiness of wit applied into thinking, saying or making anything."[98] The capacity of *agudeza* to accommodate these fluctuations is further reinforced by other definitions in the *Diccionario de autoridades*, as in the case of the entries *viveza*, *discrecion*, *penetracion*, *sutileza* or *perspicacia*, which all register the notion of "*agudeza de ingenio*."

Additionally, it is worth mentioning the association of *agudeza* with the terms *capricho*, as in the expression *hombre de capricho*—that is, one who possesses the *agudeza* to come up with new, singular, and successful ideas—and *espíritu*, defined as vivacity and promptness in thought and action.[99] Directly connected with these definitions is the idea of *agudeza* as the product of a verbal action: a witty saying, a quick and sharp remark. Its occurrence in the lexicographical sources is difficult to establish, given the economical format of some dictionaries. The use of the plural is sometimes indicative of this meaning, as in the early examples of Sánchez de la Ballesta 1587 (*Agudeza. Subtilezas. Argutiae*), Navarro 1599 (*Argutiae, arum, las agudezas*), or Bravo 1599 (*Agudezas de palabras*). There are references to it in Covarrubias's *Tesoro*, such as in the entry for *chiste* (joke), where it is claimed that for a joke to be funny it needs to be told with "much sharpness, and a few words"; or in the entries on *donaire* and *facecia*. The expression appears much later in Stevens 1706 as part of the definition of "witty speech," together with the terms *donayre* and *gracia*. And in similar terms it is recorded, in Portuguese, in Bluteau 1712: "*Agudezas, Chistes, & ditos engenhosos.*"

Interestingly, the *Diccionario de autoridades* does not register this sense of *agudeza* in its definitions of this word. Only a sub-entry for *agudo* points to the fact that certain forms of speech, allusive and equivocal, may be described as such.[100] But there are enough examples in other entries to confirm the consolidation of *agudeza* as the witty outcome of a speech act (e.g. *viveza*) or at least as a key ingredient to it (e.g. salt, *sal*).[101]

Along these lines, it is worth noting the association of this sense of *agudeza* with a series of entries in the *Diccionario de autoridades* dedicated to the important term *concepto*.[102] In a manner that resembles aspects of the English *conceit* or the Italian *concetto*, *concepto* is defined as a pithy or pointed saying, a sententious witticism (*agudeza sentenciosa*).[103] Stemming from this definition, cognates such as *conceptuar* or *conceptuoso* point to the erudition, discretion, and sententiousness that characterize ingenious discourse.[104] These entries are particularly important for two main reasons. On the one hand, they offer a glimpse of what in other contexts, literary in particular, had been and still was a central ingenuity-related motif: the technical use of *concepto* within the conceptual framework that sustained the theoretical discourse on ingenious language. In other words, these entries record instances of that complex set of notions and practices that would be considered under the term *conceptismo*.[105]

On the other hand, these entries, *conceptuar* and *conceptuoso* more specifically, are noteworthy in that they constitute two of the relatively few instances in which Baltasar Gracián is quoted as an *autoridad* in the *Diccionario de autoridades*.[106] This is a feature that has intrigued scholars: the elusive presence of Gracián, especially in those entries devoted to key *ingenio-* and *agudeza*-related terms where his work—*Arte de ingenio, tratado de la agudeza* (1642) and *Agudeza y arte de ingenio* (1648), in particular—would have indeed been authoritative.[107] It is an aspect of the *Diccionario de autoridades* that is likely to have been determined by its ideological underpinnings above anything else. In the context of early eighteenth-century debates on literary theory, ingenious language was a point of contention, since much of the linguistic "excesses" of the previous century—so heavily criticized by eighteenth-century Neoclassical scholars—had been driven by an obsession with it. Among other means, the *Diccionario de autoridades* incorporated these concerns by carefully selecting its choice of *autoridades*, therefore putting into practice the purifying tenets of its motto. However, as Aurora Egido has argued, the absence of Gracián in the *Diccionario de autoridades* does not imply that the language of *ingenio* and *agudeza* is missing too.[108] The eighteenth-century academicians relied on the *ingenios* of the previous century for their lexicographical task, and as a result a number of masters of witty language ended up featuring as *autoridades*, with Lope de Vega, Quevedo, Cervantes, and Fray Luis de Granada among the most often cited authors. Still, from the perspective of our study, it is significant that one of the most eminent theorists and practitioners of the language of ingenuity in Spanish features so poorly in such an important source as the *Diccionario de autoridades*.

Conclusion

Gabriel García Márquez tells the story of his grandfather unsuccessfully trying to find the entry for the word *mar* (sea) in his old, torn dictionary, and lapidarily concluding: "Some words are not here, because everybody knows what they mean."[109] To refer to our ingenuity keywords in these terms is, obviously, an exaggeration. The anecdote, however, serves to highlight at least two important aspects of the treatment that *ingenio* and *agudeza* receive in most of the lexicographical sources in our survey. First, early modern lexicographers appeared to assume that their readers already knew the meaning, use, and context of these terms. This is particularly so in the case of dictionaries that tend to provide very short definitions, but is also apparent in some of the more copious sources, such as Covarrubias's *Tesoro*, where some sub-definitions can be too general, and therefore less precise. Related to this, a second, important aspect of the way the language of ingenuity in Spanish is handled in the lexica is the sense of mismatch between the range of meanings and uses registered by the dictionaries, and those recorded in non-lexicographical sources. That is, there is a remarkable contrast between the rich language of ingenuity that is exercized and theorized in the works of such authors as Lope, Cervantes, Góngora, Quevedo, Gracián, or Sor Juana, and the language of ingenuity that features in the dictionaries. Much of this has to do with the idiosyncrasies of Spanish lexicography in this period. Except for a few, key lexica, neatly distributed across our timeline, the number of sources, compared to, for example, the French or English contexts, is relatively low. Ingenuity did inform the making of these dictionaries, as Quevedo wittily noted. But it could also be a point of contention, as we saw in the case of the *Diccionario de autoridades* of the Real Academia, where a concern with linguistic purity implied the purge of certain forms of ingenuity-driven language. In sum, *ingenio* and *agudeza* seem to be not so much a form of *no sé qué* / *je-ne-sais-quoi* but an expression of a tacitly assumed culture—a culture of ingenuity so pervasive and presumed that it did not need to be exhaustively recorded and explained in the dictionaries. Paraphrasing Gracián, *ingenio* and *agudeza* lend themselves to appreciation, not definition.[110]

4 French

The editor of the fifth, 1740 edition of Pierre Bayle's *Dictionnaire historique et critique* (first complete edition printed in 1697) sums up a century and a half of polemic over the possibility of any form of German wit and about the respective merits of German and French ingenuity by contrasting their defining features:

> Ce que les François appellent *de l'Esprit,* est un certain talent pour la bagatelle, ou tout au plus une je ne sais quelle vivacité généralement peu compatible avec la gravité Allemande, & avec le caractère sérieux de cette Nation. . . . J'en dis autant de la question, *Si un Allemand peut étre Bel-Esprit?* On ne dispute pas à la Nation Allemande, le plus pur Bon-Sens, la plus fine sagacité, les plus nobles saillies de l'Esprit; & contente de ce partage, elle ne regarde pas comme une grande prérogative le Bel-Esprit François.

> What the French call *wit*, is some talent for trifles, or, at best a liveliness of sorts, which is usually not suited to German gravity and to this nation's serious character. . . . I will say the same about the question, *Whether Germans can be Fine Wits, or not?* No one denies to the German nation the purest good sense,

the subtlest sagacity, the noblest traits of wit—satisfied with their share, they may not deem French wit such a great prerogative.[1]

This editorial remark draws together the threads of meanings associated with the French cognates of the Latin *ingenium* and *genius*. Its concision assumes a deeply rooted, multifaceted, and distinctive culture of ingenuity in early modern France, mostly centered on the notion of *esprit* in the seventeenth century and shifting toward *génie* by the middle of the eighteenth. In the editor's remark, *esprit* first denotes the temperament at the heart of national identity or character, then a special ability or talent, and finally a set of intellectual traits, be they quick thinking in invention (*vivacité* and *sagacité*), or ponderous, solid judgment (*bon sens*). These definitions emerged progressively from *longue durée* cultural polemics, themselves the echoes of France's military and diplomatic hegemonic attempts on the European stage: *l'esprit* and *le bon sens français* erected naturalness (*le naturel*) as a new, Gallican norm of good taste that openly opposed excessive, virtuosic expressions of ingenuity characterizing the culture of the Habsburg Empire, but also against the supposed lack of polish of German nations.[2] Yet Bayle's editorial remark is not only a semantic *précis* of ingenuity, it also sketches its history in the French context, hinting at the trajectory that led to the degradation of *esprit* into *bel esprit* or *esprit français*, both of which amounted to little more than a frivolous performance confined to the ephemera of conversation.

This derogatory outline of French ingenuity contrasted with its sturdier German counterpart in a late edition of Bayle's dictionary is one instance of the Enlightenment critique of the French *Grand Siècle*. Stemming from the seventeenth-century ideological enterprise of national self-definition and eulogy, early modern French ingenuity shaped the linguistic culture of the long seventeenth century: dictionaries, grammars, and the emerging genre of grammatical *Remarques* strove to instil the purest form of the *génie de la langue* as a cultural norm of the *esprit français*, both nationally and internationally.[3] The Jesuit Dominique Bouhours epitomizes this entanglement between *esprit français* and *génie de la langue*: the dialogue dedicated to *esprit*

Fig. 4.1. Frontispiece of the first edition of the *Dictionnaire de l'Académie françoise* (Paris, 1694).

in his 1671 *Entretiens d'Ariste et d'Eugène* sparked the polemic on German wit that Bayle's editor revives almost a century later, while his 1675 *Remarques nouvelles sur la langue* and their 1692 *Suite* turned him into a lexicographical authority for the rest of the century and beyond. This chapter unravels this entanglement between French ingenuity and lexicography: the keywords of early modern French ingenuity as they feature in period dictionaries reveal the ideological and political underpinnings of this culture, its very specific response to the question of identity, and its role in the emergence of the modern notions of style, author, and critic.[4]

Keywords

These keywords are *engin*, *esprit*, *naturel*, and *génie*: they all translate the Latin terms *ingenium* and *genius* in the first landmark of our corpus, Jean Nicot's *Thresor de la langue françoyse* (1606), either in the definition proper, or in the selection of examples.

The word history of *engin* remains a remarkably stable, antiquarian one that interprets machines as the reification of one's wit. Meanwhile *naturel*, like *esprit* and *génie*, can be used to denote a distinctively individual set of dispositions increasingly defined as a specific natural aptitude or talent, as the site of good education, and as the norm of good (self-) representation. The presentational meanings of ingenuity also prove central to the word history of *esprit*, the main keyword of French ingenuity. While *esprit* originally denotes individual identity grounded in both temperament and intellectual faculties, it progressively stands for the social and artistic representation and performance of such identity (the performative, or presentational meaning of the term), which then becomes the object of interpretation and assessment by its audience: the lexicographical fortune of the pair *esprit/bel esprit* is particularly representative of this historical trajectory. Corollary to *esprit*'s eventual loss of cultural capital (as described by Bayle's 1740 editor) is the rise of *génie*. Against the Neoplatonic story that identifies genius with the *furor* or divine inspiration, the word history of *génie* runs parallel to the one of *genius* in the Latin and polyglot dictionaries of our corpus, and suggests first a potentially libertine account of genius naturalized by classicizing dictionaries, then the antiquarian attempt to counter this naturalization.⁵ In contrast with *esprit*, however, the silences in the lexicography of *génie* point to the limits of our corpus: nothing in the dictionaries foreshadows the *Encyclopédie*'s sophisticated, pre-Romantic account of genius as the sublime, law-breaking expression of one's outstanding and sympathetic sensibility.

Lexicographical Landscape

Three lexicographical traits recur in the ten dictionaries of our corpus in varying degrees. First, some dictionaries have retained the word-for-word translation of bilingual and polyglot dictionaries. This is the case in the first dictionary of the corpus, Jean Nicot's *Thrésor de la langue françoyse* (1606), which marks the endpoint of the Renaissance bilingual tradition of the *Grand dictionnaire francois-latin* and the beginning of French lexicography.⁶ In Nicot 1606, the French word entry leads into short, sometimes tautological definitions in French, concluded by the Latin word equivalent, then followed by examples in classical Latin translated into French. At the other end of the corpus, the 1740 Nancy edition of the *Dictionnaire Français-Latin* known as the *Dictionnaire de Trévoux* is an encyclopaedic dictionary mostly written in vernacular French, which records antiquarian meanings and good usage and quotes extensively from seventeenth-century French authors and critics, but also provides Latin equivalents after headwords or quotations within brackets, and uses Latin to differentiate the various meanings of the same French word.

Second, Latin words feature as etymons rather than translations in antiquarian and technical dictionaries. The derivation of *engin* from *ingenium* is thus glossed in Gilles Ménage's *Les origines de la langue francoise* (1650) and its much expanded 1750 edition, the *Dictionnaire étymologique ou origines de la langue française*. The same etymological gloss on *engin* and comment on its obsolescence feature in the three "technical" dictionaries of the corpus. These are Antoine Furetière's *Dictionaire universel* (1690), a monolingual dictionary claiming to focus on the technical vocabularies of specific arts, sciences, and trades rather than on common usage; Thomas Corneille's *Dictionnaire des arts et des sciences* (1694), straightforwardly plagiarized from Furetière's *Dictionaire universel*; and Savary Des Bruslons's *Dictionnaire universel de commerce* (1726), a monolingual dictionary of the technical vocabularies of various trades. Etymologies also pepper Pierre Bayle's *Dictionnaire historique et critique* (1697), an erudite, antiquarian encyclopaedia of names: we have focused on its fourth (1730) and fifth (1740) editions.[7]

Third, defining good usage and lexical propriety is a prominent feature of another set of dictionaries. In fact, the main dictionaries of the end of the century (Richelet 1680, Furetière 1690, Académie 1694) all participate in this normative enterprise: good usage, institutionally defined by the state-sponsored *Académie*, thus became the linguistic expression of the rise of political absolutism. Its paradigm is the first edition of the *Dictionnaire de l'Académie françoise*, which defines lexicographical purity in common usage in the vernacular.[8] The academic normative stance had been partly preempted by César-Pierre Richelet, whose monolingual *Dictionnaire des mots et des choses* recorded a wide range of uses while quoting abundantly from the authors of the *Académie*, and partly by Furetière, whose own dictionary included a wider lexical range than the jargons with which he purports to be exclusively concerned. Trévoux 1740 crowns this normative enterprise, while the abbé Girard's *La justesse de la langue françoise* (1718, reedited in 1736 and 1740), a "distinctionary" of clusters of words wrongly assumed to be synonyms, exemplifies its pedantic excesses.[9] These dictionaries are the richest for the lexicography of ingenuity, because they display the very entanglement between the linguistic form of French ingenuity that the *génie de la langue* represents, and its political, artistic, and especially literary expressions.

Finally, at the end of our period, Diderot and d'Alembert's *Encyclopédie ou Dictionnaire raisonné des arts et des sciences* went far beyond the lexicographical ambitions of the other texts in our corpus. No fewer than four specialist entries were dedicated to *engin*, three to *génie*, and seven to *esprit*. The general entry for *engin* in mechanics was written by d'Alembert, the other three—on the mechanical arts, architecture, and nail- and pin-making—are by anonymous authors: all feature in the fifth volume (1755). Voltaire wrote the entry for *esprit-as-ingenium* or wit in that same volume. As for *génie*, the Chevalier de Jaucourt supplied the antiquarian account of the ancient tutelary god; Saint Lambert wrote the entry on *génie* in philosophy and literature; Guillaume Le Blond dealt with *génie* in fortifications; and an anonymous writer with its meaning in architecture. All can be found in the seventh volume (1757).

• • •

The question of French linguistic identity remains ideologically loaded to this day. Echoing the preface to the *Dictionnaire de l'Académie* and Bouhours's *Entretiens*, Marc Fumaroli states that the King's eloquence defines the linguistic norm—that of the *naturel* as inborn linguistic mastery equated with an innate politeness or sense of decorum—and celebrates the lexicographical and literary promotion of the *génie de la langue* as the cultural program of an absolute monarchy that effectively disciplines (polices and polishes) the nation into a purist monolingualism.[10] This position has been increasingly contested. For Hélène Merlin-Kajman, the national, political identity that lexicographical purism promoted was one of politeness as civility; the cool, more dispassionate, and rational language of the *âge classique* thus consciously broke loose with the vociferous eloquence of the Wars of Religion and its dangerous political consequences.[11] In Merlin-Kajman's narrative, the *génie de la langue* as *esprit français* is no longer merely the polished *naturel* of the king; it identifies as "polite" a vernacular salon culture of courtiers in contradistinction to the caste of erudite, Latinate *robins* whose deliberative eloquence had fueled the civil wars of the previous century. Thus Furetière noted the absence of legal and technical jargon in the *Dictionnaire de l'Académie*, whose aim was to polish the "beau style" (the language of the court) rather than the "style de palais" (the language of the Parlement).[12] Some argued that ultimately linguistic purism (the *génie de la langue*) could well be an illusion, which might in fact impoverish literature by establishing a rarefied paradigm—the *ouvrages de l'esprit* defined by Bouhours and quoted by most normative dictionaries. Linguists from

Dauzat onward have pointed at the hybridity of classical French and nuanced the depiction of enforced monolingualism, thus highlighting the inherent tensions and *bigarrures* in the *génie de la langue*.[13] Following Marie de Gournay's warning in her 1641 *Advis* that linguistic purism could amount to poetic poverty, contemporary specialists in linguistics and poetics have argued that literature—the expression of one's *naturel* as an irreducible singularity—stemmed on the contrary from a linguistic resistance to purism.[14]

The scholarly battle still rages regarding the political and cultural stakes of the lexicographical expression of French ingenuity—be they the ordering and disciplining of individual specificities and emerging social categories, or the forceful definition of the unity of a nation. In the *âge classique* itself, the French translations of the Latin *ingenium*—in particular *esprit*—were incessantly contested and debated terms in the political and cultural arenas. Even more than in Latin and polyglot dictionaries, the lexical effort at defining French monolingual propriety outlined a clear social and cultural order predicated on the identification and assessment of the variety of human *esprits* and *génies* according to their temperaments and cognitive idiosyncrasies, tempered by an increasingly normative recourse to the notion of *naturel*. *Engin* strikes a discordant, antiquarian note in this story.

Engin

The word history of *engin* involves three main meanings:

Engin, n.
    1. Cognitive power of invention:
        a. Legal cunning and trickery.
        Nicot 1606.
        b. Intellectual resourcefulness.
        Nicot 1606 > Académie 1694 (as "Industrie") > Trévoux 1740 (as "subtilité").
    2. Machines:
    Nicot 1606 > Richelet 1680 (devices for lifting weights) > Furetière 1690: specified as military siege devices > Trévoux 1740 (as legal

instruments, the sugarmill) > Anonymous 1755A and B (as lifting weights in architecture, a wire-pulling bench for nail- and pin-making) > d'Alembert 1755 (as complex machines with several parts).
3. Penis:
Richelet 1680 >.

The word history of *engin* is either a suppressed or an antiquarian one in the corpus—the word is an obscene hapax in Richelet 1680—or repeatedly described as an archaism.[15]

Nicot retrieves the medieval Latin meanings of *ingenium*, that of the hidden (criminal) intention or motive to be assessed in court, and that of the machine (usually used in sieges). Most of his successors, from Ménage 1650 onward, note that the term is obsolete, especially in its cognitive sense: Richelet 1680, Furetière 1690, and Corneille 1694 use the spelling *engien* to denote its semantic obsolescence, thus distinguishing it from the mechanical meaning of *engin*—although the mechanical *engin* itself becomes obsolete in the *Encyclopédie*.

To trace the first, cognitive meaning of *engin* in the corpus is to attend the birth of a proverb: its form is set, its author dissolves into anonymity, its original meaning becomes increasingly diffuse. While Nicot 1606 is the first one to note that *mal engin* translates *dolus malus*, Ménage 1650 first features the expression "engin vaut mieux que force," and ascribes it to Rabelais.[16] It then appears in Académie 1694 and Trévoux 1740 as an anonymous proverb. The meaning of *engin* as cunning in this proverb is progressively lost: thus the *Académie* equates *engin* with industry,[17] while Trévoux translates *engin* quite loosely as craftiness and polite, amicable manners before quoting Voiture's modern recycling of the pairing "wit and strength": Rabelais has become anonymous, proverbial, and obsolete, and must be replaced by the more elegant and up-to-date Voiture.[18]

Ménage 1650 is the first to connect etymologically the intellectual and mechanical meanings of *engin*, left disjointed in Nicot 1606: machines are wit objectified, hence their labeling as *engin*. Furetière 1690, Corneille 1694, and Trévoux 1740 all repeat this explanation of the semantic derivation from wit to machines. Mechanical *engins*

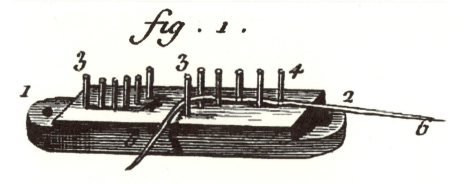

Fig. 4.2. "Engin," fig.1 in the "Aiguillier/ bonnetier" set of figures of the *Encyclopédie, ou dictionnaire raisonné des arts et des sciences* (Paris, 1762).

are primarily designed to lift heavy weights. The detailed description of their constitutive parts fills the pages of technical dictionaries such as Furetière 1690, Corneille 1694, and Des Bruslons 1726. In the *Encyclopédie* especially, the referential illusion they intend to summon soon turns into a self-referential maze of jargon: without the *planches* of supplementary volumes, the reader does not "see" the machine, but is confronted with yet more technical words in need of definitions. Trévoux's sugarmill—like the *ingenio del azucar* of the Spanish Empire—is a notable example.[19] As for the three different entries dedicated to various types of mechanical *engins* in the *Encyclopédie*—in general, in architecture, and in nail- and pin-making—they highlight hair-splitting semantic differentiation in the lexicography of the already obsolete mechanical *engin*, while reflecting the growing cultural importance of engineering and the humbler mechanical arts.[20] These technical definitions make very little sense to the lay reader on their own, and demand the supplement of technical drawings: entries on *engin* in architecture and pin-making thus ultimately send their readers back to the relevant *planches* in figure 4.2.

Richelet 1680 stands out in reminding the reader that the generative etymon of *ingenium* (*gigno*) is alive and well in *engin* as penis, yet the word operates in the satirical and burlesque registers repudiated by other dictionaries for being either unworthy of elegant conversational French or irrelevant to specialist, technical jargons.[21] Richelet is a healthy reminder of the censorship involved in the normative lexicographical disciplining of the *génie de la langue*. Already *passé* in the seventeenth century, the medieval *engin* has thus been replaced either by *machine* or, as Nicot notes, by *esprit*.

Esprit

The term *esprit* single-handedly translates several Latin terms: *spiritus*, *mens*, *anima*, as well as *ingenium*. In the corpus, *esprit* generates five semantic clusters; four of these will *not* be our main focus since they do not relate to ingenuity (they do not translate either *ingenium* or *genius* in Nicot 1606).[22]

Esprit-as-*ingenium* or *genius* can be summarized thus:

Esprit, n.
   1. Cognitive faculties of the soul.
   Nicot 1606 ("l'entendement que l'homme a de nature") > Richelet 1680 ("Substance qui pense. Partie de l'ame qui juge, comprend, raisonne, & invente ce qu'on peut s'imaginer.") > Académie 1694 ("les facultez de l'ame raisonnable") > Voltaire 1755 ("raison ingénieuse").
   2. Particular disposition or ability to succeed in a specific activity.
   Académie 1694 >.
   3. Metonymic: person. (Calls for a qualifier)
   Richelet 1680 >.
      a. *(Bel) esprit*.
      Richelet 1680 > Académie 1694 ("On appelle, *Beaux esprits*, Ceux qui se distinguent du commun par la politesse de leurs discours & de leurs ouvrages.") > Trévoux 1740 (from Bouhours 1671; positive and negative assessment of the person) >.
   4. Meaning, character of a thing or person.
   Richelet 1680 ("Caractere qui fait voir l'ame, le coeur, & la conduite d'une personne") > Académie 1694 (adds "le caractère, le style d'un auteur") >.
   5. Design, intention.
   Richelet 1680 >.

The history of *esprit*-as-*ingenium* is characterized by growing analytical precision in the definitions, and by the emergence of two related semantic traits in the presentational meaning of ingenuity. First, broadly construed, *esprit* is the faculty or tendency that drives specific, productive performances of identity. Second, its counterpart is a critical exercise intent on deciphering such a performance. *Esprit* implies a social and ultimately literary performance of identity and singularity subjected to the critical and "tasteful" reading of a community of *honnêtes gens*. The very exercise of such critical wit that drives the choice of dictionary examples, as well as the lexicographical vicissitudes of the *bel esprit*, illustrate this last point.

Indeed the definitions of *esprit* demonstrate increasing analytical rigor. Nicot 1606 provides a single, tautological, and yet confusing definition of *esprit*; it alludes to one Latin morphological etymology (*spiritus*) before concluding the definition with another Latin semantic equivalent (*ingenium*).[23] While the definition itself emphasizes the cognitive meaning of *esprit* (understanding), the translated Latin examples also allude to its temperamental meaning, and suggest yet other Latin semantic equivalences with *mens* and *anima*. By contrast, Trévoux 1740 neatly delineates the cognitive meaning of *esprit* as *ingenium*, before dedicating lengthy sub-entries to specific instances of such *esprit*: for example *bel esprit*, *esprit fort*, and *esprit particulier*.

This quest for analytical precision regarding esprit as *ingenium* starts to emerge in Richelet 1680 (four sub-entries); it becomes more pronounced in Académie 1694 and culminates in Voltaire 1755. Voltaire explicitly makes wit his focus and discards other meanings in his introduction:[24]

> Quand on dit qu'un homme a un esprit judicieux, on entend moins qu'il a ce qu'on appelle de l'esprit, qu'une raison épurée. Un esprit ferme, mâle, courageux, grand, petit, foible, leger, doux, emporté, &c. signifie le caractere & la trempe de l'ame, & n'a point de rapport à ce qu'on entend dans la société par cette expression, avoir de l'esprit. L'esprit, dans l'acception ordinaire de ce mot, tient beaucoup du bel-esprit.

> When a man is said to have a judicious mind, it does not so much mean that
> he is what we call a wit than that his reason is neat. A firm, virile, courageous,
> great, small, feeble, light, sweet, quick-tempered mind means the character and
> make of one's soul; it is not related to what is meant by wittiness in good society. Being witty, according to its usual meaning, is very much like being a wit.

This exclusion testifies to the progressive emergence of highly specific meanings of *esprit*.

The evolution of the cognitive meaning of *esprit*, from the "natural understanding" of Nicot 1606 to the detailed faculty psychology of Académie 1694, reflects the natural philosophical debates of the period on the nature of the soul.[25] The Cartesian definition of the self in Meditation 2 echoes in the initial "thinking substance" of Richelet 1680: "Substance qui pense. Partie de l'ame qui juge, comprend, raisonne, & invente ce qu'on peut s'imaginer" (Thinking substance. Part of the soul which judges, understands, reasons, and invents what can be imagined). The reference to imaginative invention as one of the "parts" of the rational soul might also relate to the precise definition of the *ingenium* in Descartes's *Regulae ad directionem ingenii*.[26] In Richelet 1680, Académie 1694, and Trévoux 1740, concept formation, imaginative invention, and judgment, to which Trévoux adds memory, are consistently ascribed to *esprit* as *ingenium*.[27] As in Latin, *esprit* lends itself to cognitive and moral qualifiers: one's *esprit* can be fast or slow, thick or thin, great or small, etc.—a point noted by Voltaire.[28] In this qualified configuration, *esprit* is particularly prone to metonymic use and denotes the individual as a whole (see 3, 131). This use is recorded in Richelet 1680 and Académie 1694. *Bel esprit* and *esprit fort* are particularly significant instances of such metonymic use. By granting these terms discreet sub-entries in Académie 1694 and Trévoux 1740, French lexicography records the emergence of increasingly prominent and contested social types in the period.

The metonymic use of *esprit* to designate a person makes the term well suited to denoting identity. The remaining two strands in the word history of *esprit* as *ingenium* provide a complementary semantic perspective on *esprit* as an identity marker. The presentational, performative meaning of *esprit* denotes the social and cultural construction and performance of identity, the display of one's abilities and one's nature, whether in economic terms (see 2, p. 131) or in artistic ones (see 3.a, 131). *Esprit* can also denote the collective interpretation and assessment of these performances of identity: this is its critical meaning (see 5, p. 131), which captures the social reception and normalization of its performative counterpart. In the dictionaries, both the performative and critical meanings disclose a recurring anxiety with sophistication as a distorted performance of identity, be it social duplicity or cultural affectation. Both also outline the reaction against such risks by promoting the social norms of politeness and *honnêteté*, and by defining the *naturel* as a cultural ideal. This anxiety can be traced in the shift from things to signs in the definitions of *esprit*: its ontological meanings—one's abilities, one's nature and temperament are increasingly taken on by the rival translations of *ingenium* in the period, namely *génie* and *nature/naturel* respectively. *Esprit* proper becomes confined to the performance of these and to the social and cultural assessment of such performance. One's nature is not readily accessible to others; it is displayed as a semiotic system in need of "reading" by others. The code of that social *semiosis* is fixed by the community of *honnêtes gens*, or *gens d'esprit*—Voltaire's "good society" (*en société*). The rise of the notion of character over the period denotes the omnipresence of this semiotic account of identity: the term *caractère* already features in Richelet 1680, eight years before the first edition of La Bruyère's *Caractères*, and in Académie 1694, where it denotes authorial intention and manner, or style:[29] "Esprit: Caractère qui fait voir l'ame, le coeur, & la conduite d'une personne" (Character which discloses the soul, heart, and moral compass of a person).[30] "Le Caractère d'un auteur, sa manière de concevoir et d'énoncer les choses" (The character of an author, his manner of conceiving things and putting them into words).[31]

According to the performative meaning of *esprit*, what one does (see 2, p. 131), the ways in which one behaves (see 4, p. 131), and what one writes (see 3.a, p. 131) all disclose one's *esprit*, that is, an identity performed in terms of practical abilities, social attitudes and, ultimately, style. Thus *esprit* denotes the specific ability to carry out a particular task or business: "Il a l'esprit des affaires, du procès" (the flair for business, or legal matters),[32] or behavioral stances in social interactions: "Un esprit commode, un esprit fascheux" (an easygoing character, an intractable character); "Les petits esprits ont l'air de parler beaucoup et de ne rien dire" (petty wits seem to speak much and say little).[33] *Esprit* as social and cultural performance of identity thus suggests in the dictionary examples the delineation—*à la* Huarte de San Juan, yet in a stridently critical fashion—of a social or national economy of wits in which individuals can be ordered, and sociocultural and national groups defined in exclusive and normative terms. Thus the libertine or *esprit fort* as an increasingly defined and confined sociocultural type deserves its own sub-entry in Académie 1694 and Trévoux 1740; its implicit negative assessment as an arrogant and antisocial stance in Académie 1694 is made explicit in Trévoux 1740: "Esprits forts. . . . Ceux qui se mettent au dessus des opinions & des maximes communes. . . . il fait l'esprit fort, il ne croit rien, il ne tombe d'accord de rien" (Strong wits. . . . Those who deem themselves above commonly received opinions. . . . He plays the part of the strong wit, he believes in nothing, he never agrees with anything).[34] "Esprit fort. . . . est une espèce d'injure qu'on dit à ces libertins & ces incrédules qui se mettent au dessus des la créance et des opinions les plus reçuës. La plupart des beaux esprits sont des esprits forts" (Strong wit. . . . is a type of insult addressed to those libertines and faithless people who deem themselves above the most commonly accepted beliefs and opinions. Most people deemed to be wits are strong wits).[35] In a similar fashion, Trévoux 1740 often denies wit to women, the elderly, and Germans, while in Richelet 1680 the unattributed reference to Pascal's *Provinciales* suggests the satirical use of *esprit* to deride the sophistication of the jesuitical style by playing on the ambiguous meaning of "l'esprit de la société": the spirit of the Society of Jesus *and* the mundane wit of *bel esprit*, Richelet seems to suggest, are not that different from each other.[36]

That same *bel esprit* was one of the most contested terms of the period.[37] The expression telescoped both the cognitive meaning of *esprit* and its metonymic use, and became the very site where the reaction against sophistication—of specific cultural stances, social types, and literary styles—was voiced more and more forcefully in terms of opposition between *bel esprit* and *esprit* "tout court" or *naturel*. Thus *bel esprit* features as a sub-entry in Académie 1694; Trévoux 1740 dedicates to it an intricate patchwork of lengthy quotations from Dominique Bouhours's dialogue on this topic; finally, it is the main focus to the exclusion of all others in Voltaire 1755. From the Académie to Voltaire, the lexicographical account of this term is increasingly ambivalent: the polite elite that the wits ("Messieurs les beaux esprits") stand for in the *Académie* has become a clique and almost a profession—the salon wit evolving into the professional writer—which can prompt envy but also, and for Voltaire it seems mostly, mockery:

> L'esprit, dans l'acception ordinaire de ce mot, tient beaucoup du bel-esprit, & cependant ne signifie pas précisément la même chose: car jamais ce terme homme d'esprit ne peut être pris en mauvaise part, & bel-esprit est quelquefois prononcé ironiquement. D'où vient cette différence? c'est qu'homme d'esprit ne signifie pas esprit supérieur, talent marqué, & que bel-esprit le signifie. Ce mot homme d'esprit n'annonce point de prétention, & le bel-esprit est une affiche; c'est un art qui demande de la culture, c'est une espece de profession, & qui par-là expose à l'envie & au ridicule.
>
> Wittiness, in common parlance, is very akin to wit, yet does not mean precisely the same; for the expression "a witty man" can not be construed negatively, whereas "a wit" can be uttered ironically. Whence such a difference? Well, "a witty man" does not mean a superior mind and marked talent, whereas "a wit" does. This expression "a witty man" is devoid of any claim, whereas "a wit" blatantly displays some: it is an art which requires cultivation, a sort of profession which, for this very reason, is exposed to both envy and ridicule.

This ambivalence is perfectly captured by Trévoux's selection from Bouhours. *Bel esprit* can be construed both positively and negatively. Bouhours summarizes the positive account of *bel esprit* as "the shine of good sense" (*le bon sens qui brille*), that is, the social display and play that the good mind of a generally well-tempered individual can put on—a mind characterized by quick yet solid invention and good judgment.[38] The performance of such *bel esprit* can be reified into ingenious works (*ouvrages d'esprit*) that Trévoux, following Andry de Boisregard, distinguishes from the products of the human mind (*ouvrages de l'esprit*).[39] Such ingenious thoughts feed lively conversation; their paradigm is literature understood as the written display of wit by men of letters.[40] However, Trévoux's entry on *bel esprit* also includes the negative account of the flashiness that is the trademark of the false *bel esprit* (*faux brillants*): a conceited wit, full of affectation (*un esprit de pointes, plein d'affèterie*). While the true *bel esprit* explores the nature of things—the *res* or referent of language—the false *bel esprit* remains trapped in mere wordplay.[41] The positive account of *bel esprit* in Trévoux 1740 asserts the paradigm of a good social and cultural performance of wit: the true *bel esprit* is the polite display of one's *naturel* encapsulated in conversation.[42] Trevoux 1740 summarized over a century of debates on the role of conversational genres that had risen to prominence over the seventeenth century.[43]

The negative account of *bel esprit* in the dictionaries captures a poor or subversive social and cultural performance, one contrived by dissimulation or sophistication.[44] Such sophistication involves an overly pronounced concern with words, that is, with style understood as pure rhetorical ornament, which has the false *bel esprit* excel in the mundane forms of epigrams, madrigals, impromptu rhymes, and other improvisations: salon games, but also minor and lapidary literary genres. Thus the academician Jean Ogier de Gombauld (1576–1666), regularly quoted from Richelet onward, makes the lowly genre of the epigram a symptom of cultural decadence: "Nos moeurs, nos actions infames/ m'ont reduit a des Epigrames."[45] The account of the false *bel esprit* in Trévoux is also, in this respect, a critical genealogy of the modern notion of writer (*auteur*) born from salon culture: "un auteur," and not "un homme" as Pascal would have it: the distinction in Trévoux carries the same pejorative baggage.[46] Indeed Trévoux notes that the term *bel esprit* has been usurped by mediocre, mundane writers to such an extent that true literary talents consider it an insult: "Vous êtes un bel esprit, disoit un Provincial à M. Racine: Bel esprit vous même, répondit brusquement M. Racine; comme si on lui eût dit une injure" (A provincial told M. Racine: "You are a wit." As if insulted, M. Racine snapped back: "wit yourself"). Fifteen years later, Voltaire's entry for *esprit* confirms the trajectory identified in Trévoux 1740: Voltaire re-labels the true *bel esprit* as *esprit* tout court, and contrasts it with *faux esprit*. The symptom of *faux esprit* for Voltaire is a *sophisticated* false thought instantiated in a conceit that violates verisimilitude: he gives the example of one such violation in a translation of Homer's *Iliad* in French: "[il] crut embellir ce poëte dont la simplicité fait le *caractere*, en lui prêtant des ornemens. Il dit au sujet de la réconciliation d'Achille: "Tout le camp s'écria dans une joie extrême, Que ne vaincra-t-il point? Il s'est vaincu lui-même." . . . toute une armée peut-elle s'accorder par une inspiration soudaine à dire une *pointe*?" (He thought he was beautifying Homer, whose *character* resides in his simplicity, by adorning him. Thus he said, about Achilles's reconciliation: "The whole camp shouted in an outburst of extreme joy: what will he not conquer? He conquered himself." . . . A whole army would agree, in a sudden fit of inspiration, to utter a *conceit* all at once? [Emphases ours]). Unlike Trévoux, however, Voltaire situates the whole discussion of *esprit* within the remits of style; by the time of the *Encyclopédie*, the conceptual inventiveness and sturdiness that still defined the true *bel esprit* in Trévoux 1740 have become the prerogative of *génie*: for the *encyclopédistes*, philosophers have or lack genius, writers have or lack wit. Voltaire reads and assesses the semiosis of *esprit* in oratorical and literary performances only: aptly enough, he quotes mostly madrigals. The debates surrounding the performative definition and disciplining of *esprit* as *ingenium* have become exclusively enmeshed with the polemic surrounding the notion of literature by the middle of the seventeenth century.

The critical meaning of *esprit* (see 5, p. 131) is the expected counterpart to the performative one we have just outlined. As a social and cultural performance of signs of one's identity—characters and styles—*esprit* demands the interpretation of a community of critical "readers." Thus, the *honnêtes gens* define the semiosis of ingenuity, assess the propriety of its performance against the paradigmatic touchstone of *esprit-tout-court* as polite *naturel,* and label accordingly the performer as a "bel esprit" (a wit), "esprit fort" (a libertine), "esprit (trop) fin" (a conceited wit), or "esprit pesant et ennuyeux" (a sluggish and boring wit). In some entries, this critical meaning of *esprit* sits alongside the medieval, legal meaning of *ingenium* as the motives or intentions of a person to be assessed by a judge in trial, but also as the spirit—by opposition to the letter—of a legal text.[47] Seventeenth-century dictionaries progressively widen this purely legal context to encompass the interpretation and assessment of intentions, whether of texts or of people, in the same entries. This medieval meaning is, as expected, missing from the classicizing French of Nicot 1606; it features explicitly in Richelet 1680, which in the same sub-entry conflates two types of legal interpretation: the motive driven by one's character with the spirit of a legal text.[48] The legal context has disappeared in Académie 1694, which juxtaposes in two continuous sub-entries the interpretation of one's motives in action reliant on character, and the interpretation of *authorial* intentions in texts. This latter instance echoes the emergence of the critic as one who interprets and evaluates such authorial *esprit*. Trévoux 1740 provides the most interesting account of this critical meaning by embedding the definition of *esprit* as "meaning, character, understanding of thing, intention, motive, design, sentiments which drive one's action" within legal, diplomatic, and finally literary and social contexts. As in Richelet 1680, Trévoux mentions the spirit of a contract; this legal context also underpins the reference to the assessment of intention in a trial for homicide, where it prefigures the distinction between manslaughter and premeditated murder: "Quand on fait le procès à un homicide, on regarde s'il l'a fait innocemment, ou par un esprit de vengeance" (When trying someone accused of homicide, attention should be paid to whether or not one did it innocently, or in a revengeful spirit).

Esprit then denotes the rationale of Spanish foreign policy, driven by a spirit of universal dominion (*un esprit de domination universelle*). Finally, it refers to the intended meaning, authorial intention, and character one should capture in a good translation, put on a par with the ability to sense the sort of polite wit that implicitly defines good society: "Il faut, en traduisant, prendre bien l'esprit de son Auteur, son sens, son caractère. Il règne toujours un esprit de politesse dans la société des honnêtes gens" (In translating, one should absorb well the spirit of the author, his meaning and character. Good society is always suffused with a sort of polite wit). Even more than the invention of the modern author defined by a singular style, the word history of *esprit* testifies to the rise of the modern critic as the referee of social and literary propriety, whose acumen and judgment increasingly become displays of wit in their own right. Thus, while satirical puns and conceits are absent from Nicot 1606 and explicitly banned from Académie 1694 (which only records the calcified form of wit that proverbs stand for), they already feature in the examples of Richelet 1680, and are omnipresent in Trévoux's mesh of quotations.[49] As for Voltaire 1755, the whole entry functions as a display of critical wit in its own right. The social and cultural disciplining at stake in this critical exercise of wit also proves increasingly violent. Praise of great wits grows rarer in the examples. By contrast, the epigrammatic or conceited attack castigates a misfit through the sort of "scapegoating laughter" whose anthropological mechanism underpins the effectiveness of *comédie de caractère*, and assumes shared social and cultural norms with the laughing audience. In this case, critical wit fosters the exclusive social logic of cliques, coteries, and classes through the merciless skewering of identities.[50] This social disciplining upholds the *naturel* as a new ideal: ease of manners, unaffected style, and effortless display of one's singularity. Paradoxically, this new social norm was an artfully constructed nature requiring education, polish, and control.[51] The *naturel* is therefore central to the early modern French culture of ingenuity, and is one of its keywords.

Naturel

Like *esprit*, the words *naturel* but also *nature*, which both translate *ingenium* in Nicot 1606, also stand for a whole raft of other Latin terms, which can be ordered in seven semantic clusters.[52] The semantic evolution of *naturel*-as-*ingenium* can be summarized as follows:

Naturel (*nature*), n.

1. Innate characters and properties that define the nature of a specific being.

Nicot 1606 >.

2. Innate dispositions from birth defining one's temperament, mode of actions, abilities.

Nicot 1606 >.

Unlike *esprit*, some overlaps between these semantic clusters are significant for the word histories of *nature* and *le naturel*, in particular the one between the mimetic, or paradigmatic, meaning of *nature* as the model of all arts (meaning F; see note 52) and *le naturel*.

The humoural meaning of *nature/naturel* related to temperament remains stable in all dictionaries: *ingenium* as *nature* and *naturel* denotes the native and bodily determinations of identity. Human singularity is only one instance of this form of corporeal specification described in terms of humours or temperaments: the dictionary examples mention the varying natures of the soil (Nicot 1606, Trévoux 1740), or the determinations of temperaments in animal species alongside individuals: lions are savage and cruel just like Nero in Furetière 1690 and in Trévoux 1740.[53] This definition of *nature/naturel* as temperament progressively leads to the definition of one's abilities: one's *naturel* also shapes what one does, one's mode of action—what one enjoys doing, and especially what one is particularly good at. This semantic trajectory from temperament to aptitudes is also that of *génie*.[54] It is encapsulated in the expression *avoir du naturel pour* (to be naturally gifted for).[55]

As embodied principle of identity and source of one's abilities, one's *nature* and *naturel* proves a favored site of social, pedagogical, and artistic disciplining but also corruption. Ongoing debates about nature versus culture and its shaping of individual identities reverberate in dictionary examples. Furetière 1690 warns that a young gentleman's good nature will be spoiled by the commerce of bad company: "Les mauvaises compagnies ont gasté tout le bon naturel de ce jeune homme" (Bad company has spoiled this young man's good nature).[56] Copious excerpts also suggest an embryonic disputation on what is more important, a good education or a good nature, in Furetière 1690, taken up by Trévoux 1740:

> "Il faut cultiver le naturel quand on l'a beau" (one must cultivate one's nature when it is a beautiful one).[57]

> "Une éducation excellente, avec un naturel médiocre, est préférable au plus riche naturel du monde, avec l'éducation ordinaire" (the excellent education of a mediocre nature is preferable to the richest nature in the world left to ordinary education).[58]

> "Bien des gens s'éloignent de leur naturel, et se defigurent pour plaire" (many stray away from their own nature, and disfigure themselves in order to please).[59]

The artistic expression and disciplining of this *nature* and *naturel* is particularly worthy of attention, as its lexicographical history in the contrasting pair *naturel/artifice* matches that of *esprit/bel esprit*. This history nuances the extremely rational views on mimesis supposedly emphasized during the *âge classique*—the "age of reason" promoted art as a ruled practice that required industry and labour, *à la* Boileau.[60] Since Richelet, the products of one's wit are *par excellence* the place where one's *nature* and *naturel*-as-abilities are not only displayed, but assessed. Seventeenth-century lexicographers take for granted that one should polish one's *naturel*: Theophile de Viau, who neglected his natural talent for poetry, is paradigmatic of the artistic failure stemming from an unpolished *naturel*.[61] While, paradoxically, the *naturel* is meant to be the expression of irreducible singularity ("what is not general in man," Furetière reminds us), it increasingly becomes a norm of good social and artistic performance.[62] This is apparent in the meanings of the adjective *naturel*: truthfulness (*genuinus*), sincerity, and polite ease of manners (*ingenuus*), also present in the definitions of *nature* and *le naturel* in some dictionaries, where these qualities are always contrasted with artifice and contrived affectation:

Naturel, adj.

Richelet 1680: "Ce mot se dit des vers et de la prose" (This word qualifies verse and prose).

Furetière 1690: "... ce qui est libre, qui ne parait point forcé. Cet Orateur a l'action belle, le geste *naturel*; il a un stile fort naturel, fort coulant qui n'est point enflé ni affecté" (what is free, does not come across as contrived. This Orator's acting is beautiful, he has a natural sense of gesture; his style is most natural, it flows nicely, and displays neither affectation nor excess).

Académie 1694: "Facile, sans contrainte. *Il a un air aisé & naturel. Il se dit aussi en ce dernier sens, des ouvrages d'esprit & de l'esprit mesme. Les vers qu'il fait sont naturels. son style n'est pas naturel. il a l'esprit naturel*" (Easygoing, unconstrained. He looks natural and at ease. Used in this latter sense about the products of one's wit, and about wit itself. His verses are natural, his style is not natural. He has an unaffected manner).[63]

Like *esprit*, *naturel* epitomizes a new cultural and artistic paradigm of polite artlessness against *artifice* and *bel esprit*. Trévoux makes this final leap from cultural norms of self-discipline and representation to artistic norms of mimetic representation by conflating *nature* as the paradigm of all arts (meaning F; see note 52) with the *naturel*-as-*ingenium* in a single entry: "se dit des moeurs, du génie, des manières particulières à chaque personne: ou de l'ordre & de la conduite de la nature dans la production et dans l'arrangement des choses (*indoles*)" (The morals, genius, and behaviors specific to every person: or about the order and discipline of [one's] nature in the production and disposition of things).[64] His examples amalgamate *nature* as the objective order that the artist attempts to capture mimetically (*natura*) and *nature* as the subjective ability and dynamism the artist needs to discipline in order to achieve good mimesis (*ingenium*, by opposition to *ars*, inherited from the rhetorical account of invention).[65] Trévoux's entry thus puts the *naturel*-as-*ingenium* squarely at the center of the mimetic poetics of French classicism; the faculty or rather the ability at stake in good artistic representation is not so much the universal reason that supposedly typified the age, but one's singular *naturel* as equated by both Nicot and Trévoux with one's *génie*:

> Nature: La nature et manière de faire que l'homme a de nature (*ingenium*) . . . ne defraudare genium: ne contenter pas nature . . . suopte ingenio: de son naturel (Nature: the nature and manner of action man gets from nature . . . defraudare genium: to rob one's nature of its due . . . suopte ingenio: out of one's own nature).
>
> Génie: Est le naturel et inclination de chacun (One's nature and inclination).[66]

Génie

The interplay between the naturalist and antiquarian meanings of *génie* in French—respectively, one's natural bent and wit and one's tutelary spirit(s)—echoes the Latinate word history of *genius*. It highlights the ways in which the temperament and moral meanings of ingenuity as markers of identity were embedded within the early modern handling of the "problem of paganism" in particular, and the question of religion in general. As in Latin, lexicographical omissions and silences about *génie* point to the limits of word histories in the investigation of early modern French ingenuity. The semantic range of *génie* in the corpus is as follows:

Génie, n.
    1. Moral character and temperament.
        a. Natural inclination or dispositions (including passions and virtues).
        Nicot 1606 >.
        b. Natural talent.
        Nicot 1606 >.
    2. Wit.
    Richelet 1680 >.
    3. Pagan tutelary spirit(s), good and bad angel.
    Richelet 1680 >.

4. Metonymic uses: people, person.
Trévoux 1740.
5. Chorography: nature, natural property.
Trévoux 1740.
6. Military engineering.
Trevoux 1740 >.
7. Architecture: winged allegories of passions and virtues.
Furetière 1690 >.

Like *esprit*, the definition of *génie* expands significantly between Nicot 1606 and Jaucourt, Saint Lambert, and Le Blond 1757. As for *esprit*, lexicographers handle this expansion analytically. *Génie* is a single entry in Nicot 1606, whereas Trévoux 1740 dedicates seven separate entries to all six senses listed above and distinguishes between various senses of *génie*: as natural talent, disposition, humour, taste, and manners (see 1, p. 144); as wit (see 2, p. 144); and as the various instances of the tutelary spirit that include the God of Nature in fables, the spirit providing moral guidance in poetry, the Christian good and bad angels (see 3, p. 144).[67] Finally, the *Encyclopédie* deals with *génie / genie / génies* in the lengthy, separate entries for *génie* (see 1–3, 6, and 7, p. 144–45).

Yet the dictionaries barely echo a booming "culture of genius." Definitions are overall stable and examples scarce. While Richelet 1680 and Trévoux 1740 indicate the semantic overlap between *esprit* and *génie*, there is nothing among the lexicographical predecessors of Saint Lambert 1757 that foretells his sophisticated, pre-Romantic account of the man of genius.[68] Dictionaries are mute as to the cultural importance of the concept of *génie* that Saint Lambert 1757 reveals: its history lies beyond the lexicographical threshold of this book.

Three main definitions dominate the lexicographical history of *génie*: the identification of *génie* as one's natural tendency (see 1, p. 144) and natural wit (see 2, p. 144), and its antiquarian definition as tutelary spirit (see 3, p. 144). Their problematic relationship mediates the question of identity through contemporary debates on the anthropology of religion. This suggests another, antiquarian rather than Neoplatonic route into the divine quality of one's *génie*; yet it also deflates this divine quality by naturalizing *génie*—no longer a tutelary god, but one's embodied soul.

Génie as tutelary spirit is missing in Nicot 1606. It appears in Richelet 1680, is the dominant meaning in Furetière 1690 and becomes increasingly autonomous from Académie 1694 onward. Richelet and the Académie list this antiquarian meaning first, and it has its own separate entry in Trévoux 1740 and in the *Encyclopédie* (Jaucourt 1757). These entries provide elaborate historical excavations of the religious culture of paganism. Jaucourt 1757 echoes its naturalization and anthropological handling, and locates *génie* within a historical anatomy of religion as superstition, in line with the agenda of the *Encyclopédie*. By contrast, Trévoux 1740 marks an antiquarian reaction to these libertine uses of pagan *génies*.

Indeed, in line with classicizing Latin dictionaries, seventeenth-century dictionaries tend to explain away the antiquarian meaning of tutelary being by appealing to humoural (temperaments), moral, or cognitive ones: the ancient tutelary god is one's nature allegorized. Furetière 1690 is paradigmatic in this respect: "Sorte de divinité chez les Anciens, laquelle il faisoit presider à toutes choses. . . . Ce mot n'estant plus employé parmy les Chrestiens, que pour signifier un certain esprit naturel qui nous donne de la pente à une chose, on represente les Genies dans les ornements d'Architecture, sous la figure d'enfans ailez, à qui on donne des attributs qui marquent les vertus & les passions" (Type of god among the Ancients, which they had preside over all things. . . . This word being only used among Christians to mean a certain natural bent that inclines us toward a given thing, *genii* are represented in architectural ornaments as winged children, who are given attributes that denote virtues and passions). First a tutelary being, *génie* is then internalized and "Christianized" as one's natural bent; a final move interprets architectural *génies* as personifications of one's dispositions, that is, passions and virtues. Jaucourt 1757 takes the same antiquarian interpretation further: the meaning of *génie* as tutelary spirit discloses the superstitious hypostasis or deification of one's nature. Thus Jaucourt inscribes the cult of the tutelary *génie*, alongside the elf and the fairy, within a long genealogy of human superstitions.[69] His final, translated quotation from Apuleius suggests that such superstition amounts to a deification of the human soul itself: Apuleius notes that the ancient genius was progressively identified with the Manes, the souls of the deceased, and concludes: "Ce fut une plaisante imagination des philosophes, d'avoir fait de leur *génie* un dieu qu'il falloit honorer" (it was a pleasing fancy of the philosophers to turn their genius into a god that had to be honored).[70] The last occurrence of the term *génie* in this sentence refers to one's nature rather than to one's tutelary spirit. Jaucourt thus accounts for the divine nature of *génie* as one's identity from within a larger, naturalist discourse on the anthropological origins of religion as superstition: one's *genius* as tutelary spirit is made up from oneself, as are any other gods. The subversive potential of that argument already lingered in the antiquarian containment and anthropological explanation of paganism from which it originates.[71] Trévoux 1740 thus forcefully denies that *génie* as a tutelary spirit ever denotes the soul of man; in doing so, the entry discloses an uneasy awareness of the possible extremes that the naturalization of the pagan *génie* could lead to in the hands of free thinkers.[72] However, the range of allegorical meanings identified by Trévoux in poetry—from *génie* as bountiful Nature to *génie* as the allegory of one's moral compass in fiction—also belongs within this explanatory dynamic that binds together the naturalistic and antiquarian meanings of *génie*.

The naturalist account of the divine quality of *génie*-as-identity also amounts to its deflation within an anthropological explanation of religious belief devoid of any (Neoplatonic) transcendence. Yet the definition of a singular, exceptional, and praiseworthy inventiveness slowly emerges in the lexicographical trajectory of the temperamental and cognitive meanings of *génie*. Corollary to it is a greater emphasis on the cognitive meaning of *génie* as one's wit. In Nicot 1606, the one and only meaning of *génie* is one's *naturel*, without any example. This meaning also features in Richelet 1680, Furetière 1690, Académie 1694, and Trévoux 1740. The Richelet 1680 entry is the first one to note that, when qualified, *génie* can mean the same as *esprit* and be used metonymically to denote and assess—mostly negatively in the examples selected—a human type,[73] whereas the Académie 1694 entry is the first one to associate *génie* with one's natural talent—*génie* as inclination thus merges one's ingrained desires with one's inborn abilities. All the dictionaries mention the set phrase "travailler de génie," which denotes an easy and natural creative process, implicitly contrasted with a contrived and industrious one smacking of effort, along the dividing lines between *ingenium* and *industria* or *ars*. Trévoux 1740 dedicates a separate entry to both the temperament and cognitive accounts of *génie*. Thus he equates *génie* as natural talent or disposition toward a specific activity with the Latin *indoles* and *natura*, which leads to yet another entry about the metonymic use of *génie* to denote the whole person.[74] The lengthy entry dedicated to *génie* as *esprit* in Trévoux 1740 notes that it is the most common use of the term, and equates it with *ingenium*. Its examples would instantiate just as well the poetic and critical meanings of *esprit* and, in particular, *bel esprit:* they all provide critical assessments of others' wits and normative statements about the role of *génie* in literature. Like *esprit*, *génie* is necessary for good writing and good interpretation, whether such interpretation takes the form of translation or criticism: Saint Évremond and Bouhours, omnipresent in the *esprit* entries, take center stage again here: "Il faut un génie vif et brillant pour la Poësie. Marot avait un génie facile, aisé, plein de délicatesse et de naïveté" (Poetry requires a bright and lively genius. Marot's genius was fluent, easy, full of delicate ingenuousness). "Il faut pénétrer le génie qui anime les Ouvrages des Anciens" (One must grasp the spirit that animates the works of the Ancients).[75] Trévoux's entries themselves are witty displays of conceits in their own right: his critique of the bad critic, lifted from the abbé de Bellegarde, thus reads: "Vous avez détrompé le public, qui vous regardait comme un génie de premier ordre" (You have disabused the public, who held you to be a first-rate genius).[76] Trévoux's entry on *génie* as *ingenium* equates it with *esprit* and testifies to the emergence of a more cognitive meaning of the term.

In striking contrast, *génie* in Saint Lambert 1757 is the very opposite of *esprit*. Saint Lambert first defines *génie* as the creative imagination stemming from a particularly developed, sympathetic sensibility, before examining such genius at work in the arts (epic poetry and tragedy, bucolic poetry and comedy), the sciences (experimental philosophy), and practical knowledge (state governance and military leadership). In each of these, Saint Lambert opposes the manifestations of *génie* to those of *esprit*—his psychology reworks commonplace attributes of the *ingenium* accordingly. Thus, Saint Lambert redefines one of the cognitive abilities of the good *ingenium*, namely reminiscence: it is no longer equated with the analytical ability to reconstruct a chain of inferences in a process of discovery, but with the ability to "relive" or experience a past (or imaginary) event to the fullest of its sensible potentialities. In the arts, good taste is the manifestation of *esprit* and defines a poetics that meets the demands of *decorum*, that is, a social definition of norms. In epic and bucolic poetry and in tragedies, the work of *esprit* is a fine, elegant, and polished one; in comedy, it contributes to enforcing those social norms. By contrast, genius manifests itself in the sublime and defines a poetics of irregularity, which breaks all rules—the rhetorical or poetic rules of a given genre or those of linguistic propriety—and prompts astonishment. Thus Virgil and Racine are elegant wits, and Milton and Homer are geniuses. In comedies, the sublime consists in pinpointing deviation not so much from the social norm as from a cosmic, natural, and universal order.

In the sciences, Saint Lambert notes that experimental philosophy, by its very nature, requires more *esprit* than *génie*, because it involves method, that is, careful analytical progress in abstract reasoning, and *slow*, systematic confrontation of observations in order to identify similarities and differences. The good philosopher, concerned with truth, is therefore driven by a curious intellect, whereas the philosophical genius is driven by a fiery imagination—his thinking is decidedly unmethodical and lateral; swayed by circumstances and his passions, he quickly jumps to conclusions in sweeping moves; he is prone to building systems whose beauty he mistakes for truth. Locke is a good philosopher in that he coldly observed and deduced a number of truths; Shaftesbury is a philosophical genius whose fantastical systems, often built without foundations, contain many errors and occasional sublime truths: "Il y bien peu d'erreurs dans Locke & trop peu de vérités dans milord Shafsterbury [*sic*]: le premier cependant n'est qu'un esprit étendu, pénétrant, & juste; & le second est un génie du premier ordre.

... nous devons à Locke de grandes vérités froidement apperçûes, méthodiquement suivies, séchement annoncées; & à Shafsterbury des systèmes brillans souvent peu fondés, pleins pourtant de vérités sublimes" (There are very few errors in Locke and too few truths in milord Shaftesbury: yet the former was only a wide-reaching, sharp, and accurate mind whereas the latter was a genius of the first order.... We owe Locke great truths that were coldly perceived, methodically tracked, dryly expressed; we owe Shaftesbury brilliant systems, usually devoid of solid foundations, yet full of sublime truths).[77] While too much imagination would in principle hinder good philosophical practice, Saint Lambert also acknowledges that the quick leaps and bounds of the philosophical genius have contributed more to the advancement of learning than the systematic findings of the pedestrian, methodical philosopher—*pace* Descartes, who remains listed among the philosophical geniuses who sparked the progress of the discipline alongside Bacon, Malebranche, and Leibniz. As for practical knowledge—statesmanship, counsel, and military leadership—the man of genius might well be able to find out the rules of good government; yet his heated, passionate temperament and his love of fanciful, grand systems mean that he risks lacking the cold blood, steadiness, discretion and realism required of a good prince, counselor, or minister. For the same reason, the man of genius is a great tactician but not a great strategist in war; his ability to take in, assess, and react to occasion is a marked advantage in the heat of battle, but he neglects the rational reality checks needed to plan properly a whole military campaign.

Saint Lambert 1757 assumes a rich and continuous intellectual history of the notion of ingenuity and its expressions—rhetorical invention, reminiscence, or quick thinking—at the crossroads of *esprit* and *génie* in early modern France: nothing in the entries dedicated to *génie* of his predecessors suggests such a history. Saint Lambert's definition and assessment of *génie* assumes that, while it is a universal capacity, it is better suited to the arts than to any other intellectual pursuits. The whole entry, however, is suffused with its author's partiality for the pre-Romantic anthropological ideal of "l'homme de génie," a hot-blooded character governed by his imagination and his sensibility, who heralds the demise of "l'homme d'esprit," the fine, elegant, and polished wit of the preceding century.

Conclusion

Opening with the redefinition of *esprit* against Habsburg *ingenio*, and ending with new meanings of *génie* that owe much to the promotion of sensibility in both the German and English contexts, the early modern French lexicographical history of ingenuity echoes from within the pages of dictionaries the role played by European cultural debates, diplomatic alliances, and military conflicts in the articulation of French national identity.[78] By highlighting the peculiarities of early modern French ingenuity and its importance in the fabric of the *âge classique*, the word histories of *esprit*, *génie*, and *naturel* confirm yet also nuance two Foucauldian intuitions. First, French classicism was the age of representation—understood, however, not as the transparent, inner reflection of a rational consciousness, but as the presentational, social, and ultimately artistic performance of one's individual cognitive idiosyncrasies and temperament. These were tempered by the expectations of propriety and taste of the audience, those cultural expressions of social order that disciplined the *naturel* and made it a norm.[79] Second, the modern subject emerging from this age of representation was indeed a site for ideological, political, and cultural policing; paradoxically, however, such policing promoted singularity and was instrumental in the emergence of the modern notions of author, style, and critic. Envisaged from the perspective of ingenuity, the emergence of the modern subject fulfils a disciplinary function in the *âge classique*, yet also promotes the very means for its contestation, especially in upholding the *naturel* as the new norm of good social and artistic performance. The eighteenth-century lexicographical account of *génie* in the *Encyclopédie* thus redefines the *naturel* in terms of sensibility and as the higher-order dismissal of the social rules of taste. At the end of the lexicographical trajectory of the keywords of French ingenuity, the pre-Romantic *génie* of Saint Lambert has become the polar opposite of the discriminating social and artistic intelligence so cherished by Voltaire in his entry on *esprit*.

5 German and Dutch

The Germanic language of ingenuity is related to *ingenium* by lexicographical artifice since, unlike the Romance languages and even English, it has no roots in Latin. The late eighteenth-century German philologist Johann Christoph Adelung was more sensitive than most to the challenge of defining German apart from the other European languages. It was one thing that people *did* in fact borrow Latinate words. It was another to *explain* what those words meant in German. Adelung gave full credit to currency: in his own two-volume treatise on German style, he defended the French loanword *Genie* as useful for celebrating genius in modern literature.[1] The word *Genie* had become a keyword in the German literary culture of his time. He therefore gave it a significant entry in another work, his *Grammatisch-kritisches Wörterbuch der hochdeutschen Mundart* (first edition 1774–1786), the culmination of over two centuries of German lexicography. The entry's first line alerted readers to the novelty of the word, "which was recently taken from France into German." Adelung defined the word, however, with older terms that drew on a distinctive German vocabulary: *Art* or *Natur* describe one's inborn nature; *Sinn* or *Gemüthsfähigkeit* refers to the higher capacities of soul that characterize *Genie*; meanwhile, "*Kopf* is perhaps the only German word which in time the French term might displace."[2] There is not a complete match between the new German *Genie* and the old language of ingenuity: Adelung does not use *Verstand* or *Vernunft* (intellect or understanding) to define it, even though these would have been obvious German stand-ins for *ingenium* in earlier times. Elsewhere in the dictionary, Adelung argued that the older German word *Gemüthsfähigkeit* did not capture the newer meanings. "Some would like to introduce this expression for the French *Genie*, but [*Gemüthsfähigkeit*] does not entirely exhaust the meaning, since *Genie* deals mainly with the higher powers [of the soul]."[3] *Genie* was offset from the older language of ingenuity.

As a late-comer to German, Adelung's *Genie* directs us to the longer history of ingenuity in the Germanic languages. This longer history is unique among the languages studied in this book. Although the Germanic languages had always translated and incorporated pan-European discourse, they boasted origins quite different than Latin. Julius Caesar, Tacitus, and other ancient ethnographers had already enshrined the northern barbarians as the noble "other," a theme that was constantly replayed in German histories written after Boccaccio and Bracciolini recovered manuscripts of Tacitus's *Germania*.[4] Some German and Dutch writers found irresistible the notion that their language had a purer source in antiquity. But the idea of a pristine Gothic language constantly foundered on the historical experience of flux. The diversity already began when just naming the various Germanic strands. Sixteenth-century lexicographers such as Josua Maaler and Cornelis Kiliaan spoke of "Teutsch" or the "lingua teutonica," to mean "Hochteutsche" (High German) and "Nederduytsch" (Low German, or Dutch and Flemish). In the Germanic-speaking countries, dictionaries enforced new order, bringing discipline to the languages as they sprawled across multiple cultural hubs—unlike English and French lexica, which bowed to the dominance of London and Paris. It is important, therefore, to compare both Dutch and German in this chapter, because they encoded ingenuity in related but different ways.

Keywords

Early modern dictionary-makers found equivalents for *ingenium* quite distant from Adelung's Romantic notions. The sixteenth-century polyglot dictionaries discussed in Chapter 1 offer a set of German and Dutch words that cluster around the Latin terms *genius*, *ingenium*, and their cognates. In particular, they give the following words: *Art, Gemüt*, and words formed on the stem *Sinn-*. For Dutch, these polyglots offer only slightly different terms: *aard/aerdt, geest*, and *sinrijck*, and other words based on the stem *sin-*.[5] Roughly plotting their neighbors now will help us examine them more closely in the remainder of the chapter.

Historians of philosophy and aesthetics have focused on other keywords relating to some aspects of ingenuity. Very common translations for *ingenium* were *Vernunft* and *Verstand* in German, and their Dutch relatives *vernuft* and *verstand*.[6] The semantic fields of these words are easy to grasp. *Vernunft/vernuft* (distantly from the word *nehmen*, "to take") translates the faculty of understanding, *ingenium* or *intellectus* as the capacity for "taking" mental hold—much like Latin *apprehensio* or *comprehensio*.[7] *Verstand* (from "to stand") does the work of *ingenium* or *intelligentia*, describing what "stands behind" the senses.[8] Predictably, related adjectives and adverbs describe qualities of mental reliability, industry, capacity, etc., following a well-worn vocabulary in philosophical psychology and practical wisdom literature. But this translation is relatively straightforward, well studied, and entirely addresses head and not hand—it lacks the moral and artful axes of ingenuity—so this chapter will not address these terms directly.

Indeed, the history of German aesthetics and philosophy has long centered on words that have an apparently straightforward but misleading relation to the history of ingenuity. Two key instances are *Geist* and *Witz*, which before the late eighteenth century are only marginal to the language of ingenuity in these lexica. In the *Dictionary of Untranslatables*, Alain Pons quite rightly pointed out that *Witz* emerged in Kant's writing as an important technical response to *ingenium*. But unlike the other words explored in this chapter, the word *Witz* never bore the cultural weight of *ingenium* before Kant; just as Adelung's *Genie* was a late import from French, this late development probably evolved as a roundabout response to the philosophically loaded English *wit*. Indeed, as we show below, in his anthropology Kant uses the term *Sinnlichkeit* to do the work of *ingenium*. The historiography of aesthetics has too often been preoccupied with *Genie* and the Romantic vocabulary of genius that emerged in the late eighteenth century.[9] But this newer vocabulary—words that we might consider neighbors of early modern ingenuity rather than family members—hides two things. First, it effaces a well-developed language of ingenuity that the Germanic languages had developed a century earlier. Second, it hides the extent to which the newer aesthetic vocabulary shifted this older one.[10] As a case in point, *Sinnlichkeit* barely appears in modern surveys of aesthetic vocabulary.[11]

Thus, by foregrounding "family" terms such as *Art, Gemüt, Sinn-* (in German) and *aard, geest*, and *sinrijk* (in Dutch), our study counterbalances the usual story of German keywords in aesthetic theory and philosophy. In both languages, *Art/aard* (kind, species) deeply resonates with the various axes of the language of ingenuity. As we shall see, *Art* was thought to stem from "earth" and its gift of fertile production; it addresses the moral and the cognitive status one is born with, but is also twinned as *art oder neigung* (kind or tendency), keeping in view how these innate qualities might unfold in practice and various modes of display. *Art/aard* is thus often closely linked to neighbors such as *Natur/natuur* and *Geist/geest* (spirit).

What the two languages share can be seen in their use of *Sinn-*. By itself, the basic meaning of *Sinn* (sense) is not central to the word family of ingenuity. Yet it is the stem for important words that are: the adjectives and adverbs formed on the stem *sinn-*, especially the various related words in Dutch such as *sinrijk* (intelligent, comparable to *bon sens*). From the earliest printed lexica throughout the period, words such as *Sinnreich* and *Scharfsinnig* (sharp-witted or perceptive) translate *ingeniosus*, only becoming more and more closely related to the larger meanings of ingenuity. The link intensified, as we will see, around 1700 when German writers required an analogue to the

French culture of *esprit* and the English culture of *wit*. By that point, words based on *sinn-* describe cognitive abilities and also account for witty comportment and modes of presentation. A survey of German and Dutch lexica used in this chapter gives many close neighbors: words such as "artful" (*Kunstlich, constigh*), "trick" (German: *Klug, List*; Dutch: *truc*), or "clever" (Dutch *subtijl*, often found paired with *geestig*).[12]

While the first two keyword sections of this chapter focus on shared domains, the final sections look to keywords where German and Dutch diverge. The distinctive paths of Dutch and German are seen in their use of *Gemüt* and *geest*. *Gemüt* has a Dutch relative (*gemoed*), but in German the word develops into a central touchstone for artistic theory, creativity, spiritual sensitivity, and even emotional forms of knowing, from Dürer's writings on art to Goethe's Enlightenment.[13] As a result, it deserves particular attention. Meanwhile, Dutch *geest* plays both an aesthetic role and a social one, closer to the French *esprit*; German *Geist*, in contrast, rarely refers to ingenuity, retaining a close link to the "spiritual" rather than the "spirited." One reason for focusing on *geest* in Dutch therefore is the heroic work it did in theology, in aesthetic theory, and even in humbler domains such as matter theory.

Lexicographical Landscape

Precisely because Germanic ingenuity links to *ingenium* only by lexicographical artifice, we must examine the Germanic origins of the lexicographical art a little longer than we have done in other chapters. The histories of German and Dutch lexicography share the roughest of outlines. At the beginning of the period, since dictionaries were made by scholars who hoped to introduce their countrymen to the Latin of the Republic of Letters, they were largely intended to explain Latin words. By the end, following the early seventeenth-century French, Italian, and Spanish academic dictionaries, German and Dutch lexicographers were lured by the search for the roots of their beloved vernaculars to insist that their languages held special history, beauty, and power.

Not that the earlier lexicographers lacked love for their mother tongues. Their goals were just more modest. The schoolmaster Petrus

Dasypodius stands at the origin of a larger project in the early sixteenth century to invent a German lexicography to match Nebrija in Spanish, Estienne in French, or Elyot in English, and more generally the work of Calepino.[14] This project took on momentum with the great encyclopaedist Conrad Gessner, a Basel-based polymath who had studied with Dasypodius in Strassburg and was enthused by the growing self-assurance of Germanic erudition in the previous generation of Conrad Celtis.[15] Gessner himself wrote polyglot dictionaries of plants and other zoological encyclopaedias, as well as a *Mithridates, de differentiis linguarum* (1555), a comparison of languages and cultures.[16] Although Gessner never published a dictionary himself, he contributed to Johannes Frisius's *Dictionarium latinogermanicum* (1541; based on Robert Estienne's *Dictionarium latinogallicum*, 1538). Here lie the origins of German lexicography, and it is tempting to find in this circle the beginnings of a nationalist Germanic program, especially in its greatest achievement, the *Die Teütsch Spraach* (1561) of Josua Maaler (Pictorius), the first major dictionary to be constructed around German headwords. Gessner contributed a long letter to the reader, recounting conversations with Frisius in which they agreed that it was high time that Germans had tools equal to those already available to the Spanish, Italian, French, and English. "It is appropriate for learned men, that one should have also—with other things by which he might adorn his homeland—a means for cultivating his maternal tongue (a necessary tool in all the arts and sciences, in all of civic life, and not least in religion and piety)."[17]

The key phrase is "for learned men." The German word family of ingenuity that emerged in the sixteenth century does not branch far off the Latinate trunk of the family tree. None of these works report German street usage; rather, their social world is that of the schoolroom and the study. These dictionaries provide Latin equivalents to students and scholars for writing—for dealing with Latin culture in German.

This early stage in Germanic lexicography supplies the reason we have selected *Art, Gemüt, Sinn-* (in German) and *aard, geest*, and *sinrijk* (in Dutch) as our Germanic keywords. Three important Latin-German dictionaries exemplify how these words come to hand as equivalents for Latin *ingenium*. To see how German meanings sit uneasily with the Latin, consider the first such dictionary, Petrus Dasypodius's *Dictionarium latinogermanicum* (1535). The entry for *gigno*, the root of so many Latin terms related to ingenuity, takes several directions in German. After addressing meanings of *gigno* to do with conceiving and bearing fruit, the entry turns to ingenuity: "Ingenita vis, *Die angeborn naturlich krafft*. Et ingenium, *die angeborn und naturlich art, verstendtlicheyt, synreyche*. Ingenium soli, *Die art des bodens,* Ingeniosus, a, um, *Synreych, klüg*. Item, Ingenuus, a, um, *frey geborn, naturlich*. Ingenue loqui, *dapfferlich, warlich reden*. Ingenuitas, *freyheyt, edelkeyt*" (Ingenita vis, *the inborn natural power.* Et ingenium, *the inborn and natural kind, knowledgeability, cunning*. Ingenium soli, *The nature of the soil,* Ingeniosus, a, um, *Cunning, skilfull.* Item, Ingenuus, a, um, *born free, by nature*. Ingenue loqui, *to speak bravely and truly*. Ingenuitas, *liberality, nobility*). The entry fragments into the vernacular language, shooting off in every direction; no single set of words share a common root the way that *ingeniosus* and *ingenuitas* both stem from *gigno*. Rather, each German term selects only one branch on the family tree of ingenuity, without echoing the others. *Synreych* (sensible, crafty) does not remind the reader of *angeborn art* (inborn nature) the way that *ingeniosus* brings to mind *ingenium*. Not only does the translation lack etymological or morphological echo, but the German equivalent inescapably narrows the definition: "inborn natural power" chooses only one of the several meanings of *ingenium*.

In our second case, Johannes Frisius (1541), another of Gessner's protégés, sets *Art* and *Sinn* within the cross-section of Latin ingenuity. For *ingenium*, we find words that explicitly overlap with the semantic range of Latin *natura*: "die art und natur, oder natürliche neygung" (the kind and nature, or natural tendency). When the Latin example relates to pleasure, *Art* is used to give the sense of inborn nature (*Ich bin also geartet*, I am made of such a nature). In the long lists that characterize cognitive capacities, Frisius offers *Verstand* as a main equivalent—but elsewhere in the dictionary, *Verstand* also stands for *intellectus*, revealing a narrower cognitive meaning than *ingenium*. The moral meanings of the Latin *ingenuus* generally correspond to phrases with the word *frey* (free), so that *ingenuitas* is primarily translated with a phrase rather than a directly equivalent word: e.g. *angeborne freyheit* (inborn liberality). Meanwhile, *Sinn* emerges in the adjectival and adverbial forms, in a cohesive group:

> Ingeniosus, adiectivum, Verstendig, sinnreych, scharpffsinnig.
> Quo quisque ingeniosior et solertior, hoc docet iracundius et laboriosius.
> Ad causas ingeniosa fuit, Spitzfündig.
> In alieno libro ingeniosus.
> Ingeniosa miracula fecit natura.
> Ingeniose, adverbium, Mit grossem verstand, künstlich, artlich.

> (Ingeniosus, adjective, Intelligent, clever, sharpwitted.
> The more ingenious and skillful a man is, the more angrily and laboriously he teaches [Cicero, QRosc. 31].
> She was ingenious with reasons [Ovid, *Heroides* II.22], Persnickety.
> [To be] ingenious on another's book [Martial, *Epigrammata*, 1.10].
> Nature made ingenious marvels [Pliny, *Naturalis historia*, 7.32].
> Ingeniose, adverb, With great intelligence, artful, naturally.)

Here the direct translation of Latin ingenuity results in several fields of meaning beyond the axes of identity (*Art*) and mental capacities (*Verstand*): *Sinn* implies an ingenious capacity for display of inner wit (*Art*), perhaps skillfully deploying craft (*Kunst*). *Art* and *Sinn* turn out to be durable members of the ingenuity family; if we turn to the first major dictionary with German headwords, Josua Maaler's *Die Teütsch Spraach* (1561), we find that the one additional word that closely matches this topography of ingenuity is *Gemüt*.[18]

As our third starting point, the *Dictionarium tetraglotton* (1562) of Cornelis Kiliaan (Kiel) gives us a comparable case for Dutch. This work emerged in the richly erudite print shop of Christoph Plantin, who would soon undertake the famous polyglot Bible of Antwerp, finally printed in 1574. Like the dictionaries from Gessner's circle, the *Tetraglotton* conveniently circumscribes meanings through equivalents, allowing us to discern a distinctively Dutch branch of the family tree of ingenuity in Greek, French, and Dutch: "Ingenium, ij, n.g. φύσις, εὐφυΐα. *La nature qu'un chacun a. Esprit et entendement qu'on a de nature.* De nature die een iegelick heeft. Gheest ende verstant dat men van naturen heeft" (Ingenium, *physis, euphuia.* The nature that each one has. The spirit and understanding that one has by nature). Here, the Dutch language of ingenuity is established not by direct translation from Latin, but by way of French. *Nature*, *esprit*, and *entendement* were already the focus of the French vocabulary of ingenuity, established in Estienne's *Dictionarium latino-gallicum* (1538). Rendering the French phrase in Dutch, Kiliaan already aligns the two vocabularies. The effect is only stronger in the adjectives: the *ingeniosus* is "Verstandich, Die goeden geest ende goet verstant heeft (bon esprit et entendement)" (*Intelligent, one who has good spirit and understanding*). This French dependence reflects the social and cultural proximity of Plantin's Antwerp to France—Dutch subtly veers away from German in Latin-to-vernacular polyglot dictionaries. The cognitive words *Verstand* and *verstand* differ very little. Just as for German, in Dutch the moral connotations of *ingenuitas* are explained with phrases concerning the "edelheyt de vrije persoonen" (*nobility of free persons*). But it was another matter in the embodied language of identity and display: where German tended to reach for *Art*, *Sinn*, and *Kunst*, in Dutch Kiliaan instead turned to *natuur* and *geest*.

At this early stage of printed lexica, Frisius and others in Gessner's circle were not nationalist in the sense of being motivated to find a pure Ur-language for the Germanic peoples. To be sure, some earlier German and Dutch erudites hoped that if they traced the roots of their mother tongue far enough into the past, it would prove to have at least the antiquity and thus the dignity of Greek and Latin. The outsized example of this impulse in Christoph Plantin's circle was Johannes Goropius's *Origines Antwerpianae* (1569), a wonderful magpie collection of etymologies intended to reveal Dutch as the original language.[19] Yet most Germanic dictionary writers had the much more modest goal of helping their countrymen access the Republic of Letters, not supplant it. Even the earlier pioneering dictionaries of Josua Maaler for German (1561) and André Madoets for Dutch (1573), despite being organized around vernacular headwords, were intended not to transform the vernacular language into a real alternative to Latin, but to set compatriot scholars on an equal footing with peers abroad.[20]

These aims changed in the late seventeenth century. While Dasypodius, Maaler, and Kiliaan constructed a set of terms in High and Low Dutch on the Latinate model of *ingenium*, later Germanic lexicographers grew deeply conscious of differences between their native tongues and cultural rivals in the Romance languages. Already in the sixteenth century, Dutch and German scholars had written grammars for their languages.[21] French and Italian had their own linguistic academies working on national dictionaries by the early seventeenth century. For decades, without a unified political patron, the Germanic languages lagged. The closest such effort was Henisch's *Teütsche Sprach und Weißheit* (1616), but Henisch inconveniently died after the first volume was published, so the project foundered after the letter "G." A circle of German scholars started up the *Fruchtbringende Gesellschaft* in 1617, which nurtured a generation of scholars determined to see German recognized as a literary language. They wrote a flood of theoretical and orthographical works on the proper identity, dignity, and history of High German, attacking the "Tyranney der Lateinischen Sprach" as Joachim Junge (1587–1657) put it.[22] Despite seventy years of false starts, and a great deal of theorizing about how best to write a German dictionary, Caspar Stieler finally published the first complete, great equivalent of the academy dictionaries in German: *Der Teutschen Sprache Stammbaum und Fortwachs* (1691). This was the first large-scale monolingual dictionary of German based on the distinctively Teutonic etymological notion of *Stammwörter*, the irreducible monosyllables theorized as the ancient roots of the Germanic tongues.[23] The first academy-influenced single-language dictionary in Dutch was produced even later. John Considine suggests that the strong polyglot tradition of Kiliaan and Plantin, and the prominent place of classical scholarship in Dutch linguistics, are important reasons for why a Dutch-only dictionary was not produced until after 1723, when Lambert ten Kate published his two-volume *Aenleiding tot de kennisse van het verhevene deel der Nederduitsche sprake*, a work of etymological linguistics intended to support a new etymological dictionary.[24] Ten

Kate was a well-respected *liefhebber*, collector, and amateur mathematician, and he brought important theoretical underpinnings to his lexicographical work. He believed in an original *gemeenlands*, a "common speech" that had characterized an age of perfection in Dutch, which existed in the pre-Roman era.[25] This original speech could be traced through vowel shifts that had crept in over time. Thus, ten Kate was the first to describe the "sound laws" (or what he called *streekhoudende dialect-regel*) of Germanic root vowel changes, which the philologist Jacob Grimm adopted under the term *ablaut* in the nineteenth century.[26]

These Dutch and German grammarians repeatedly reconstructed their languages without Latinate origins. Yet their dictionaries persistently dropped hints leading back to Latinate definitions of ingenuity. Beginning with French refugees from religious violence and enduring well after the Francophilic Enlightenment court of Frederick the Great, French was the language of courts in German-speaking lands. People often complained about the popularity of French words. In one satirical work on lexicography, the *Alamodischer Brief* of Karl Gustav (in *Der Teutsche Palmbaum*, 1647), as much as a third of the vocabulary is French. And the dual-language dictionaries between Dutch, German, and the Romance languages shared an emerging language of ingenuity. Dutch especially is a crucial case, because while German language theorists struggled to free their language from Latinity, Dutch was more open to influence.[27] Though a small region, the Low Countries became a center of global trade and therefore a go-between for warring states all across Europe. For native speakers, therefore, translation dictionaries from Dutch to other languages became enormously popular.[28] French-German dictionaries reveal similar trends in the eighteenth century, as the phrase *lingua franca* became meaningful. The dominant resource for German speakers was Matthias Kramer's grand dictionaries for Italian (2 vols. 1676 and 1678) and French (4 vols. 1712–15).[29]

The dictionaries of Stieler and ten Kate from this period exemplify why it was so difficult to eradicate the Romance languages from German or Dutch, even when they were devoted to purifying the language. In his dictionary of 1691, Stieler reimagined a world of etymologies to show that the word *Natur* was indigenous to German. Clearly he found the word so important that he could not ignore it. He explained its semantic range in terms from the Latin notion of *ingenium* as one's identity or potential: *natura, indoles, genius, inclinatio*. But to avoid admitting its ultimate origins in Latin, Stieler took drastic measures. He conjectured that *Natur* was formed on the Germanic stem *Ur* (i.e. *Nat-ur*), having to do with origins and generation. *Ur*, in turn, he saw as formed from the family of words relating to *Alt*, or "ancient." In other words, unwilling to deny the well-worn word a place in the German language, Stieler substituted an appropriate pedigree by conjecturing a new etymology for *Natur*. This was not the only option. He could have ignored it, as ten Kate did a few decades later. Even though the word was omnipresent in Dutch culture, ten Kate simply left *natuur* out of his list of Dutch headwords, treating it as a foreign word.

Perhaps Stieler found himself in this predicament because of his choice of metalanguage. Even though he structured his dictionary around German *Stammwörter*, he explained his entries in Latin. These Latin equivalents pushed against his effort to retain the specificity of the German, linking *Natur* to the temperamental, pedagogical connotations of *indoles* and the divinity of *genius*; the new etymology retained the hues of generation and genesis that colored the orthodox Latin etymology. This irrepressible drift toward a European-wide language of ingenuity will occur again and again as we address the keywords in this chapter.

Art

Germanic ingenuity could depend too much on Latin, to the point of falsity. For instance, the German word *Art* (kind, nature, manner) appears to support the same meanings as Latin *ars*, falsely implying a shared etymology. Such problems were especially knotty for Gessner and Maaler, who believed dictionaries should be organized by etymology. For this reason, Maaler (1561) presented definitions in an order that allowed readers to infer etymologies. Thus his presentation of the word *Art* runs perilously close to the Latin *ingenium*, beginning with the first example of *die Art unnd eigenschafft der erden* as simply *ingenium soli*. This phrase, "the nature and quality of the earth," immediately draws the reader into Latinate connections between earth, fertility, and the range of meanings implicit in *gigno*. *Art* is therefore constructed on a Latin model, a construction Maaler repeats in many other entries.[30] The etymological impulse does not verge into illegitimate territory until a couple of entries further down the page:

> Artist (der) Artificiosus, Solers.
> Artlich / artist. Argutus, Artificiosus.
> Dise auslegung ist gantz Artlich und subateyl. Habet acumen haec interpretatio.
> Artliche brieff / unnd geleertlich beschriben. Scitè et literatè perscriptae literae.
> Artliche unnd listige mätz. Meretrix arguta.
> Artlich / vast artlich. Affabrè, docte, Industrie, Ingeniose, Argute, Solerter, Artificiose, Dextere.
>
> Artist (the) An artful, skillful man.
> Artlich / artist. Sharp, artful.
> This interpretation is entirely artful and subtle. This interpretation is sharp.
> An artful letter and written learnedly. Knowingly and learnedly written letters.
> An artful and crafty maiden. A skilled courtesan.

Thus *Art* transmogrifies into *Artist* and *artificiosus*, completely effacing the distinct histories of the German and the Latin words. For *artist* comes from the Latin *ars* > *artista*; yet here it develops the German *Art*, which in fact originates in words relating to earth and ploughing.[31] The order and content of the entries sews together a false etymology; the material and verbal structure of the dictionary actually fabricates meaning. With *Art* set alongside *artificiosus*, the Germanic family of ingenuity takes on echoes of the Latinate family:

> *Art, aerd*, n.
>   1. Kind or species.
>   a. Nature.
>   German: Dasypodius 1536 (as "ut, Ingenium soli") >; Maaler 1561.
>   Dutch: Goropius 1569 >, Madoets 1573.

b. Gender.

German: Dasypodius 1536.

Dutch: —

c. Genius of a language.

German: Kramer 1676 (under *genio*, as "il genio della lingua Italiana, die Art und Eigenschafft der Italiänischen Sprache, etc.").

Dutch: —

d. Race or nation.

German: Kramer 1712–15 (as part of the first definition: "Art, f. Espece, race, f. Genre, m. it. Generation, nation, f. V. Geschlecht").

Dutch: Madoets 1573 >.

2. Manner

a. Character.

German: Dasypodius 1536 (as "Art oder weiss im schreiben, oder reden") >.

Dutch: Madoets 1573 >.

b. Tendency.

German: Dasypodius 1536 >.

Dutch: Madoets 1573 >.

c. One's *naturel*.

German: Kramer 1712–15; Gladov 1728.

Dutch: Madoets 1573 >.

d. Craft of a maker.

German: Maaler 1561 (as "Ein wäsenlicher Artiger knächt. Servus non incallidus"); Henisch 1616 (as "Ein artlicher redner / logodaedalus. Ein artlicher und kunstreicher handwercksman / mechanicus, mechanarius ingeniosus, solers fabricator operum, quae ingenio pariter et manu perficiuntur") >.

Dutch: Madoets 1573 (as "Aerdich. Ingenieux, artificiel. Ingeniosus, artificiosus, industrius, venustus -a -um, artificialis -le, elegans") >.

e. Craft of an object.

German: Maaler 1561 (as "Dise auslegung ist gantz Artlich und subateyl. Habet acumen haec interpretatio"); Henisch 1616 (as "Das ist sein art / also ist er geschaffen / das ist sein köpffsein, sic est ingenium, pro, huiusmodi") >.

Dutch: Maaler 1561 >.

3. Source

    a. An inherited quality.

    German: Kramer 1678 (as "ein Art von Pferden, hunden, etc., razza di cavalli, cani, etc.") >.

    Dutch: Goropius 1569 (as "*Ard* enim id apud nos notat, quod cuique vel a natura, vel a parentibus inest") >.

    b. Generation, to beget.

    German: Stieler 1691 (as "Arten, geartet, patrissare," though see "ab Arten" in Dasypodius 1536) >.

    Dutch: Goropius 1569, ibidem >.

    c. The land, earth.

    German: Maaler 1561 >; Wachter 1737 (as "Ab *erde* terra, quia res naturales ingenium eius soli, in quo natae sunt, referre solent, non fructus tantum, sed etiam incolae").

    Dutch: Kiliaan 1599 (as "Aerde/eerde. Terra, tellus, humus, solum: arida") >.

When, around 1600, German dictionaries began to collect and taxonomize the riches of German itself, not just facilitate access to other languages, Latin was still used as a neutral space from which to analyze other languages. The great dictionary of Henisch (1616) treated German as a cultural treasury comparable to Latin—but Henisch uses the Latin title *Thesaurus* and explains German terms with Latin equivalents, before offering a German definition: "Art/natur/gestalt/eigenschaffe/neigung/ *forma, natura, ingenium*. Von ἀρετὴ, oder *virtus*, so under andern auch bedeut ein krafft, natürliche neigung" (Kind/nature/form/identity/tendency/ *form, nature, ingenium*. From *arete*, or *virtue*, which among others things also means a power, a natural tendency). Even in this definition, *Art* is not simply an indigenous German word, but ultimately rooted in Greek *arete* (often translated as the excellence of a thing). The effect is one of slotting *Art* into a pre-existing pan-European language of ingenuity—an effect only heightened by Henisch's ensuing list of equivalents: in English (*Witt, nature*), Dutch (i.e. "Belgisch," *verstant*), French (*la nature qu'un chagun a*), Greek (*phusis*), and Italian (*natura*). As John Considine put it, Henisch and Gessner "were seeing a dictionary both as a comprehensive repository of culture, and as a comprehensive account of things."[32] But it was difficult to unimagine the picture of culture as the grand edifice of Greek and Latin antiquity that the great humanists reconstructed in books of antiquities, commonplaces, proverbs—and in dictionaries. On the same model as Estienne's *Dictionarium*, Henisch completed the entry on *Art* with German sayings about ingenuity, including several from Maaler 1561: "Art und neigung / sonderlich der Jugent zu gutem unnd bösem, indoles, natura. . . . Das ist sein art / also ist er geschaffen / das ist sein köpffsein, sic est ingenium, pro, huiusmodi" (The particular kind and tendency of a youth toward good and evil; innate disposition, nature . . . That is his sort; he is made that way; that's the way his head is; he's this sort of wit; and the like). Proverbs describing the training of youth and their characteristic docility are taken over from contexts soaked with Latin meaning, so that behind the German phrases echo Ciceronian school maxims about teachable boys and Plautan jokes about the *ingenia* of drunken soldiers.

So far, *Art* has largely shown us ingenuity as it revolves around the moral axis. German left cognitive meanings to the words *Verstand*, *Vernunft*, and especially *Sinnlichkeit*. But over the period *Art* came to be defined more and more in terms relating to making and display. Even though he avoided Maaler's spurious association between Latin *ars* and German *artlich*, Henisch nevertheless found the word a straightforward way to talk about literary and mechanical craft: "Ein artlicher redner / logodaedalus. Ein artlicher und kunstreicher handwercksman / mechanicus, mechanarius ingeniosus, solers fabricator operum, quae ingenio pariter et manu perficiuntur" (An elegant author: logodaedalus. An elegant and crafty handworker: a mechanic, ingenious mechanician, skillful maker of works done by wit and hand together). This notion of *Art* evolved in German into more than simply a quality of execution, but by the late seventeenth century was identified with "manner" (*Weiss*) or style. In his German-to-Italian dictionary of 1678, in fact, Kramer separated this sense of *Art* into a distinct lemma, giving as equivalents "Manier, Modo, Maniera," and elaborating on *Art* as *stile* in everyday life, studies, and artistic production.[33]

Dutch *Aerd* shared all of these meanings discussed so far, and it would be prolix to work through many examples.[34] A major difference, however, is that Dutch lexicographers early on admitted resonances of ingenuity in other languages, notably French. Already in 1573 Madoets attended closely to the French *naturel*. Several times he repeated the French notion of *naturel* as a translation for *aard*: "Sijnen aerdt volghen. Suivre son naturel. Ingenium suum sequi" (*To follow his nature* . . .). Perhaps even more strikingly, Madoets translates the verb *aerden* with "taking on the nature or *mode du pais*." *Aerden* becomes the equivalent of "induere genium alicuius loci" (to take on the ingenium of some other place), which firmly associates *aerd* with the *genius loci*.

Both *naturel* and *genius* of a place or nation come to be associated with German *Art* only much later. Not until Kramer's German-French dictionary of 1715 do we find *naturel* as a translation.[35] And it is via French and Italian that *Art* begins to refer to the *genius* of a language, race, or nation. In fact, it is Kramer in 1676 who makes *il genio della lingua Italiana* equate to "the kind and quality of the Italian tongue" (*die Art und Eigenschafft der Italiänischen Sprache*), and then gives as the first definition of *Art*, "species, race, kind . . . nation" (*Espece, race, genre . . . nation*).[36]

In fact, this shift explicitly followed Dutch, which was used to rationalize this late drift of *Art* further into the ingenuity family. Indeed, Wachter's German etymological dictionary of 1737 refers to Dutch in order to explain *Art*, informing readers that "Earth comes from *Erde*, since natural things usually are borne back to the *ingenium* of the soil from which they were born; not only produce, but inhabitants too."[37] The topos of the *genius loci* here comes into German, but via Dutch — under the assumption that Dutch testified to a more authentically German meaning.

There is a second wonderful irony here, for while German had been slow to incorporate more Latinate senses of ingenuity into *Art* — but at last gave in — Dutch became more cautious over time. Wachter, an etymologist well informed on the history of his language, most likely had read the entry for *Aerd* in Lambert ten Kate's Dutch etymological study of 1723. By this time, ironically, Dutch lexicographers had become sensitive to the origins of *Aerd* in the earth. Not until Binnart 1649 is *terra* set as the first meaning of *aerd*, next followed by *natuere*.[38] Wachter thus encountered the non-Latin etymology within ten Kate's pioneering effort to assemble a comprehensive lexicon on a Dutch-only etymological footing. Ten Kate's etymology reasoned out a range of early Germanic precedents, including English such as *earth* and *eortha* alongside early High-German precedents such as *Erde*. He even drew on historical ethnology, recalling from Tacitus that the ancient German pagans had a goddess *Hertha*, mother earth, "that is, the earth serves as the mother and father of all flesh, who keeps us at home and peaceful as if sitting in her lap."[39] Thus drawing on the best principles of historical philology, ten Kate reanimated *aerd* with a vibrant etymology.

However, this Dutch repatriation of *aerd* ultimately returned to the roots of *ingenium*. The very notion of a pagan deification of natural powers resonates with the language of the tutelary genius, precisely along the lines pioneered by Perotti in the fifteenth century.[40] Indeed, at the end of the entry ten Kate tipped his hand, showing his sources and mental framework: he cited Kiliaan and offered precisely the same Latin equivalents common in the sixteenth century, concluding with the observation that what *aerd* owed to *aardich(-lijk/-keit*, etc.) was "even gelijk het Lateinsche ingeniosus van ingenium" (just like the Latin *ingeniosus* from *ingenium*).

A final striking moment in *aerd*'s story concerns its adjectival form. Josue van Iperen sketched plans for a new dictionary of Dutch on native principles, which accidentally deepened the conceptual overlap of *aard* and the family of words around *ingenium*.[41] Van Iperen decided to advertise his dictionary scheme by publishing a proof of concept entry using *aard* or *aerd* as the archetypal Dutch word. After reviewing various spellings and pronunciations, including the consonant shift from *Art* in "High Dutch" or German, he supplied words based on the stem *aard*. Thus he not only supplied *aardig*—a standard Dutch translation of *ingeniosus*—and *boosaardig*, *eigenardig*, and so on; but he also noted words based on the past participle of *aarden*, such as *Wel-geaard* (good natured). He noted that *regt-aard* (right natured) described a son with respect to a property inherited from his father, unconsciously echoing a topos in Latin entries on *ingenium*.[42] As Wachter had noticed, *aard* had been an ancient suffix in many Germanic languages, which could be traced in the derogatory epithets of other languages: the English *bastard* or *drunkard*, the German *Bankart*, the French *babillard* or *louschard*, and the Flemish *dronkart*.[43] Precisely what made *ingenium* so useful to Renaissance thinkers was that it could span the tension between heredity and individuality in the richly variegated makeup of human beings. In the end, it was this tension that kept *Art* and *aerd* within the language of ingenuity.

Sinnlichkeit, sinrijk, etc.

Perhaps the most surprising words adopted into the ingenuity family, in both Dutch and German, are those terms formed on the stem *Sinn-* (sense). In his *Anthropology from a Pragmatic Point of View* (1798), Kant made *Sinnlichkeit* (sensibility) the set of senses that present the manifold of experience and submit it to the judgment of *Verstand* (understanding).[44] He also described *der Scharfsinn*, the faculty of noticing the smallest similarity or dissimilarity, a faculty he says is bound to *Witz* and to the power of judgment (*Urteilskraft*).[45] *Witz* has long been associated with English *wit*, and indeed Kant glosses it as *ingenium*. Insofar as scholars have focused on *Sinnlichkeit* and related words, it tends to be in the context of Enlightenment sentiment or even sensibility.[46] But we suggest that these words, at least in German and Dutch, drew on the language of ingenuity too, as a synoptic overview will quickly suggest:

Sinn-, adj., adv.
  1. Skill
    a. Industrious.
      German: Anonymous 1495 (under headword "*Sinnig / Sinrich* as "ingeniosus, industrius") >; Dasypodius 1536.

Dutch: Madoets 1573 (under *Sinrijck* as "verstandich ende begrijpich in den sin. Riche d'entendement, ingenieux, capable de sens. Ingeniosus, capax, docilis, sagax, solers, industrius") >.

b. Crafty prudence.

German: Dasypodius 1536 (under headword *Sinn* as "Sinnreiche, klügheit, solertia") >; Kramer 1676 (under *Sinnreich, sinnig* as "*Sensato, Prudente, it. Ingegnoso, Spiritoso, Giudicioso*. V[ide]. Scharffsinnig, Klug").

Dutch: —

2. Judgment

a. Sharp insight.

German: Dasypodius 1536 ("Scharpffsinnig, Klüg. Solers, Perspicax, Acre ingenium, Cordatus, Ingeniosus") >.

Dutch: Madoets 1573 (under lemma *Subtijl van sinnen* as "subtil de sens, fin, ingenieux") >; Kiliaan 1599 (under headword *scherp-sinnigh* as "Acutus ingenio, acer animo, solers, acri ingenio praeditus, ingeniosus").

b. Good sense.

German: —

Dutch: Madoets 1573 (under headword *Sin*, as "Den natuerlicken sin. Le sens naturel. Sensus communis"; "Sinrijck, verstandich een begrijpich in den sin. Riche d'entendement, ingenieux, capable de sens. Ingeniosus, capax, docilis, sagax, solers, industrius") >.

3. Invention

German: Stieler 1691 (under *Aussinnen* as "Vernünftig aussinnen / ratiocinari, ingenio eruere"); Kramer 1700 (under headword *Sinnig* as "eigen-sinnig / capriccioso").

Dutch: —

4. Tendency

a. One's true nature.

German: Maaler 1561 (under headword *Sinn* as "Nach seinem Sinn, Fantasey, art und lust, läben. Ingenio suo vivere"; "Sinn / m. Gedancken / Meinung / it. Verstand / it. Gemüt / Neigung / Senso, Pensiere, Opinione, etc. . . . das Ding ist recht nach meinem Sinn / questa cosa mi da batte vivamente nel genio") >.

Dutch: —

b. Appetite for pleasure.

German: Kramer 1676 (headword *sinnlich* as "id est, fleischlich, sensuale"); Kramer 1712.

Dutch: —

c. Purity.

German: —

Dutch: Ten Kate 1723 (*zindelijk / zinlyk* as "singulari cura mundatus" [polished with singular care]).

5. Faculty of mind
   a. Understanding.

German: Maaler 1561 (under headword *Sinn* as "(der) Verstand, Das fassen unnd erwütschung unsers verstands. Sensus. Vernunft, Gemüt, Meinung. Mens, Comprehensio"; under headword *Sinnreyche* as "(die) Ein güter verstand. Facultas ingenii") >.

Dutch: —

Before moving to *Sinnlich*, etc., we should observe the enduring resonance of the neighboring terms *Kunst* in German, and *conste* in Dutch. In his German > Latin section, Dasypodius sets *Kunst* very close to the Latin *ars*: "Eruditio, Scientia, Literae, arum, Disciplina, Ars."[47] The notions of craft, skill, and artifice seem stable throughout the period. As with *Sinnreich, Scharpfsinnig*, etc., the adjectival forms explicitly fall into the word family of ingenuity: *Kunstreich* is very close to *ingeniosus*. For example, the word also has the additional valence of craftiness and cunning, setting *Kunst* more with worldly comportment (e.g. *Klug*) and even deceit (e.g. *List*). For example, Kramer 1712 even supplies as his French translation "secret, artifice, etc." The "etc." suggests that he expects his reader can supply further French examples without difficulty. Would he have included *engin*, where artifice quickly has negative meanings? We may also trace intriguing connections between *Kunst* and words in Dutch and English, which, while etymologically related, follow trajectories veering slightly away from ingenuity. For instance, the Dutch *kunnen* (to be able) relates to "I can" in English; Kiliaan 1599 points out the relationship to "cunning" in English (or *konningh*, as he spells it). Yet the word only relates to negative abilities for Sewel 1691, where English "cunning" is only translated with words for devious trickery (*loos, listig, behendig*). In these cases, the word has lost some of the moral reversibility that characterizes ingenuity family terms.[48]

Sinn is not central to the ingenuity family of terms, taken entirely on its own. In the early sixteenth century Dasypodius found the cognitive meanings pendant on *Sinn* to be wide and many. He especially focuses on mental capacities for insight, such as *perspicuitas*, and for the moral strength of a *sinnreich* person to be *industrius* and *solers*. Nevertheless, he goes no further than cognition. Kiliaan's Dutch dictionary (1599) offered it as a translation for senses both internal and external to the mind, arguing that it is based on the Dutch verb "to see" (*sien*).[49] But Kiliaan's colleague at Plantin's press, Josua Maaler, suggested that in German the notion of "sense" could stand in for one's mental power, an alternative to both *Vernunft* and to *Gemüt*. Maaler expands the German word. It certainly includes *Verstand*, where it represents aspects of the moral axis of ingenuity. For example, he describes one as "living according to one's *Sinn*, fantasy, kind, and desires," to capture the Latin *ingenio suo vivere*.[50] *Sinn* therefore not only is cognitive; it describes temperament.

Maaler also added adjectives; to be *Sinnreyche* was to have "a good understanding; a faculty of the *ingenium*."[51] It is in this form—the adjectives and adverbs formed on *Sinn*- such as *Sinnlich* and *sinrijk*—that these words became an especially rich vocabulary of ingenuity in both German and Dutch. A subtle transformation takes place when the suffix *-reich* or the prefix *Scharff-* is added to the stem *Sinn*. For example, one tends to associate these words with ingenuity when describing someone is "rich" with an abundance of good sense (*Sinnreich*) or characterized by "sharp sense" (*Scharffsinnigkeyt*). Already the earliest printed dictionaries of both languages show this; an anonymous German wordlist from 1495 simply translates *sinnig, sinrich* with the stock Latin twin "ingeniosus, industrius."[52] As Madoets defined the Dutch *sinrijck* as being "understanding and comprehending in the senses," he offered French and Latin equivalents for defining skillful *ingenia*.[53] In both German and Dutch, as *Sinn* shape-shifts between combinations of suffixes and prefixes, its semantic topology stretches across many of the same domains as does *wit* in English, *esprit* in French, and *ingenio* in Spanish.

In German lexical works, these meanings ca. 1700 furl ever tighter around themes of pleasure and wit that we quickly recognize from other discourses of ingenuity. This shift marks various editions of Kramer. In his first German-Italian dictionary of 1676, Kramer repeats the cognitive meanings of *Sinnreich* and *Sinnlich* just mentioned, and for the first time in a lexicon adds what has become the modern primary meaning of *Sinnlich* as "sensuality."[54] In the greatly revised edition of 1700, Kramer linked the terms much more closely with *ingegno*. To be *Sinn-reich* is "to be rich in sense and sentiments, namely sensitive, sharp, subtle, spirited, judicious, ingenious."[55] So it is not surprising when one of the rich terms of ingenious invention in Italian, *capriccioso*, is used to translate *eigensinnig*.[56]

Kramer's account of *Sinn-* therefore resonated closely with ingenuity. This was partly because he tried to bridge to Latinate languages, but another reason was that German lexica—in the effort to identify a self-sufficient, authentic German linguistic history—actually constructed alternative analogues to Latinate word families of ingenuity. Kramer had followed the lead of Caspar Stieler, whose 1691 dictionary was the first to follow distinctively German ordering around singlesyllable stem-words (*Stammwörter*). Under Stieler's gaze, *Sinn-* was revealed as a thick forest of terms, an example of German's lexical riches. Thus the terms characterize a whole range of individuals:

> Sinnlich et Sinnig / it. Sinnisch / adj. et adv. animo affectus, ingeniosus, et ingeniose, sensilis, sensilibus, *et vulgo sensibiliter,* solers, et solerter. Unsinnlich, vecors, furiosus, rabidus, insaniens, et Unsinnlich / furenter, rabiose, furialiter. . . . Rechtsinnig / et Redlichsinnig / cordatus, ingenuus. Scharf- *sive* Tiefsinnig / ingeniosus, cogitabundus. Schwersinnig / melancholicus, subtristis. Starrsinnig / rigidus. Mansinnig / opiniosus, prava sentiens.

Sinnlichkeit / die / et Sinnigkeit / solertia, acumen ingenii, docilitas, indoles, et vigor naturalis. *Inde:* Eigensinnigkeit, pervicacia, obstinatio. Hochsinnigkeit, fastus, ferocitas, sublatio animi. Kaltsinnigkeit / frigus amorsi. Leichtsinnigikeit / futilitas, levitas animi. Stumpf *sive* Tummsinnigkeit / stupiditas ingenii. Unsinnigkeit / insania, dementia, rabies, *dicitur etiam* Mansinnigkeit.

All terms share the fundamental task of cognates to ingenium: to qualify the range of human temperaments. Many of these are words rarely found in other dictionaries, such as *Rechtsinnig*, here given as *ingenuus*, or *Schwersinnig* for the melancholic. Despite his distinctively German principle of organizing the dictionary, Stieler borrows heavily from Latinate ingenuity to explain his abundant German vocabulary.

Dutch moves in a slightly different direction. While German words to do with *Sinn-* are heavily laden with emotional, moral, and even passionate meanings, the Dutch counterparts remain firmly in cognitive territory; not a single Dutch dictionary indulges in examples about appetites or pleasure. In contrast, German lexica are bereft of clear examples relating to common sense or "good sense." In particular, although there exist many examples of related words (*onzinnig, sinneloos, krankzinnig, uitzinnig...*), in contexts where it translates ingenuity Dutch is preoccupied with the sharpness of ingenuity, as in *scherpsinnigh.*[57] Conceptually, this might explain why Dutch ingenuity is so closely connected with the word *subtijl*. Hexham defines *scherpsinnigh* in his Dutch-English dictionary of 1647 as "Subtill, Suttlewitted, Acute."[58] The notion of "subtlety" has its own deep history as a Latin term of praise and blame; *subtilitas* characterized one capable of fine reasoning—which could be all too fine when it turned sophistic. But Dutch linked *sinnig*, *sinnerijk* and *sinnlijk* so strongly to *subtijl* that Martin Binnart defined one with the other: "Subtijl/scherpsinnigh/ gheestigh, acutus, argutus, ingeniosus."[59]

So far, we have encountered no applications of *Sinn-* to objects. In Dutch lexica, this corner of ingenuity remains purely about people—though examples may easily be found in other genres. But by the late seventeenth century, German lexicographers such as Stieler and

spectio. Obsehen/ der/ & poët. præfectus, præfes, præpositus, custos.

Quersehen/ obliqvè intueri, limis oculis aspicere, luscum esse, per transennam videre. Quergesicht/ das/ imago catagrapha, aliàs Seitengesicht.

Seitab- & Seitwärtssehen/ it. Nebenhinsehen/ obliqvô vultu respicere, à conspectu alicujus se avertere, declinare vultum, deflectere oculos, & detorqvere, alterô oculô contueri.

Schälsehen/ non soliùm est limis oculis intueri, sed etiam vultum adducere, censorium supercilium facere, inimicô vultu intueri.

Versehen/ propr. visu falli, aberrare oculis, deinde autem providere, instruere, parare, procurare, & tandem peccare, delinqvere. Inde Versehung/ die/ & das Versehen/ non solùm est lapsio, delictum, & peccatum; sed etiam providentia, spes, ac fiducia. Sich des besten versehen/ bene ominari. Versihe dich mit Gelte/ provide argentum! Die Jungfer hat es versehen/ virgo vitiata est. Ein kleines Versehen/ erratum tolerabile. Großes/ & Haubtversehen/ lapsus summus. Verseher/ errans, provisor. Zuversichtlich/ certus, indubitatus, &: certè, eqvidem, sine dubio. Zuversichtlicher Freund/ amicus spectatus, exploratus, perspectus, fide dignus.

Umsehen/ propr. circumspicere, circumspectare, collustrare, huc atqve illuc intueri, qvod etiam dicitur Umhersehen/ deinde est sollicitum esse, prospicere sibi. Es ist nach guten Gründen sich umzusehen/ reqvirendæ erunt rationes firmissimæ. Ich will mich nach einer eigenen Wohnung umsehen/ habitationem mihi prospiciam. Herrliches Umsehen/ prospectus præclarus. Umsehung/ die/ provisio, cura, sollicitudo. Umsehung nach Getreide/ rei frumentariæ cura. Umsehung nach der Flucht/ meditatio, sive consilium fugæ. Ohne Umsehen/ haut respiciendo.

Wiedersehen/ revisere, intervisere, it. redire, recipere, & referre se, ad aliqvem remigrare. Wir wollen uns bald wiedersehen/ protinus me tibi reddam. Wiedersehung/ die/ & das Wiedersehen/ reditus, reversio, recursus, vulgò revisio. Wiedersehen bringt Freude/ reditio aspectu reficit.

Zersehen/ Sich zersehen/ sæpiùs revisere, oculos pascere, animum levare adspectu, figere

obtutum in aliqvam rem, acerrimè contemplari.

Zuesehen/ idem qvod Aufsehen/ lustrare, inspectare. Er hat der Komödie bis ans Ende zugesehen/ ludum scenicum perspectavit, i. e. uq, ad finem stetavit. Seht mir doch nur dem Handel zu/ egregium spectaculum capessite oculis. Zuesehung/ die/ & das Zuesehen/ spectatio. Zuescher/ spectator. Sed Zuesehen/ etiam sumitur pro tolerare, & connivere. GOtt kan nicht länger zuesehen/ Deus hæc amplius dissimulare, sustinere, sufferre non poterit. Zuesehens/ coram, in præsentia.

Zurücksehen/ retrò videre, circumvidere, retorqvere oculos, aciem oculorum reflectere. Refertur etiam ad animum, estq, in memoriam revocare, meminisse. Siehe zurück auf deine Jugend/ subeat recordatio hæc eadem, te fuisse juvenem. Zurücksehung/ die/ & das Zurücksehen/ recordatio, reminiscentia, vulgò reflexio.

Zwischen- & Darzwischensehen/ introspicere, it. interea cogitare, interim providere. Zwischen die Spalten durchsehen/ per rimas spectare.

Sims/ der/ plur. Simse/ podium, projectura, it. basis transversa fenestrarum, aliàs Vordächlein. Seulensims/ sive, qvod idem est, Gesims/ seu dimin. Gesimslein/ cymatium, triglyphi. Fußgesims/ basis columnarum. Haubtgesims/ hypotrachelium. Rund Gesims/ corona. Ofengesims/ lorica testacea. Simse etiam dicuntur repositoria parietibus in conclavibus affixa, qvibus imponuntur canthari, vitra & urcei, aliàs mutuli. Hinc venit verb. Simsen/ i. e. Simse machen/ construere & striare mutulos, vel triglyphos. Simsen etiam dicuntur junci, aliàs Semden/ sed hoc Sims venit à Seim.

Sing/ singen/ Ich sang/ & sung/ du sungest/ er sung/ & sang/ wir sungen/ ꝛc. Ich sünge/ gesungen/ canere, psallere, modulari, cantare. Nach dem Takt singen/ ad modos numerosqve artis vocem accommodare. Hoch singen/ pæana citare, voce alta, elevatâ canere. Mit halber Stimme singen/ voce submissâ, vel depressâ canere, aliàs Kleinsingen/ & fistuliren/ minurire, inde Kleinsingung/ sive Fistulirung/ die/ minuritio. Sehr lieblich singen/ aures svavissimô cantu permulcere. Alleine und ohne Instrument singen/ asâ voce canere. Elend sive abgeschmackt singen/ insulsè, vel absurdè canere. Einen Meistergesang

Fig. 5.1a. The entry for *Sinn-* in Stieler, *Der Teutschen Sprache* (Nuremberg, 1691) sets out a forest of terms translated by the Latinate family of ingenuity.

gesang singen / rudiorem symphoniam efficere. Den Diskant singen / acutum canere in musico concentu. *Inde* Diskantist / der / *oxyphonus.* Den Alt singen / alterum ab acuto canere. Den Tenor singen / voce subgravi canere. Den Baß singen / grave canere. Ein geistlich Lied singen / psallere hymnum. *Compos. sunt :* Absingen / decantare. Ein Gesetzlein absingen / inchoatum absolvere nomon. Ansingen / accinere. Ein Haus ansingen / cantu gratulatorio familiam qvandam & carmine laudare. Aussingen / certis voculis & recto cursu cantitare, *it.* finem canendi facere. Auffsingen / incantare. Die Nachtigall singet auf dem Baum / philomela incantat arbusculis. Besingen / consalurare, proseqvi, concelebrare cantu. Tapfferer Leute Tahten besingen / de clarorum virorum laudibus carmina cantare. Durchsingen / cantiunculam voce percurrere. Er hat das ganze Lied durchgesungen / integram cantilenam sono reddidit. Ein- & Dreinsingen / ad harmoniam canere. In die Orgel singen / jungere vocem organô. Ein Kind einsingen / infantem cantu mulcere, excomponere somnum modulatione. Fortsingen / cantitare, usqve & porrò canere. Für / *sive* Vorsingen / præcinere, cantum præire. Etwas hersingen / cantum edere, & exercere, vocem explicare. Mitsingen / compositâ modulatione canere. Gemein hin *sive* wegsingen / simplices planosqve modos canere. Nachsingen / succinere. Ubersingen / explorare canendi facultatem. Zierlich singen / vocem canendo crispare. Zusammen singen / concinere, concentum *vel* harmoniam efficere. Mit Koloraturen singen / *aliàs* Koloriren / chromatibus uti.

Singer / & Sänger / der / cantor, modulator, cantator. Singerinn / *sive* Sängerinn / die / cantrix, cantatrix. Gute Singer / gute Schlinger / cantores amant humores. Der Schwan ist sein Todteusänger / cantator cygnus funeris ipse sui. Alte Sänger singen was gutes / vetulus cantor canit memoriâ dignum. Obersinger / phonascus, chorostates. Vorsänger / præcentor. Liedersänger / rhythmorum modulator. Meistersänger / cantionum magistralium cantator. Todtensänger / siticen. Todtensängerinn / præfica. Hochzeitsänger / hymenæi psaltes. Schnapperliedersängerinn / cantionum cinædicarum psaltria. Vorsinger / choricus.

Singung / die / das Singen / & der Gesang / cantratio, cantus, melos, oda, cantio, canticum, cantilena, modulamen. Bulengesang / cantilena amatoria. Kindergesang / lallus. Todten- *sive* Trauergesang / næniæ, cantus lugubris. Meistergesang / cantio magistralis. Geistlicher Gesang / hymnus, psalmus. Schambar Gesang / *aliàs* Schnapperlied / cantio cinædica. Weihnachtgesang / canricum natale, &c. Gesänglein / das / *dimin.* cantiuncula. Dem Singen zuhören / aures præbere cantibus. Besingung / & Besingnüß / *it.* Ansingung / cantatio. Besingnüß der Todten / exeqviæ, inferiæ, parentalia. Zusammensingung / symphonia ; cantio ternione, qvaternione , qvinione, ogdoade musicorum constans. Koralgesang / cantus Gregorianus & vocibus assis.

Singhaft / Singicht / & Singerlich / *adj. & adverb.* canens, cantans, cantatus, psallens, & modulatè. *aliàs* Gesangsweise. Es ist mir gar nicht singerlich / à cantando prorsus abhorret animus, fastidiosus sum cantilenarum. Vorsingerliche Pfeife / incentiva.

Sinn / der / *plur.* die Sinnen / sensus, & qvidem propr. qvinqve sensus exteriores, die fünf euserliche Sinnen / *aliàs* mens , animus, ingenium. Innerliche Sinnen / sensus interiores, qvales sunt : Intellectus, memoria & voluntas, seu potiùs : Sensus communis, phantasia, & memoria. Harter Sinn / durum caput, homo pertinax, atq; hujus significationis est : Das Sinnlein / *sive* Sinnchen / ingenium contumax , mens indomita, ferox, propr. sensiculus. Sinn *etiam qvandoq; est* sententia, argumentum, scopus, intentio, *ita :* Der Sinn des Gesetzes / sententia legis, & verbis, *sive* literis legis opponitur. Sein Sinn stehet nach Ehren / gloriæ cupidus est. Mein Sinn ist darauf nicht gericht / consilia mea eò non referuntur, non collineat ad hanc rem intentio mea : *appellant etiam das* Datum. Es gehet ihm nach seinem Sinne / ex voto res succedit. Er hat nichts Gutes im Sinn / meditatur mala, volutat in animo astutias. Die Sinnen sind Boten / Dolmetscher und Fenster des Gemütes / sensus sunt qvinqve nuncii ac interpretes rerum animiqve fenestræ. Bey Sinnen seyn / seine fünf Sinnen haben / propr. sensibus præditum ac instructum esse, *communius autem :* habere sensum incorruptum atqve integrum, intelligere, animadvertere. Seiner Sinnen gebrau-
chen /

chen / mit den Sinnen begreifen / sensu capere, comprehendere, judicare, sensibus uti. Seiner Sinnen beraubt werden / sensu carere, vacare. obstupescere, sine sensu esse. Die Sinne betriegen nicht / sensus vera nunciant. Stumpfer Sinn / sensus hebes, tardus, stupidus. Scharfer Sinn / acutissimus, acerrimus. Schlimmer Sinn / animus malevolus, mens prava. Beständigen Sinn haben / animô stare. Seinen gantzen Sinn worauf legen / mente totâ, omniqve animi impetu in aliqvid incumbere. Es kommet mir in Sinn / in mentem venit, illius rei memini, in memoriam redeo, recordatio subit. Viel Köpfe / viel Sinne / qvot capita, tot sensus. Man siehets ihm an Augen / was er im Sinn hat / animus ejus in oculis habitat. Abersinn / pertinacia. Andachtssinn / mens devota. Bauersinn / intelligentia vulgaris, popularis. Blutsinn / mens barbara, sæva, atrox animi. Buben sive Diebs Laster, Schand- & Schelmsinn / furens audacia, scelus anhelans, omnium scelerum flagitiorumqve documentum. Christensinn / mens pia, innocens, Christiana. Edelsinn / ingenium ingenuum, nobile, liberale, præclarum. Ehrensinn / animus masculus, generosus, pectus stabile. Eigensinn / pervicacia, obstinatio, & refractariolus. Erdensinn / animus imbecillus, & angustus. Flattersinn / & Wankelsinn / inconstans, mollis, natura desultoria. Fleischessinn / homo carnalis. Freselsinn / mens temeraria, audax. Friedenssinn / pacifica. Gemeinsinn / sensus communis. Glaubenssinn / ratio fidei. Heldensinn / altitudo, magnitudo animi, mens heroica. Hertzenssinn / pectus fidele, sanctum, ex animo. Himmelssinn / mens cœlô nata, dedita. Hui sive Schnellsinn / animus irritabilis, præceps. Hurensinn / libidinosa mens. Jugendsinn / juvenilis. Kaltsinn / tepidus. Leicht sive liederlicher Sinn / mens frivola, futilis. Liebessinn / gratiosa. Mittelsinn / mediocris. Mutter & Vatersinn / storge. Narrensinn / sententia vaga, absurda, stulta. Poltersinn / mens tumultuans, inqvieta. Rabensinn / obscena, fœda, immanis. Rachsinn / vindictæ cupida, sangvinaria. Rätzel sive Fabelsinn / epimythium, sensus ænigmatis. Rede- Wort- & Schriftsinn / sensus literarum. Schadsinn / damnosus, damnificus. Schwindelsinn / fanaticus. Soldatensinn / ingenium militare. Teufelssinn / toller Sinn / & Unsinn / mania, phrenesis, delirium. Tugendsinn / animus magnus, excelsus. Wansinn / opinio. Weibersinn / mens abjecta. Weltsinn / levis. Wandel- Wetter- & Zweyfelsinn / errans, anceps, vaga, dubia. Wundersinn / homo qverulus. Sinnen / Ich sinne / sonne / & san / du sinntest / sonnest / & sanst / er sinnte / sonne / & san / ꝛc. Ich sönne / gesonnen / & gesinnet cogitare, animô volvere, secum commentari, reputare, retractare: It. fingere, meditari. Was sinnestu? qvid tecum commentaris? Lieder sinnen / carmina fingere. Scharfsinnen / in acerrima cogitatione versari. So ist er gesinnet / sic animatus est.

Sinner / der / & Sinnersinn / die / mas, & fœmina meditabundi, spectantes, contemplantes.

Sinnung / die / meditatio, meditamentum. Das Sinnen / id. & cogitatio, fictio.

Sinnlich / & Sinnig / it. Sinnisch / adj. & adv. animô affectus, ingeniosus, & ingeniose, sensilis, sensibilis, & vulgò sensibiliter, solers, & solerter. Unsinnig / vecors, furiosus, rabidus, insanens, & Unsinnlich / furenter, rabiose, furialiter. Bössinnig / malignus. Tummsinnig / stupidus. Frechsinnig / temerarius. Gegen- sive Wiedersinnig / discrepans, contrariæ opinionis. Eigensinnig / contumax, cerebrosus, singularis, & obstinatè. Gutsinnig / benignus. Hochsinnig / arrogans, fastuosus. Kaltsinnig / tepidus, frigidulus, & negligenter. Leichtsinnig / levis, futilis. Rechtsinnig / & Redlichsinnig / cordatus, ingenuus. Scharf- sive Tiefsinnig / ingeniosus, cogitabundus. Schwersinnig / melancholicus, subtristis. Starrsinnig / rigidus. Wansinnig / opiniosus, prava sentiens.

Sinnlichkeit / die / & Sinnigkeit / solertia, acumen ingenii, docilitas, indoles, & vigor naturalis. Inde: Eigensinnigkeit / pervicacia, obstinatio. Hochsinnigkeit / fastus, ferocitas, sublatio animi. Kaltsinnigkeit / frigus amoris. Leichtsinnigkeit / futilitas, levitas animi. Stumpf- sive Tummsinnigkeit / stupiditas ingenii. Unsinnigkeit / insania, dementia, rabies, dicitur etiam Wansinnigkeit.

Absinnen / disponere rem, & intra se formare, acie mentis dispicere, animô concipere, propr. animadvertere, cogitatione comprehendere, & conseqvi. Ich kan das Ding nicht genug absinnen / ad hujus rei intelligentiam satis penetrare neqveo,

Kramer (and the eighteenth-century authors who follow them) begin to connect the lexis of *Sinn* with broader European-wide cultures of mechanical and especially presentational ingenuity. Kramer suggests that textual objects can be *sinnreich*, referring to a *sinnreicher* saying as "an ingenious phrase," and to a *sinnreicher* verse as "a spirited poem."[60] He also glosses *Sinnreich* as an "invention" (*Erfindung*). It is possible that Kramer is just following the lexis of his target language. But it is also possible that here again he follows the lead of Stieler, who offers a striking connection between *sinn* and creativity. For Stieler, the prefix *aus-* makes *Sinn* about original creation, invention: an *Aussinner* is literally an "inventor." The term includes precisely the overtones of deceitful trickery so often found in the Romance languages: *Ein Schelmstück aussinnen* is about "machinating some trickery in one's heart" or to "make up a deceitful scheme."[61]

By this point in our study, it is clear that German writers are reaching for the broader European culture of ingenuity. Wachter presents the very precise example of a *Sinnbild*. This sort of image is not a vague sort of *figura ingeniosa*, but belongs precisely to the genre of emblems, *pictura loquens*, that comprised such an important part of the culture of ingenuity throughout Europe. In fact, Wachter writes, this is "properly called an 'ingenious image,' made of a figure and theme, and is so called 'ingenious' from *Sinn* or *ingenium* . . . but few *ingenia* rise so high that they can construe the whole symbol."[62] The first emblematists to use the term *Sinnbilder* explained that these emblems hid meanings produced out of their authors' *Sinnen*, meanings that in turn demanded readers to deploy their *Sinnen* to interpret.[63] By defining the German word in terms so familiar to the European culture of ingenious images, Wachter sets the concrete shape of the close link that he and Stieler are forging between *Sinn* and *ingenium*. Indeed, looking away from lexica to the broader German literary discourse around 1700, we see how the *Sinn* branch of ingenuity was growing as a distinctive element in new German literary movements. Drawing in part on the notion of *acutezza* in Tesauro's *Cannocchiale Aristotelico* (1654) and the writings of Gracián, German authors such as the dramatist and pedagogue Christian Weise turned the "penetrating style" (*stilus argutus*) into a German genre of *scharfsinnigen Inschriften*—a German response to the baroque literature of wit.[64]

Geest (Dutch)

As we have seen, German agonized over finding a ready counterpart for Latinate ingenuity. Dutch was less conflicted, at least at first. Already in the late sixteenth century, Dutch found an easy equivalent for the French in *geest*: "Esprit *ou* Esperit, m. Geest, sin ende verstant" (sense or understanding).[65] In fact, whereas the German *Geist* remained largely a matter of "spirituality," part of the lexicon of devotion and ghosts, the Dutch *geest* (or *gheest*) edged ever closer to the Latin *genius* and *ingenium*, nestling ambivalently between the two.

> *Geest*
> 1. God or spirit
>   a. Daemon, good or bad spirit. Madoets 1573 >
>   b. Holy Spirit. Madoets 1573 >
> 2. Soul
>   a. Mind. Madoets 1573 (as "mens"); Sasbout 1576 (as "Esprit *ou* Esperit, m. Geest, Sin ende verstant") >;
>   b. Soul. Madoets 1573 (as "animus") >.
>   c. Religiously minded. Madoets 1573 >.
>   d. Liveliness. Madoets 1573 (as "Dat is zeer geestlich gedaen. Cela est for ingenieusement faict. Affabre factum est, eleganter, ingeniose ac scitissime factum," "geestlicheyt. Vivacité d'esprit. Vivacitas, dexteritas") >.

Already in 1573, André Madoets matched the Dutch lexis of *geest* to Latin, most likely using Robert Estienne's *Dictionarium latinogallicum* of 1538. The entry for *Geest* stands out in Madoets's dictionary, with an unusually long list of phrases that build on the preliminary definition itself (*esprit*, *spiritus*, *animus*, *daemon*). It would be no surprise that Madoets drew on Estienne—elsewhere in this book we have seen the influence of those lexica. It is perhaps more interesting that most of the French and Latin phrases translating *geest* are drawn not from Estienne's lemma for *ingenium*, or from *spiritus*, but from *animus*, which

Estienne chiefly rendered in French as *esprit*. Therefore "my *geest* [inclination] is to pursue that" is rendered as "*mon esprit* and *animus* are so inclined."[66] Thus, even in a dictionary heavily dependent on Latin, the word that became central to the culture of ingenuity in Dutch was mediated by the growing French culture of *esprit* rather than directly by Latin *ingenium*.[67]

However, non-lexical sources indicate the extent to which *geest* was situated somewhere between French *esprit* and Latin *ingenium*. The painter Guillem van Haecht played with the relationship between these cognate words by placing the motto "Vive L'Esprit" upon the portal that dominates his painting *The Gallery of Cornelis van der Geest* (1628).[68] This motto translates the first half of a famous phrase from the pseudo-Virgilian *Elegiae in Maecenatem*: "Vivitur ingenio, caetera mortis erunt" (He lives by the spirit, the rest belongs to death).[69] In the painting, the French words pun obviously on the surname of renowned *liefhebber* Van der Geest, implying that his fame will endure beyond the grave through his patronage of contemporary artists—a group of whom (Rubens among them) are gathered in the gallery depicted. Yet the phrase invokes more broadly the spiritual and cognitive aspects of *ingenuity* that *geest* could denote, for it is situated immediately beneath a dove—standing for the *Heilige Geest* (the Holy Spirit)—perched atop a helm sporting outstretched wings: a standard iconographical attribute of ingenuity.[70] This pairing of verbal and visual motifs succinctly indicates the semantic range of *geest* in the cultural milieu of Antwerp's *liefhebbers*, among whom ingenuity was a prominent theme. For example, in the second half of the sixteenth century the city's Chamber of Rhetoric (the *Violieren*) staged several performances addressing processes of intellection, which featured personifications of *vernuft*, *verstand*, and *geest*.[71] Elsewhere, the painter and theorist Karel Van Mander used the term *geest* repeatedly in his *Schilder-boek* (1604) to refer to certain types of pictorial subject matter that, as Melion explains, "must be pictured *uyt den gheest*, since they are either too fleeting or too multifarious to be captured *nae t'leven* (from the life)."[72] As Van Mander writes, "Leaves, hair, air and fabric all are *gheest* and *gheest* alone teaches how to fashion them."[73] These, then, are parts of

Fig. 5.2. The armorial and inscription "vive l'esprit" (long live wit), of the *Cabinet of Cornelis van der Geest* (1628).

a painting in which ingenuity-as-character, close, even, to individual style or the *non-so-che* of Italian artistic theory, may be observed.[74] It is no coincidence that such qualities are abundantly present in the varied paintings lining the walls of Van der Geest's gallery, which—in displays of personal and collective ingenuity—are scrutinized discriminatingly by the collector's *geestige* friends.

The dictionaries confirm the word's semantic range and its centrality to the Dutch lexis of ingenuity. *Geest* tracks quite closely the ambivalent moral value of ingenuity. Martin Binnart gives "the good or bad genius" as his first translations of the word.[75] Already in the first major Dutch dictionary of Madoets (1573), *geest* is matched with both positive and negative adjectives. As Hexham (1647) notes, not every geest is light and quick; one can be "*swaremoedigh van geest*, Of a dull and heavie Spirit."

Like *esprit* and *ingenium*, one's *geest* could be shaped with practice. Far from identifying with Romantic genius, *geest* described the niceties of acquired comportment. Binnart included the word in a classic example of honing one's *ingenium* in rhetorical disputation. "To make subtle" or "to sharpen one's wits" in the cut and thrust of school argument, he wrote, was the equivalent of "sharpening one's senses and *geest*."[76] A particularly vivid example is Sewel's English-Dutch dictionary of 1691, which supplies a series of reflections on the word:

> Genius, Aardt, inborst, geest.
> Genteel, Aardig, net, hupsch, geestig.
> Genteelness, Aardigheyd, geestigheyd.
> Genteely, Aardiglyk, op een geestige wyze.
> . . .
> Ingenious, Zinryk, vernuftig, scherpzinnig, verstandig, geestig, aardig.
> Ingeniously, Vernuftiglyk, geestiglyk, aardiglyk, op een zinryke wyze.
> Ingeniousness, Zinrykheyd, vernuftigheyd, geestigheyd, aardigheyd.

In each line, *geest(ig)* emerges alongside *aard(ig)*, reflecting the extent to which both had sedimented into a coherent vocabulary of ingenuity. In fact, *geest* with its variants is the one word here that emerges in *every* definition. It certainly translates one's genius (which here must mean one's "nature"), but also picks out a rich social habitus: the elegant, gentle display of character, as well as the "sensible wits" of one behaving "ingeniously."

As for German *Geist* and French *esprit*, by contrast, the connotations around *geest* continue to touch on the "spiritual," the saintly, and the devout, rather than ingenuity. To be *geestelijk* could be to display upturned eyes and a hymnbook close to hand. But to be *geestig* is quite different. Its lexis moves firmly toward some senses of ingenuity.[77] De la Porta's translations between Dutch and Spanish (1659) is a typical case, which sets *geestig* squarely within the Spanish culture of *ingenio*: *ingenioso, industrioso, agudo, diestro, agudo de ingenio . . . constelijck, artificiosamente, industriosamente, sutilmente*. Perhaps because of its alliance with French *esprit*, encouraged certainly by a seventeenth-century culture of Dutch dictionaries that focused less than German on difference and more on connecting to the Romance languages, *geest* sits evenly across the various axes of ingenuity.

Gemüt

When Adelung used *Gemüthsfähigkeit* (an ability of the soul) to explain the new Romantic *Genie*, he was reaching for a word intriguing because of its mismatch with *ingenium*. *Gemüt* has a deep history in the German vocabulary of the soul, and became particularly associated with the soul's artistic, aesthetic powers. The history of *Gemüt* found a high point with Johann Wolfgang von Goethe, who focused his account of artistic capacity on that word.[78] Lexica only hint at such grand stakes; therefore as a keyword *Gemüt* takes us to the limits of our focus on ingenuity in German. Its main meanings are as follows:

   1. Spirit
      a. Mind. Dasypodius 1535 (as "mens") >.
      b. Soul. Dasypodius 1535 (as "animus") >.
      c. Religiously minded. Maaler 1561 (as "Götelich Gemüt") >.
      d. Characteristic spirit. Kramer 1676 ("eine weibisches, feiges Gemüth, talento feminile") >; Kramer 1712.
      e. Imagination. Kramer 1700 (as "Gemüts-bildung / f. imaginatione, Facoltà imaginativa").
   2. Affective faculty
      a. Heart. Henisch 1616 (as "cor, Hertz") >; Kramer 1712 (as "Gemüt, n. ame, esprit, coeur; naturel, genie. V. Mut, Natur, Art").
      b. Sensitivity. Henisch 1616 >.
      c. Will, desire. Stieler 1691 (as "*Hestig gemüt,* animus inflamma-

tus, fervens, ardens . . . Er had ein leichfertig Gemüt, ingenii ancipitis, vani et mutabilis") >. (Cf. also Wachter 1737, "Gemüt, animus et omnes animi facultates et inclinationes, bonae vel malae").

The first German dictionaries offer short but messy definitions. For Dasypodius (1536), *Gemüt* is *Verstand, Will, animus*, and *cor*. Understanding (*Verstand*) and mind (*animus*) are easily recognizable as neighbors of ingenuity. But will (*Will*) and heart (*cor*) pull dramatically in another direction, suggesting perhaps emotional, affective experiences as well as more intellectual ones.[79] In fact, these rough definitions probably did not fit usage very well. When Henisch wrote his larger dictionary of 1616, the first focusing on German itself, he repeated definitions for *Gemüt* that linked heart, mind, and even sensation.[80] But he also collected various proverbs that clarified the word. In particular, these proverbs make *Gemüt* do the kind of work that *ingenium* performed in Latin. Not only did it stand in for one's mental material and abilities, but it tempered the products of that mind. Thus one could be *Wolgemutet* (well-natured). "Speech is an image of the mind" (*Die Red ist dess gemüts Bildnuss*); "As the word is, so is the mind" (*Wie die rede ist, so ist das gemüt*); and "nobility, not blood, makes the mind" (*Edel macht das gemüt, nicht das geblüt*).

In these lexica, *Gemüt* shares much with *ingenium*, representing mental ability, piercing perception, the sum of mental faculties, as well as the soulish principle that makes one person different from another. But there are also big differences. The word never stands in for an object: one might recognize someone's *Gemüt* in a text, but the text can never *be* the *Gemüt*. And the notion of intuition, of immediate and reflexive response, allows *Gemüt* to be related to the heart so often; this is unusual in the language of ingenuity. It therefore seems to be a deeply inward sense of self.

It might be tempting, therefore, to see the Romantic *Gemüt* of Goethe as part of the eighteenth-century rise of genius. But that would entirely miss the longer history—only just visible in these lexica—that links *Gemüt* to ingenuity. The term's history weighs on works such as the *Künstlerroman* of Johann Tieck, *Franz Sternbalds Wanderungen* (1798), which tells the story of one of Dürer's apprentices, who leaves his master's workshop for Rome. On the journey, the young man struggles with the manner in which Renaissance masters such as Dürer and Raphael had described beauty, and he complains that his experience leaves him unable to paint. "My inner images multiply with every step I take, each tree, each landscape, each wanderer, the rising and the setting of the sun, the churches that I visit. . . . My soul [*Gemüt*] is so bewildered that I cannot even dare to set myself to work."[81] Even as he makes *Gemüt* the source of deep human feeling—literally, of being *moved beyond*—Tieck also signals the longer history of *Gemüt* as bound to ingenuity. Dürer had famously used this language for artistic productivity, the site of invention and imagination; Dürer's heirs turned the language inside out, so that images overwhelm.

For the overflowing "inner images" that cloud Sternbald's *Gemüt* may well refer to Dürer's own aesthetic theory, with his "Gemüt voller Bildnuss," usually translated as "mind full of figures."[82] The phrase was published posthumously in Dürer's *Vier bücher von menschlicher Proportion* (1528), in a long passage traditionally dubbed the "aesthetic excursus."[83] The "aesthetic excursus"—so-called because in it Dürer grappled with notions of beauty and the ends of art—is suffused with the German word family of ingenuity. In each of his drafts for the text, Dürer describes learning as a process of filling the *Gemüt*. He asserts that Nature "pours in" (*eingossen*) knowledge of the "truth of each thing" (Wahrheit aller Ding), but "our weak *Gemüt* cannot contain the bounty of all arts, truth, and knowledge." Thus, we must "sharpen our *Vernunft* with learning," and "no man is so thick that he if does not apply his *Gemüt* to the utmost, he will have no *Vernunft*." This difficult-to-attain skill of artists, which Dürer compares to "foreign speech" (fremde Sprach), is what kings have properly valued in the past. At its highest levels, "they valued such *Sinnreichigkeit* as a creator alike to God." These and similar passages enunciate a German version of ingenuity, with *Gemüt* at its core. At one point Dürer even borrows Latin, a little uncertainly. He surmises how much art theory has been lost to time, and cannot restrain himself from complaining that "it often happens that, because of coarse printers, the noble *Ingeni* [sic] are lost." Venting frustration with his own printers, Dürer aligns himself with the plight of ancient *ingenia*.

Dürer's use of *Gemüt* resonates especially with Middle High German, where it was connected with the "ground" (*grund*) of the soul, which marks one's deepest identity. Meister Eckhart, the mystical theologian of the thirteenth century, had described the *Gemüt* as the portion of the soul "which [Augustine] calls an enclosure or container of spiritual shapes or of formal images," namely the memory.[84] Here Eckhart explained the German word with reference to Latin theology. By Dürer's time, a distinctively German late medieval theology drew on meanings of *Gemüt* as more than the sum of intellectual operations. Just as those "poor in spirit" inherit the kingdom of heaven (Matt. 5.3), only those with humble selves (*demütigen gemüt*) can welcome Christ within: one must strive to discipline one's *Gemüt*.[85]

Out of this longer history, *Gemüt* came to describe the seat of Romantic affect, having moved from religious feeling to aesthetic response. But the change is not so great. The word never had resonance on the mechanical or performative axes; its domain remained mildly cognitive and, above all, moral, a matter of one's nature, identity, and character. What one loves is who one is. It is this aspect of ingenuity that lexicographers emphasized when defining the word. Stieler (1691) explains *Gemüt* as one's whole character, so that the notion of having "a flighty *Gemüt*" (*er had ein leichtfertig Gemüt*) is interpreted as being "double-minded" (literally, "two-headed wit," *ingenii ancipitis*). *Gemüt* lays open the inclinations of the heart, an angle also hinted at in Kramer's French-German dictionary (1712–15), which offers a string of equivalents: *ame, esprit, coeur, naturel, genie*. Three of Kramer's equivalents come directly out of the French language of ingenuity.[86] But *ame* and *coeur* set *Gemüt* slightly apart. No other keyword in this book says so much of the heart.

Conclusion

Within Dutch and German lexica, the trajectories of ingenuity certainly differed from trajectories of usage outside the lexica, as examples repeatedly show. But these trajectories were not entirely independent. We can see this in the Dutch and German reception of Huarte de San Juan's *Examen de ingenios para las ciencias*, a deeply set cornerstone of the early modern culture of ingenuity.[87] As a physician, Huarte mobilized a tremendous amount of classical lore to support a determinist account of how physiology and geography determine ability, with an eye to classing exceptional, heroic *ingenios*; he began with Cicero and the (pseudo)Aristotelian framework widely available in Latin. His Dutch and German translators made these powerful ideas available in terms that we can recognize from our lexicographical sources.

The word *ingenio* clearly gave pause to the Dutch translator of the *Examen*, Henryk Takama (Huarte 1659), who began his running commentary with several notes on terminology. The choice of title, *Onderzoek der byzondere vernuftens*, set ingenuity as the human capacity for abstract reasoning. Takama explained by first comparing *vernuft* with *ratio* and *mens*. He then expended two pages on the usual scholarly divisions of the soul's vegetative, sensitive, and rational powers.[88] The whole discussion aimed at the cognitive qualities that individuate one person from another—as the next note succinctly put it: "Aardt Lat. *natura*, wat eygentlijk dat is wordt zeer treffelijk verklaardt in het 2. Hoofdsts. § 10" (*Aardt* is in Latin *natura*, and what in fact that is, is explained very well in chapter 2, § 10).[89] This flattened some meanings; the *ingenio* of Hercules and Demosthenes, for example, could possibly refer to their *daemones*, while the *vernuft* of Demosthenes could only be about understanding, not inspiration. The tension between abstract cognition and its embodied ground persists throughout the work, so that *vernuft* bears the direct load of *ingenio*, but the weight is dispersed through explanations using *ingebooren aardt* and words related to *zinrijk*.

In choosing *vernuft*, Takama's *Onderzoek* typifies the medical and natural philosophical assumptions that underlay the moral and cognitive axes of ingenuity. The translation also typifies the Netherlands in the seventeenth century: soaked with Spanish influence and the wider waters of European culture, while firmly determined to translate them readily and flexibly into Dutch. Nevertheless—perhaps because Dutch scholars had easy access to Spanish, English, Italian, and Latin translations of Huarte—the *Onderzoek* was not reprinted and seems to have left little impression on Dutch culture or scholarship.

The German case is somewhat more complicated. Huarte was widely read by German encyclopaedists and polyhistors; but the first German translation was not published until 1752, the end of the period we cover in this book. Gotthold Ephraim Lessing undertook the translation as part of his MA studies at Wittenberg, publishing it as *Prüfung der Köpfe zu den Wissenschaften*. Lessing's translation itself hints strongly at the Romantic genius that he would later espouse in his *Hamburgisches Dramaturgie* (part 34). Without fail, he translated *ingenio* as *Genie*: the term is directly borrowed from the rising French (and English) usage. In the *Prüfung*, Huarte's text keeps Lessing a breath away from Romanticism: one *has* a *Genie*, one cannot *be* a *Genie*. As in the lexica studied in this chapter, German resisted Latinate forms of ingenuity until late.

But the choice of *Kopf* (literally, "head") in the title indicates the tensions that characterized the earlier, Germanic culture of ingenuity found in our lexicographical sources. *Kopf* was the sum of one's mental powers; metonymically, it could reflect the whole person.[90] Lessing used the word to refer to exceptional individuals. One could be a *Kopf* like Xenocrates, while one might only possess *Genie* like Democritus. But Lessing here echoed earlier usage; he likely took his title from Johann Christoph Gottsched's German rendition of Pierre Bayle's *Dictionnaire*.[91] When Gottsched himself reflected on *Genie* elsewhere, he dismissed it as un-German.[92] In contrast, *Kopf* was a distinctively German choice, one faithful to the language while also denoting the cognitive abilities of an exceptional individual. Adelung, we saw at the beginning of this chapter, would claim that "*Kopf* is perhaps the only German word which might displace the French [*génie*]."[93] Gottsched's disapproval notwithstanding, Lessing's translation of Huarte de San Juan therefore pulled in both directions at once. By adopting a newly fashionable French word within the text, Lessing remained part of the narrowing focus of ingenuity to inspired genius. At the same time, in the *Kopf* of his title Lessing reprised the embodied, pedagogical, and even rhetorical aspects of the older pan-European culture of ingenuity that Huarte had helped to shape. This tension rendered Lessing a faithful emblem of German ingenuity.

6 English

In the opening pages of *A Discourse of Wit* (1685), the physician and translator David Abercromby neatly summarized the linguistic and conceptual challenge of his subject:

> Yet shall I not say with a great Man of this Age, that *Wit* is, *un je ne scay quoy, I know not what*: For this would be to say nothing at all, and an easie answer to all difficulties, and no solution to any. Neither shall I call it a certain Liveliness, or Vivacity of the Mind inbred, or radicated in its Nature, which the *Latines* seem to insinuate by the word *Ingenium*; nor the subtlest operation of the Soul above the reach of meer matter, which perhaps is mean't by the *French*, who concieve *Wit* to be a Spiritual thing, or a Spirit *L'esprit*. Nor with others, that 'tis a certain acuteness of Undestanding, some men possess in a higher degree, the Life of discourse, as Salt, without which nothing is relished, a Celestial Fire, a Spiritual Light, and what not. Such and the like Expressions contain more of Pomp than of Truth, and are fitter to make us talkative on this Subject, than to enlighten our Understandings. But what then is *Wit*?[1]

Rejecting the puffed up and imprecise meanings implied by Latin and Romance-language words, Abercromby offered instead what he clearly considered to be a sensible English alternative. "Wit," he opined, "is either a senceful discourse, word, or Sentence, or a skilful Action." This combination of "sense" and "skill" is, he claimed, necessary for "meriting the Honourable Name of a *Virtuoso*, and a true *Wit*." Doubtless he had in mind the most renowned "Christian" virtuoso of later seventeenth-century England, his friend Robert Boyle, several of whose works he translated.

Abercromby's easy equation of *wit* with *sense* and *skill* aptly reflects the fledgling Royal Society's promotion of reasoned industry, in which judgment and talent (both intellectual and mechanical) were harnessed for the general improvement of knowledge and the commonweal. It belies, however, the already strongly contested nature of the lexis of ingenuity in early modern England. By the mid-eighteenth century, Johnson 1755 could list no fewer than eight different senses of *wit*, ranging from "the powers of the mind" (sense 1) through Abercromby's "Sense; judgment" (sense 6) to "a man of genius" (sense 5). Notably, Johnson's entry for *ingenious* opened with the synonyms "witty, inventive, possessed of genius." This diversity—not to say ambiguity—of meaning had arisen in part through the special lexicographical circumstances of English (a mongrel language bred from Teutonic and Romance stock), in part from vigorous philosophical and literary debate over the nature of mental faculties and creative talent. Against the backdrop of growing national self-confidence—in particular the formation of a robust literary canon in the later seventeenth and early eighteenth centuries—Anglo-Saxon *wit* morphed gradually into English *genius*. This transformation was neither smooth nor straightforward, yet it was decisive. It paved the way for a grander inflation of *genius* in Romanticism and the (still) pervasive association of that word with the artistic and scientific heroes of early modern England: Shakespeare and Newton.[2]

Keywords

The word family of early modern English ingenuity clusters around four terms: *genius*, *ingenuity*, *wit*, and *cunning*. The first two are etymologically related, deriving from the Latin *genius* and *ingenium*; the second two are etymologically unrelated, deriving from Anglo-Saxon words that denote knowing: *gewit*, *wit*, *witan*, *cunnan*, and *cunnung*.[3] The four keywords' closest semantic neighbors are *conceit*, *inclination*, *nature*, and *subtlety*, but they also nudge up against words such as *art*, *artifice*, *deceit*, *fancy*, *freedom*, *guile*, *humour*, *imagination*, *invention*, *judgment*, *pregnant*, *reason*, *sense*, and *spirit*. While *engine* is etymologically closest to the Latin-derived words and (via Old French) informed their meaning, it is rarely related semantically to the group as a whole in the dictionary definitions.[4] This is a good example of the extent to which the lexica sometimes fail to reflect period usage. For example, in his *Entertainment for Britain's Burse* (1609), performed to mark the opening of London's *New Exchange*, Ben Jonson punned on the relationship between *engine*, *wit*, and *cunning*. The Master of the *New Exchange*, introducing his audience to the exquisite commodities offered there for sale, draws attention to a particularly ingenious object: "Here's a second rarity, a conceited saltcellar, an elephant with a castle on his back, where beside the art of the artificer in the whole dimensions, the spreading of the ear, winding the proboscis, mounting of the tusks, and architecture of the castle, do but observe his engine! Why an elephant more than any other creature? He might have made it a mule, a camel, or a dromedary, but the elephant, being the wisest beast, it was fit he should carry the salt from 'em all, for by salt is understood wisdom: *sal sapit omnia*."[5]

Here, Jonson displays his own wit in a clever ekphrasis that plays on the lexis of ingenuity. The saltcellar is *conceited* (clever, amusing), not only because it displays the *art* (craft skill) of its maker (*artificer*), but because it displays the goldsmith's *engine* (ingenuity, cunning) by punning on the relationship between wisdom and salt. The conclusion is especially apt, since salt carried strong connotations not only of *wis-*

Fig. 6.1. Elephant salt-cellar (ca. 1550).

dom, but also of *wit*.[6] For Thomas 1587, the Latin *persalse* is "verie salt, verie wittily," while Florio 1598 offers for the Italian *salse*: "salt or seasoning. Also mirth, pleasant wittines in wordes, merie conceites or wittie grace in speaking. Wit, conceit, invention, pleasantnes. . . ." In Jonson's scene, then, the *conceit* consists in rebus-like double meanings, which in early modern English could have been called a *device* or, figuratively, an *engine*.[7] Connections such as these should be borne in mind as we trace our four keywords' trajectories, for the terms often carried a greater number of connotations and associations than may be glimpsed in the dictionary definitions.

None of the keywords have benefitted from a full investigation in the lexical sources, and they have rarely been examined in a comparative context.[8] Even C. S. Lewis's fairly expansive study of *wit* seems limited when set within the compass of our dictionaries. Lewis recognized correctly that an important early sense of *wit* is as the translation of *ingenium*.⁹ Indeed, this is how *wit* is translated in early sources, such as Geoffrey the Grammarian's *Promptorium parvulorum* (1499). Yet over the course of the following century this simple synonymy came under serious pressure. For example, in Thomas 1587 *ingenium* is defined as the "nature, inclination, or disposition of a thing: also wit, wisdome, will, or propertie, fansie, invention, cunning." Elsewhere in this text, *ingenium* is associated with such English words as *nature, pleasure, pregnant, sharpness, spirit*, and *will*, and—often via *wit*—with a wide variety of Latin words, including *animus, argutus, genius, mens, natura, sagacitas, solertia*, and *spiritus*. As Ruthven aptly remarked of the early modern *conceit*, "it was all very confusing."¹⁰

The four keywords are clearly united, however, in denoting the qualities possessed of an individual. Yet even here, *genius* differs from *ingenuity, cunning*, and *wit* in referring (especially in early sources, which are closest to the Latin sense) to an angel, spiritual being, or spirit of a place. Each keyword may refer to the properties of a thing rather than a person, although the lexical category used in this case varies across the words. The "inborn" aspect of the definitions derives, as we shall see, largely from the Latin *ingenium*, their explicit association with knowledge from the Anglo-Saxon sources.¹¹ The history of the word family in the early modern period is essentially the intermingling of these two strands (called by C. S. Lewis *wit-ingenium*), injected with connotations of liberal high-mindedness from the *ingenuity* strand and of deviousness from the "machinating" that stems from the Old French *engin*. The frankness associated early on with *ingenuity* (stemming in part from confusion between *ingenious* and *ingenuous*) is first intermingled with, and then distinguished from, the devious aspects of *cunning* as the period progresses.¹² The "special talent" associated in modern usage with *genius* tends, before the eighteenth century, to be denoted by *ingenuity* or *wit* (normally in their adjectival forms), but also occasionally by *cunning*. In the first half of the eighteenth century, *genius* underwent a semantic inflation whereby the earlier qualities associated with *wit* and *ingenuity* attached to it, such that by Johnson 1755, *genius* had come to mean not just "special" but "superior" ability.

For the most part, the words tend to indicate either neutral or positive qualities, the latter especially in relation to the capacity to invent. This sense, however, underlies their intermittent association with craftiness, which—especially in the case of *cunning*—may be unfavorable. Of the four terms, only *wit* had a technical meaning: in rhetoric, where it was associated with the orator's inventiveness, and in faculty psychology, where (in relation to the intellective soul) it was sometimes associated with reason.[13] While the latter meaning is rarely explicit in headword definitions, it is made plain in the explanation of the neologism *witcraft* in Lever's *Art of Reason* (1573): "Wit in oure mother toung is oft taken for reason." In this context, *wit* became a hotly contested term in philosophy and aesthetics ca. 1700, especially in (sometimes oppositional) relation to judgment and imagination. This derived in part from the impact of Hobbes's and Locke's views on the subject, some of the effects of which are reflected in the dictionary definitions, especially in Johnson 1755.

Lexicographical landscape

Before examining the fortunes of our words in greater detail, let us consider briefly the lexicographical contexts in which they appear. England is unusual in that, compared to other European countries, its first major lexicographical project came late: Johnson's *Dictionary* (1755).[14] The impetus behind this project—dissatisfaction with existing English dictionaries—offers some indication of the general state of English lexicography in the early modern period.[15] Indeed, the first monolingual dictionary was not published until the early seventeenth century (Cawdrey 1604).[16] As such, it is vital to include Johnson's *Dictionary* (1755), not least because it identifies obsolescent words/senses and, unlike previous English dictionaries (even including its significant precursor, Bailey 1730), plentiful examples of usage from a wide variety of literary and philosophical sources.

Prior to Johnson, the general trend in English dictionary publishing is a gradual shift from polyglot (mainly bilingual) dictionaries in the mid-to-late sixteenth century to monolingual "hard word" dictionaries in the first half of the seventeenth century, then to the "general purpose" dictionary—incorporating "canting" (slang) words—in the second half of the seventeenth century.[17] These developments were by no means smooth, but notable landmarks include Elyot's popular and much reprinted Latin > English dictionary (Elyot 1538); Cawdrey's monolingual "hard word" dictionary (Cawdrey 1604; sometimes called the "first" English dictionary); the much-expanded mid-seventeenth-century monolingual dictionaries Blount 1656 and Phillips 1658; and the somewhat more modest Coles 1676, which nevertheless incorporated new canting material and provided some etymological information from classical languages and European vernaculars.[18] Developments in the eighteenth century are of two kinds: a move toward "ordinary" (as opposed to "hard") words, and a dramatic expansion in the scale and scope of dictionaries.[19] Both trends are observable in Kersey's dictionaries of 1702 and 1706 (the latter a revision of Phillips 1658), and in those by Bailey published in the 1720s and 30s, before reaching their apogee in Johnson 1755. This period also witnessed the development of encyclopaedias, such as Chambers 1728, which provided important material for Johnson and which, because of its extensive lexicographical content, has been included in this study.

The trend from polyglot and "hard word" dictionaries to general-purpose dictionaries reflects the susceptibility of English to neologisms and loan words at a time when the language was barely understood abroad.[20] According to the OED's estimate, the number of words available to English-language speakers more than doubled in the period ca. 1500 to ca. 1650.[21] These new words—which included *ingenuity*—were derived principally from Latin and Greek, but also from the European vernaculars, especially French and Italian.[22] This influx, along with increasing travel to the Continent and growing international trade, resulted in a flourish of polyglot dictionaries in the later sixteenth and early seventeenth centuries, notably the Latin-English dictionaries Cooper 1578, Thomas 1587, and Rider 1589, along with the bilingual vernacular dictionaries Florio 1598 and Cotgrave 1611, and the multilingual Minsheu 1617.

Polyglot works such as these are especially important for our word histories, in three senses. First, because *genius* and *ingenuity* derive from the Latin *genius* and *ingenium*, both of which were extensively glossed in scholarly lexica published abroad in the sixteenth century (see chapter 1). These works, especially those by Calepino and Estienne, enjoyed widespread popularity in learned English circles and provided important source material for our dictionary makers.[23] Second, the continental dictionaries encouraged a fondness for humanist *copia* in lexicography, at least in the bilingual tradition.[24] This is especially significant given the relative smallness of English monolingual dictionaries and the striking brevity of their entries prior to the eighteenth century.[25] Third, the boundaries between polyglot and monolingual dictionaries were decidedly blurred in early modern England, such that monolingual dictionaries have even been called a "special case" of bilingual ones.[26] All this shows the necessity of incorporating bilingual dictionaries into our study, confirming C. S. Lewis's sense that a proper history of a word such as *wit* should take full account of its relationship both to *ingenium* and to the Italian *ingegno* (for the latter, see chapter 2).[27]

Equally, and as the copiousness of certain polyglot dictionaries makes plain, we will do well to avoid an overly simple reliance on headwords when tracing our terms' histories. We may rightly construe as significant the appearance of an ingenuity term in a headword position in any given dictionary. Indeed, we risk spinning out of semantic control if we stray too far from them. Yet working by headwords alone fails to capture fully the extent of pollination from the European vernaculars and Latin (and, less commonly, Greek), not least because many polyglot and "hard word" dictionaries did not use English for their organization, which was often alphabetized. By way of example, *genius* does not appear as a headword in Rider 1589, but it does appear as a descriptor in that dictionary's entries for *grace* and *nature*, both of which are important aspects of the early modern meaning of *genius*. Thus, we will track our keywords' appearance as descriptors where relevant, a task greatly facilitated by corpus-linguistic digital resources such as LEME.[28] Other resources of this kind, such as EEBO and ECCO, are beyond the scope of this study. In any case, the sheer quantity of instances in which our keywords appear in such corpora prohibits their serious consideration here.[29] We will, however, have occasional recourse to non-lexica sources for the purpose of amplifying and further contextualizing the dictionary entries where necessary. Lastly, we should note that in the OED English has an unusually sophisticated resource for historical lexicography. While not infallible, it provides a useful structure for our words' several definitions, helps establish the extent to which early modern lexica capture period meanings, and indicates whether the dictionary definitions are synchronous with other forms of period usage. Thus, in the remainder of this chapter we will situate our words in relation to the OED, identifying the senses in which they were defined and the chronology of their appearance as a headword in the dictionaries.

Genius

We will begin with *genius*, not least because its semantic history has attracted the most critical attention, although the fortunes of the word prior to the eighteenth century have been comparatively neglected (at least in relation to its companion, *wit*).[30] A notable exception to this is Wiley, who, in "Genius: A Problem in Definition" (1936), was the first to attempt a systematic examination of the word in a selection of early modern English lexica. Wiley's valuable account was notably prescient of later developments in word history, offering an early critique of the type of conceptual history later codified in *Begriffsgeschichte*.[31] Acknowledging that the semantic inflation of *genius* in the eighteenth century led eventually to a great range of definitions, Wiley nevertheless identified five "exclusive semantic classes" for the word before 1800: (1) genius as attendant spirit; (2) genius as inclination, bent, or bias; (3) genius as mental endowment; (4) genius as the person endowed with superior faculties; (5) genius as superior ability to succeed in some art. While we find all these meanings in the period lexica (indeed, they are roughly equivalent to the definitions eventually offered in Johnson 1755), the limitations of Wiley's sources led both to omissions and to errors in chronology. In particular, she was surely wrong to conclude that "there is perhaps no actual relation between any two of these five principal significations of genius."[32] A deeper and broader investigation of the sources suggests the following meanings of *genius*, organized chronologically by order of appearance as a headword:

Genius, n.
  1. A spirit or deity.
     a. The god of procreation.
     Anon. 1500, as "the god of weddynge," single instance. Not in the OED, but connected to OED, AI: "a supernatural being" (1387 >).
     b. An angel or spirit, whether good or bad, that attends each man.
     Elyot 1538 > Bailey 1730. OED, AI.2: "Either of two mutually opposed spirits imagined as accompanying a person throughout his or her life and exerting either a good or bad influence" (1572 >).
     c. The spirit or soul of man.
     Elyot 1538 >, as "the spirit of a man"; Bullokar 1616 >, as "the spirit or soul." Not in the OED, but connected to AI.1: "the tutelary god or attendant spirit allotted to every person at birth to govern his or her fortunes" (1387 >); AI.2 and AII: "character, ability, and related senses" (1586 >).
     d. The spirit of a place.
     Cooper 1584 >; sometimes conflated with (b.), e.g. Blount 1656. OED, AII.6e: "The essential character or atmosphere of a place." 1741 >; "Genius loci," 1 (1575 >) and 2 (1605 >).
     e. The spirit of a group of people, nation, or age.
     Phillips-Kersey 1706 > (but see also Miège 1677 >).[33] OED, AII.6b: "[Spirit of] a group of people, a nation, period of time, etc.: prevalent feeling, opinion, sentiment, or taste; distinctive character or spirit" (1639 >).
     f. A spiritual being; specifically an intercessor.
     Livy 1600 >. OED, AI and AI.4: "A quasi-mythological personification of something immaterial" (1600 >).
  2. Nature, fancy, or inclination of an individual.
  Elyot 1538 >. OED, AII.6a: "Characteristic disposition; natural inclination; temperament" (1586 >); OED, AI.1b: "appetite" (1607 >).

3. The appealing quality (or "grace") of a thing.
Thomas 1587 >. Not in the OED, but see AII.6d: "With reference to a material thing, a disease, etc.: natural character or constitution; inherent tendency" (1675 >).
4. Natural aptitude, talent, or inclination toward a specified thing.
Phillips-Kersey 1706 (as endowment) > Bailey 1730 >, as "natural talent or disposition to one thing more than another." OED, AII.7a: "A person's natural aptitude for, or inclination towards, a specified thing or action" (1611 >); AII.7b: "natural ability or capacity" (1649 >).
5. Mental capacity.
Chambers 1728 > as "a force or faculty of the soul by which it thinks or judges"; Johnson, 1755 > as "mental powers or faculties." Not in the OED, but connected to AII.7.
6. Superior ability.
Johnson 1755 >. OED, AII.8b: "An exceptionally intelligent or talented person" 1711 >; (?)AII.9: "Innate intellectual or creative power of an exceptional or exalted type" (1749 >).

Before exploring these meanings, we should note that the lexica omit two senses that the OED indicates came into use around the middle of the seventeenth century. They are:

AI.5: "A god, spirit, or other figure associated with the influences of an astrological body; a combination of sidereal influences represented in a person's horoscope" (1644 >).

AII.6c: "With reference to a language, law, institution, etc.: prevailing character or spirit; general intent or meaning; characteristic method or procedure" (1647 >).

While the first does not appear in relation to *genius* as a headword, we find a single instance of it in the description of *Eudemon* in Moxon's specialist *Mathematical Dictionary* (1679): "[In Greek, *Eu* signifies Good, or Well; and *Daemon* a Spirit.] The *Good Genius* or *Spirit*. The 11th House of a Celestial Figure is so called, by reason of its good and prosperous Significations. . . ."[34] Notably, this is the only instance in the English lexica in which the notion of *daimon* is connected explicitly to *genius*, although we may observe a loose connection through the association of *genius* with *spirit*, since the latter (sometimes as the near-homonym *sprite*) appears regularly in entries for *demon* (or *dae/imon*) as a headword.[35] It is this sense that gives us *genie*, a word in circulation since at least the early seventeenth century but largely absent from the dictionaries, including Johnson 1755.[36] The other absence—*genius* as the spirit of a language, etc.—is hard to explain, given its appearance in the Latin dictionaries upon which sixteenth- and seventeenth-century English lexicographers drew.

The other aspects of *genius* as "spirit" denoted by the sources remained relatively stable throughout our period, the commonest being that of the "good or evil angel," presented first in Elyot 1538 and repeated almost verbatim up to Martin 1749 (although Johnson eschewed it, perhaps because of its superstitious overtones).[37] As per the Latin sources on which he drew, Elyot 1538 noted the origins of this sense in pagan beliefs: "Amonge the Paynims [i.e. pagans] some supposed it [*genius*] to be the spirite of a man." We find no elaboration on this "antiquarian" aspect of the definition until Chambers 1728, who amplified it through sources such as Vestus, Varro, and Augustine.[38] Citing the latter's *De Civitate Dei*, Chambers is one of the few lexicographers to identify *genius* as the ancient god of generation: "*Genius* was a God who had the Power of generating all things; and presided over them when produced."[39] Drawing on Chambers, Bailey 1737 notes that *Genius* was fabled to be the son of Jupiter and Terra, goddess of the earth, and that he "was thought to be that spirit of nature which begets all things, assists at all generations, and protects whatever is produc'd; and all things were agreeable to him that tended to mirth and pleasure."[40] This is the sense in which *genius* first appears as a headword in the English sources: in the *Ortus vocabulorum* (1500) it is defined as "the god of weddynge." Derived from Roman belief, this sense was common in romance literature and familiar to English authors of the sixteenth century, such as Spenser.[41] Yet despite such prevalence it is absent from the vast majority of period dictionaries (as well as the OED), representing a good example of the potential mismatch between dictionary definitions and more widespread usage.

While not explicit in the lexica, it seems likely that the "generation" sense undergirds some of the English definitions of the Latin *genius* as "delectation" and "pleasantness."[42] The connection is certainly apparent through the adjective *genial*, which as early as Blount 1656 was defined as "full of mirth: pertaining to marriage; the marriage-bed was of old called the *Genial-bed*, *quasi* Genital-bed."[43] But this sense must have been current earlier, since in Elyot 1538 we find "Genio dare operam: to lyve voluptuously" and the notoriously unabashed Florio 1598 offers: "*Geniale*: whatsoever pertaines unto a mans *genio*. Also pleasant, blithe, bucksome, merie, given to pleasure and recreation. Also pertaining to mariage, houshould, and procreation."[44] One of his definitions for the headword *genio* is "instinct of man given him by nature," indicating that these definitions pertain equally to *genius* as "inclination," especially the freedom to act according to one's nature. It is through this "freedom," especially, that *genius* relates to *fancy*, the latter used to define the former from Coles 1676 onward.[45] With *fancy* we are thrust firmly into the realm of the creative artist, for it is a term used in English to denote not only free will, but also the imagination and its products.[46] Indeed, in defining *congeniality*, Blount 1656 offered: "A likeness of Genius or fancy with another; As Sir Henry Wotton saies Poets and Painters have alwaies had a kind of congeniality."[47] Thus, we see how the "generation" sense of *genius* leads effectively to the notions of artistic temperament and the artist's creative potency. The latter connects *genius* not only to *fancy*, but also to *conception*, and thus to its near-synonym *conceit*. As we shall see, this relationship is significant for the adjectives *witty* and *ingenious*, since both were used to denote someone "full" (i.e. pregnant) with sharp, clever, or amusing quips. Indeed, since "temperament" in the early modern period was a matter for humoural theory, we have tripped over the roots of *humour* as "mirth."[48]

Despite the "humoural" aspects we have now encountered, the English lexica never define *genius* in relation to the "genial" melancholy of inspiration.[49] Indeed, although there is undoubtedly a semantic relationship between the ingenuity family of terms and *inspiration*, the latter word is usually defined either straightforwardly as "inward breath" or in relation to divine gifts, particularly for prophesy.[50] In this sense it is connected to *enthusiasm*, which (as Mee and others have shown) has a complex history in the poetics of Romanticism.[51] While this aspect is rare before the mid-eighteenth century, we find it early on in Cotgrave, who makes plain its relationship to the *furor poeticus*: "*Enthusiasme* . . . A ravishment of the spirit; divine motion, or inspiration; poeticall furie" (Cotgrave 1611). One wonders whether the importance of poetical fury in French literary traditions (especially Rabelais and the Pléiade) had any bearing on Cotgrave's choice, although we find it simultaneously in Florio's definition of the Italian *enthusiasmo*.[52] In any event, while Cotgrave's definition is picked up in Blount 1656 and Miège 1677, it then largely disappears.[53] This must be due in part to widespread suspicion of enthusiasm as a destabilizing form of religious fanaticism, not least because of its feigned aspect. For example, Lloyd-Wilkins 1668 defined *enthusiasm* as "counterfeited inspiration," which connects it with the negative aspects of another ingenuity keyword: *cunning*.[54] These several concerns also underpinned rationalist suspicion of any claim for divinely inspired creativity. As Hobbes, whose psychological writings proved especially influential on later seventeenth-century literary criticism, wrote: "Why a Christian should think it an ornament to his poem, either to profane the true God, or invoke a false one, I can imagine no cause, but a reasonless imitation of Custom, of a foolish Custom; by which a man enabled to speak wisely from the principles of nature, and his own meditation, loves rather to be thought to speak by inspiration like a bagpipe."[55] It remains for us to note, however, a final connection of "ravishment" with *genius*, albeit an oblique one. We will recall that one of the definitions of *genius*—offered first in Cooper 1584—is "the grace and pleasantness of a thing."[56] The semantics of *grace* is especially complex in the early modern period, but one of its meanings in English (although found predominately in the polyglot dictionaries) was *beauty* or *comeliness*.[57] It is presumably this sort of thing—*grace* as *beauty*—that Henry Wotton had in mind when describing the qualities of a "sound piece of good art" in *The Elements of Architecture* (1624): "In truth a sound piece of good Art, where the *Materials* being but ordinarie stone, without any garnishment of sculpture, doe yet ravish the Beholder, (and hee knowes not how) by a secret Harmony in the Proportions."[58]

We have already seen that one of the key meanings of *genius* in English was (as in the other European vernaculars) *nature*. This meaning was stable right up to the turn of the seventeenth century, when it began to morph from "natural inclination" to "natural ability" or "natural talent" (although these three senses were often used interchangeably). For example, in *A New Dictionary of the Terms Ancient and Modern of the Canting Crew* (1699) by the anonymous "B. E.," *talent* is defined as "the same with Capacity, Genius, Inclination or Ability; . . . *His Talent does not lye that way*, he has no Genius for it, or his Head does not lean to it." Previously, *talent* had been defined either as the ancient weight of money or, more rarely, as *lust*, as in Geoffrey the Grammarian 1499: "Talent or lust. Dilectacio . . . Appetitus." It is this aspect of *talent*, as delectation or appetite, that renders it susceptible to being defined as *genius* (we will recall that as early as Elyot 1538, *genius* is "dilectation moved by nature"). Indeed, while the dictionary definitions do not indicate it, the OED shows that *talent* as "mental endowment; natural ability" was widespread in the fifteenth and sixteenth centuries, even that it could denote "special" or "superior" ability.[59]

Something of this is at work in Blount 1656: "Talent . . . also signifies a faculty or ability; as we say, a man of good *talents*, id est of good parts or abilities," although the lexis of *genius* is notably absent from this entry. For the most part, when *talent* is used to define *genius* it does so in a neutral sense. We find this in Bailey 1730, the entry from which is worth quoting in full: "Genius [among the *Antients*] was used to signify a Spirit either good or evil; which they supposed did attend upon every Person; they also allow'd *Genii* to each Province, Country, Town, etc, also a Man's natural disposition, Inclination, etc. Genius, the Force or Faculty of the Soul, considered as it thinks or judges; also a natural Talent or Disposition [to] one thing more than to another." Here we have some familiar meanings: *genius* as "spirit" (but now extended to encompass a people, place, etc.); as natural disposition; and as a talent for certain things. To these Bailey adds something approaching "mental capacity": the "Force or Faculty of the Soul, considered as it thinks or judges." This intervention was not his own, for the entry as a whole is lifted from Chambers's far more learned—but also more loquacious—

Cyclopedia (1728). There, in a long entry that lingers particularly on the ancient variations of *genius*-as-spirit, we find that "Genius is more frequently us'd for the Force or Faculty of the Soul, consider'd as it thinks, or judges. Thus we say, A happy *Genius*, a superior *Genius*, an elevated *Genius*, a narrow confin'd *Genius*, &c. In the like Sense we say, A Work of *Genius*; a Want of *Genius*, &c." Thus, the "faculty" sense of *genius* is in fact the traditional one ascribed to *ingenium* (see chapter 1). However, three aspects of Chambers's account of this sense should give us pause. First, his invocation of "judgment," which suggests he may have been responding to ongoing debates about the nature of *wit*, discussed below. Second—and for the first time—we have an objectified sense of *genius*, as in "a work of genius." Third, Chambers has introduced the notion of extraordinary ability with "superior" and "elevated" *genius*, even if these remain elements in a spectrum rather than the absolute quality of *genius* in Romanticism.

Around the time Chambers was producing his encyclopaedia, writers such as Addison and Shaftesbury were starting to endow *genius* with the "special" properties that would lead to its inflation as "exalted intellectual or creative power."[60] It has been claimed that despite his awareness of these changes, Johnson did not recognize this aspect of genius in his *Dictionary* (1755).[61] The matter is rendered debatable and complicated by Johnson's discussion of *genius* in his non-dictionary work.[62] In the *Dictionary*, Johnson offers five definitions of *genius*, four of which we have already encountered: "The protecting or ruling power of men, places, or things" (sense 1); "Mental power or faculties" (sense 3); "Disposition of nature by which any one is qualified for some peculiar employment" (sense 4); and "Nature; disposition" (sense 5). His second sense, however, is a departure from the earlier lexica: "A man endowed with *superior* faculties" (italics added), accompanied by a quote from none other than Addison: "There is no little writer of Pindarick who is not mentioned as a prodigious *genius*." As Jonathan Bate has argued, it may reasonably be assumed that Johnson expected his readers to have in mind not only these specific comments on the "wild Pindaric ode" but also the context in which they were made: Addison's famous essay no. 160 in *The Spectator* (1711), in which he offered a discursive attempt at defining *genius*.⁶³ Let us conclude, then, by noting that while modern sources may equivocate over the meaning and precision of the second sense of Johnson's definition, his peers did not. In his long review of the *Dictionary* in *The Monthly Review*, Sir Tanfield Leman singled out *genius* (along with *character* and *taste*) as "instances of his [Johnson's] judgment and success in explaining words."⁶⁴ Thus, we may be confident that by 1755 the definition of *genius* as an individual—not a faculty, thing, or quality—possessed of superior powers, accurately reflected period usage.

Ingenuity

While *genius* entered the English language unchanged from the Latin, its etymological and semantic relative *ingenium* turned into the adjective *ingenious*, but was also translated—via the Anglo-Saxon—as the noun *wit*. *Wit* was by far the more common term, widely used in many domains and, like *genius*, a capacious word, which became hotly contested in criticism and philosophy. Conversely, the word *ingenious* was far less mobile and controversial, appearing later and less frequently in the lexica than *genius*, *wit*, and *cunning*. The noun form *ingenuity* appears toward the end of the sixteenth century, but was never used as a direct translation of *ingenium*. Rather, in its early usage it denoted "ingenuousness": a frank and open nature, possessed especially of freeborn gentlemen, very different from the guile with which *ingenuity* would later be connected.⁶⁵ Indeed, the history of the words *ingenuity* and *ingenious* in early modern English is one of etymological confusion and semantic fluidity, stemming largely from the conflation or interchange of *ingeni-* and *ingenu-*stemmed Latin words as they moved into the vernacular. While this confusion passed largely unremarked by lexicographers, it was sufficiently widespread for Elisha Coles to remark bluntly that "*ingenious and ingenuous*, are too often confounded" (Coles 1676). Although the meanings of these words started to be clearly distinguished around the turn of the seventeenth century, as late as Phillips-Kersey 1720 we find *ingenuity* defined as "Ingeniousness, Quickness of Wit, Smartness; also Ingenuousness, Sincerity, Frankness in Speech or Dealing."⁶⁶

Ingenium itself first appears as a headword in Elyot 1538, where it is translated as "the proper nature of a thynge. Also wytte." This double meaning—of inborn nature and general mental capacity—remained relatively stable until the mid-eighteenth century, when Johnson abandoned the Latin term as a headword, noting that its English translation (the obscure *ingeny*) had become obsolete. However, and as several commentators have observed, the simple translation of *ingenium* as *wit* or *nature* hardly did justice to the range of its possible meanings.[67] Notably, the only lexicographers to take up the challenge of *ingenium*'s elastic meaning were those steeped in humanist learning: Thomas Thomas and John Rider. Thomas 1587 offers a suite of meanings for *ingenium*: "The nature, inclination, or disposition of a thing: also wit, wisdome, will, or propertie, fansie, invention, cunning." The examples given in Rider 1589 are if anything more expansive:

> *Nature*: Natura. Genius, Physis, endogenia. The nature of a man, or any other thing: Indoles, ingenium, Minerva, proprietas, dispositio.
>
> *Pregnant*: Pregnans. A pregnant, or sharpe witte. Acre ingenium. Acutum ingenium.
>
> *To make Proper unto*: . . . The property, or natural disposition of a thing: Ingenium, natura.
>
> *To give up the Spirit or ghost*: Exhalo, extremum halitum efflo. The spirite of a man: Animus spiritus, genius. The spirite, or wit that a man hath by nature: Ingenium.
>
> *To Wil, or command*: . . . Will, or pleasure: Libido, ingenium, arbitrium.
>
> *To be out of ones witte*: Insanio. Witie: Ingenium, mens, minerva, acumen. A smal, a prettie witte: Ingeniolum. Sharpenesse of witte: Subtilitas, solertia, sagacitas, acrimonia, acumen. . . . Wittie: Solers, scitus, ingeniosus, industrius, subtilis, cordatus, argutus, artificiosus, cautus, dædalius, dædalus, salsus, acutus, perspicax. . . . Quicke witted, or very wittie: Peringeniosus, persubtilis, peracutus, perargutus, persalsus, perspicax. . . . Wittilie: Ingeniose, scite, solerter, industrie, cordate, acute, argute, salse, sagaciter, provide. Very wittily: Peringeniose, peracute, praeacute.

These entries indicate the extreme ambivalence of the Latin *ingenium*, as well as its proximity to our other English keywords. When translated into English, *ingenium* could shift meaning from the relatively passive *inclination* to the active *will*, the acquired (*wisdom*) to the inborn (*nature*). Strikingly, it was morally ambiguous, the wiliness implicit in certain of its *wit*-inflected meanings a far cry from the plain dealing normally denoted by *ingenuous*. These "dangerous" qualities were transferred to the English *ingenious* even in instances apparently unconnected to *ingenium*. For example, in the first appearance of the adjective in the lexica, Baret offered for *very subtle* "ingenious, or wittie. Perargutus . . . ," encouraging his readers to consult also the entry for "guile, wile, deceit" (Baret 1574). A few years later, Cooper 1584 cribbed from Baret to translate *perargutus* as "very subtile, ingenuous or wittie," transposing *u* with *i* in a perfect example of the lexical confusion we have noted. Thus, the words *ingenious*, *ingenuity*, and the rare *ingeniosity* and *ingeny* spanned the broad semantic gap between devious cleverness and openhearted liberality, all the while embracing the creative potential associated with a quick wit.

The definitions of the four terms as headwords are as follows:

Ingenious, adj.

1. Witty, quick witted, mentally sharp.

Cawdrey 1604 >; suggested also in Baret 1574 >: *perargutus* (very sharp). OED, I.1.a.: "Having high intellectual capacity; able, talented, possessed of genius" (1483 >); I.2.a.: "Intelligent, discerning, sensible" (1560 >).

2. Ingenuous, freeborn, liberal.

Phillips 1658 >. Preston ca. 1674 as "good natured, well-born and bred." OED, II.4: "Used by confusion for ingenuous; Having or showing a noble disposition, high-minded; honest, candid, open, frank" (1597 >); II.5: "Well born or bred" (1638 >). See also Cockeram 1623 (as definition of *noble*).

3. Fancy.

Lloyd-Wilkins 1668. Not in OED, but presumably connected to 4, below.

4. Inventive.

Miège 1677 >; see also Cotgrave 1611 (as synonym for *ingenieux/ euse*). OED, I.3.a.: "Having an aptitude for invention or construction; clever at contriving or making things; skilful" (1576 >).

5. Cunning, shrewd.

Kersey 1702 >. Presumably OED, I.1.a., I.2.a. and I.3.a.

6. Of a thing: exquisite, excellent.

Miège 1677 >. OED, I.3.b: "Of things, actions, etc.: Showing cleverness of invention or construction; skilfully or curiously contrived or made" (1548 >).

7. Possessed of genius.

Johnson 1755 >. OED I.1.a., although not distinguished from talent in general.

Ingenuity, n.

1. As per *ingenious*, 2.

Blount 1656 >; see also Florio 1598, as translation of *ingenuità*.

Phillips-Kersey 1706 and Johnson 1755 specifying sincerity, frankness, candor.

2. As per *ingenious*, 1.

Phillips-Kersey 1706 >.

3. As per *ingenious*, 7.

Johnson 1755 >.

Ingeniosity, n.

1. As per *ingenious*, 1.

Blount 1656 >.

Ingeny, n.

1. As per *ingenious*, 1.

Coles 1676 >.

2. Natural disposition.

Phillips-Kersey 1706 >. OED, 1a: "mind, intellect, mental faculties; mental tendency, disposition" (1477 >).

As these definitions show, quickness and acuity marked the adjective *ingenious* from the outset. Its first occurrence as a headword in Cawdrey 1604, where it is defined as "wittie, quicke witted," confirms this sense. Thus, from early on the word signaled not just a general faculty, as *wit* could, but a particularly swift, subtle, or pointed cleverness—much like the Italian *ingegno* (see chapter 2).[68] This sense explains Kersey's inclusion of *ingenious* in his entry for *smart*—"sharp, brisk, witty, ingenious, &c"—and the related *shrewd*: "subtil, smart, or ingenious" (Kersey 1702). The Anglo-Saxon root of this sense is well illustrated, albeit obliquely, by Hogarth, who explains that the now obsolete first name *Kenard* may be "drawn from Keen Anglo Saxon. Cene, Sharp and Aerd, Nature, i.e. one acute and ingenious by nature" (Hogarth 1689).[69] The "acute" sense was sustained throughout the period, although eighteenth-century lexicographers tended to distinguish it from other meanings. For instance, Martin 1749 gives "*Ingenious* (of *ingeniosus*, L[atin] of *ingenium* wit) 1. sharp, witty. 2. cunning, shrewed. 3. industrious, inventive. 4. exquisite, or excellent." Sometimes, the potentially negative connotations of the word were stressed. For example Johnson 1755, who defines *ingeniousness* (from *ingenious*) as the apparently positive "wittiness, subtilty, strength of genius," quotes Robert Boyle's *Occasional Reflections* (1665): "The greater appearance of *ingeniousness* there is in the practice I am disapproving, the more dangerous it is."[70] We may note here the gendered aspect of Boyle's *ingeniousness*, which appears in the context of a discourse upon dissimulating female courtiers who decorate their closets in an overly lavish manner. This sort of gendering—rare in the lexicographical sources, but prevalent elsewhere—suggests that the ingenuity family of terms could take twists and turns similar to the equally morally ambiguous *curiosity*, a word that connects to *ingenuity* in several ways, especially through its connotations of skill.[71]

As we have seen, associations with dissimulation were always lurking beneath the surface of the ingenuity family of terms. For instance, Florio 1598 lists *wilie*, *cunning*, and *craftie* alongside *ingenious* as synonyms for the Italian *ingegnóso*. In fact, it is Florio who first introduces the noun *ingenuity* into the lexica, but in the "ingenuous" sense of the term. Translating *ingenuità*, he offers "freedome or free-state, ingenuitie, a liberall, free, or honest nature and condition." Similarly, in Blount 1656 (where the term first appears as a headword) *ingenuity* is "the state of a free and honest man, freedom, a liberal nature or condition." Bullokar, who introduces *ingenuous* as a headword, simply defines it as "gentlemanlike" (Bullokar 1616). There is a simple explanation for these apparent reversals and conflations. As C. S. Lewis observed, in antiquity the word *ingenuus* was often a stand-in for *liberalis*, meaning both the social condition of being freeborn and the ethical qualities or behavior associated with men of that position. In early modern England, the freeborn gentleman was thus *ingenuous*, his education, pursuits, and actions *liberal* (i.e. free, disinterested) in contrast to those of a servile nature or station. The latter are differentiated from the former by their self-interest and cunning. As Lewis observes: "The true servile character is cheeky, shrewd, cunning, up to every trick, always with an eye to the main chance, determined to 'look after number one.' . . . Absence of disinterestedness, lack of generosity, is the hall-mark of the servile. The typical slave always has an axe to grind. Hence the miserably betrayed Philoctetes in Sophocles' play (l. 1006) says to the cunning Odysseus 'Oh you—you who never had a sound or eleutheron thought in your mind!' Odysseus has done nothing without 'an ulterior motive.'"[72] We shall have more to say about these aspects in our discussion of *cunning*, but for now it suffices to note that in the rigidly hierarchical social world of early modern England it was natural to ascribe to a gentleman not only an open nature, but also the innate prowess (mental, athletic, etc.) granted by a noble bloodline. Hence Phillips 1658 could define *ingenuous* as "natural power," thus approaching both *ingenium* and *genius*, the latter of which was, as we have seen, associated in English with "free" fancy and creativity.[73] By contrast, socially and ethically inferior tradesmen and artisans—practitioners of the servile "mechanical" arts—are not *ingenuous* but *ingenious*. Confusion arises when naturally quick-witted gentlemen turn from patronizing to practicing these mechanical pursuits, thus becoming *virtuosi*.

The *virtuoso*—a neologism from the Italian, derived from *virtù* (power or force)—was not only *ingenuous*, as a gentleman should be, but also intellectually potent and able.[74] Blount 1656 defines the word as "vertuous, honest. It is also used substantively, for a learned or ingenious person, or one that is well qualified," while just two years later Phillips 1658 gives "a man accomplisht in vertuous Arts and Ingenuitie." Since the *virtuoso* was both generous and skilled in many (previously distant) branches of learning, the "pointed" sense of *ingenious* waned in this context. While the *virtuoso*'s wit was keen, he did not engage in sharp dealing nor was his learning acutely focused on a single domain, unlike artisans committed to the sole profession to which they are inclined by nature.[75] Thus, the freeborn gentleman usurped certain qualities of the artificer, transforming self-serving cunning into liberal ingenuity through acquired abilities vested in a broad, natural power. The social and intellectual ramifications of this shift were complex, sometimes debated, and especially evident in domains where *virtuosi* were most active, such as the visual arts, medicine, and (especially in the later seventeenth century) natural philosophy.[76] The contested status of technicians alongside the promotion of gentlemanly disinterest in the early Royal Society is a perfect example not only of *ingenuity*'s range but also of the tensions that could arise from its shifting valence.[77]

In fact, the direction of *ingenuity*'s travel went both ways. While gentlemen adopted the practical cunning of craftsmen, artisans could benefit from association with the "ingenuous" strand of *ingenuity*. An early example is the renowned portrait miniaturist Nicholas Hilliard, who, in his unpublished treatise on limning (composed ca. 1600), appealed to the word's noble aspects in a passage that knocks together gentlemanly freedom with the god-given talent associated with *genius*.

> [Limning] is for the service of noble persons very meet.... And this is a work which of necessity requireth the party's own presence for the most part of the time, and so it is convenient that they be gentlemen of good parts and ingenuity, either of ability, or made by prince's fee able to themselves as to give such seemly attendence on princes as shall not offend their royal presence. Seest thou not that these men, then, must

often in their busines stand before princes, though they be born but common people? But God, the author of wisdom and the giver of all good gifts and goodness, He giveth gentility to divers persons, and raiseth man to reputation by diverse means: we read that he called Bezaleel and Aholiab by name, and filled them with wisdom, skill and understanding, without any teaching, but only of his own gift and grace received. He taught them Himself to be cunning in all fine and curious work, in embroidering in silk, in painting, in setting of precious stones in gold, and carving of wood. The text sayeth He filled them with the spirit of God to devise cunningly in such works, being men brought up but in slavery and making of bricks in captivity, they and their ancestors for many generations.

Here is a kind of true gentility, when God calleth: and doubtless though gentlemen be the meetest for this gentle calling or practice, yet not all but natural aptness is to be chosen and preferred, for not every gentleman is so gentle spirited as some others are. Let us therefore honour and prefer the election in God in all vocations and degrees: and surely he is a very wise man that can find out the natural inclinations of his children in due time, and so apply him that way which nature most inclineth him. . . . [78]

We may note here Hilliard's claim that artistic talent is a quality granted by the grace of God. This is entirely in keeping with continental definitions of ingenuity, such as Cesare Ripa's summary of *ingegno* in his much-reprinted *Iconologia* (first edition 1593). There, creativity is a spiritual gift, making the artist not just *ingegnoso* but also *spiritoso*.[79] In this context, Lloyd-Wilkins's inclusion of *sprightly* (i.e. spirited) alongside *fancy* as a definition of *ingenious* makes perfect sense (Lloyd-Wilkins 1668).

These connections underscore the strengthening bond between *ingenuity* and inventiveness in early modern English, which by Johnson 1755 had become one of the most pronounced senses of *ingenious*.[80] The association may have first entered English via French or Italian, since certain of the synonyms for *ingenuity* in those languages—*esprit*, *engin*, and *ingegno*—were closely associated with creative tal-

Fig. 6.2. Nicholas Hilliard, *Self-portrait, aged 30* (1577).

ent, especially for crafting exquisite artefacts. For Florio 1598, the *ingegnóso* is "craftie, full of invention," while Miège 1677 offers for *ingenious, or witty*: "*ingenieux, qui a bon esprit*. Ingenious, or inventive, *ingenieux, subtil, inventif*. An ingenious piece of work, *un Ouvrage exquis, Ouvrage travaillé avec une rare industrie*. Ingeniousness, ingeniosity, *esprit, invention*. Ingeniously, *ingenieusement, avec esprit*." Such connections explain why *ingenious* was used to describe a clever or skillfully wrought object: a meaning amplified by the fact that the

Latin *ingenium* could denote a useful mechanical contrivance. Phillips-Kersey 1706 notes of *ingenium* that "In our old Records it is taken for an Engine, Instrument, or Device," but the word *engine* also had vernacular roots.[81] In his entry for *engine*, Hogarth 1689 explains that the word derives "from the French *Engin*, or the Italian *Ingegno*, ingenuity; so called, because of the curiousness of the workmanship thereof. From hence, the French *Enginier*, and our English Engineer, a maker of Engines."[82]

Jessica Wolfe and Jonathan Sawday have examined in detail the relationship in early modern England between engines, human ability (especially imaginative capacity), and social change, such that we need not dwell on it here.[83] We may, however, observe that in relation to "inventive ability" the trajectory of *ingenuity* in English not only mirrors the rise in status of mechanical artificers (engineers, especially) but of artists of all kinds. As the middle of the eighteenth century approached, an *artist* could be defined specifically as one possessing *ingenuity*. For instance, Defoe 1735 offers: "Artist, one who understands his Art, an ingenious Workman." By this date, knowledge—deriving from the *cunnan*-inflected sense of *ingenuity*—of one's art, combined with skill in working, could serve to distinguish an *artist* from a "mere" artificer. Perhaps this explains why Bailey invoked *ingenuity* as his definition of a new literary genre, the *novel*: "an ingenious and diverting story, in which the writer dresses up an invention of his own, with all the imbellishments of art; to render it both agreeable and instructive" (Bailey 1737). Indeed, while the adjective *ingenious* was often used to describe finely crafted goods, it is in the verbal arts that *ingenuity* first took flight, especially via its ties to *wit*.

Despite the late appearance of "witty" as a definition for *ingenuity* as a headword, this sense was undoubtedly in circulation much earlier. The OED identifies its first use in Ben Jonson's *Every Man out of his Humour* (1599), in a scene of wooing between affected courtiers Fastidious and Saviolina, observed by the mocking Macilente:[84]

> Fastidious: 'Fore God, sweet lady, believe it, I do honour the meanest rush in this chamber for your love.

> Saviolina: Ay, you need not tell me that, sir. I do think you do prize a rush before my love.
>
> Macilente: [Aside] Is this the wonder of nations?
>
> Fastidious: Oh, by Jesu, pardon me, I said 'for your love,' by this light; but it is the accustomed sharpness of your ingenuity, sweet mistress to—mass, your viol's new strung, methinks.
>
> *[Fastidious] takes down the viol.*
>
> Macilente: [Aside] Ingenuity? I see his ignorance will not suffer him to slander her, which he had done most notably, if he had said "wit" for "ingenuity," as he meant it.[85]

Here, Jonson plays upon the "sharp" sense of *ingenuity* we have already encountered, contrasting—through Macilente's acerbic aside—Fastidious's pretentious use of a neologism with his evidently dull wit. It is a joke that continues throughout the scene:

> Fastidious: How like you her wit? (*Tobacco.*)
>
> Macilente: Her ingenuity is excellent, sir.
>
> . . .
>
> Saviolina: Nay, I cannot stay to dance after your pipe.
>
> Fastidious: Good—nay, dear lady, stay. By this sweet smoke, I think your wit be all fire. (*Tobacco.*)
>
> Macilente: [Aside] And he's the salamander that lives by it.[86]

In these passages, Jonson was likely lampooning *ingenuity*'s currency as an inkhorn term around the turn of the century. In fact, one of its earliest appearances in print was in the "University Wit" Thomas Nashe's list of affected terms, published in *Strange Newes* (1592): a vituperative response to Gabriel Harvey's *Foure Letters and Certain Sonnets* (1592). Harvey had used *ingenious* in the "ingenuous" sense of the word, although it appears in such close proximity to the language of the humours and of *conceit* that its meaning was surely intended to be fungible: "young bloud is hot: youth hasty: ingenuity open: abuse impatient . . . the Satyricall humour, a puffinge, and swellinge humor:

Conceit penneth. . . ." Moreover, the fact that Nashe placed it in his list immediately before the phrase "Jovial mind" suggests that erudite scribblers associated it with some of the senses of *genial* that were current at the time: "The floures of your *Foure Letters* it may be I have overlookt more narrowlie, and done my best devoire to assemble them together into patheticall posie, which I will here present to Maister Orator *Edge* for a New yeares gift, leaving them to his *wordie* discretion to be censured whether they be currant in inkehornisme or no. *Conscious mind: canicular tales: egregious an argument:* when as *egregious* is never used but in the extreame ill part. Ingenuitie: Jovial mind."[87] Strikingly, however, *ingenuity* was never used to define *genius*, or vice versa, in the period lexica. The closest we get is the very rare *ingeny*, defined in Phillips-Kersey 1706 as "Genius, Natural Disposition, Parts, Humour, etc. . . ." This definition doubtless comes via the increasing proximity of *genius* to *wit*, since in its first (and only, prior to Phillips-Kersey) appearance in the lexica, *ingeny* is defined by Coles 1676 as "ingeniosity, wit, wittiness."[88]

Wit

Of all our English keywords, *wit* is the most semantically expansive and its evolution the most complex. An important and much-debated term in natural philosophy, rhetoric, and criticism, it has been treated so extensively in the secondary literature that we can offer only a summary of its lexicographical fortunes here.[89] Indeed, *wit* is a good example of the limitations of our dictionary sources, the definitions of which seem slight compared to the giddy fecundity of the term in the hands of the period's poets and critics. Lexicographers struggled to keep pace with the rapid inflation of a term that proved vexingly elusive to many who attempted to define it, and that could often be understood only by virtue of what it was not. As Cowley put it in his *Ode: Of Wit* (1650): "What is it then, which like the Power Divine/ We only can by Negatives define?" Indeed, *wit* in English came very close to the *je-ne-sais-quoi*: something recognizable but ultimately indefinable.[90] For instance, Richard Flecknoe celebrated *wit* in *A Short Discourse of the English Stage* (1664) as the "spirit and quintessence of speech," but acknowledged "'Tis in vain to say any more of it; for if I could tell you what it were, it would not be what it is; being somewhat above expression, and such a volatil thing, as 'tis altogether as volatil to describe."[91] Lexicographers were thus presented with formidable challenges as the word morphed, Proteus-like, from its original Anglo-Saxon sense as "understanding" to become a term of almost overwhelming variety, associated not just with the powers of the mind in general, but with imagination, facetious speech, humour, literary brilliance, and philosophical judgment. As we have seen, by 1755, Johnson could offer no fewer than eight senses of the term:

1. The powers of the mind.
2. Imagination; quickness of fancy.
3. Sentiments produced by quickness of fancy.
4. A man of fancy.
5. A man of genius.
6. Sense; judgment.
7. In the plural. Sound mind.
8. Contrivance; stratagem; power of expedients.

While this judicious attempt at definition hardly does justice to *wit*'s full range, it at least indicates some of the key senses the term had acquired by the mid-eighteenth century, when it started to be associated explicitly with *genius*. Notably, however, Johnson omits the "humorous" or "facetious" sense of *wit*, in part owing to the denigration of this aspect as "false wit" in Augustan poetics and his own critique of "metaphysical wit" (i.e. *discordia concors*) in the poetry of Cowley and others.[92] Johnson's definitions do, however, capture one of the key transformations of *wit* in early modern English: from a word meaning, generically, "the powers of the soul" (especially in relation to the mind) to one specifically denoting the capacity to invent (hence *wit*'s frequent conflation with *imagination*) or the capacity to judge. The latter capacity was keenly contested, however, especially by Hobbes and Locke, who distinguished sharply between *wit* as the inferior ability to "observe similitudes" (which Hobbes associated with a "good fancy," i.e. a good imagination) and the superior *judgment*: the ability to discern the differences between things.[93] Johnson's willingness to identify *wit* as both *judgment* and *fancy* (or *imagination*) reflects the successful attempt by literary figures, Pope and Addison especially, to rehabilitate *wit* and imaginative power more broadly in their accounts of creativity and sentiment. These efforts, in which *wit* was deftly associated with the truth of nature, underpinned its later significance in aesthetics, especially the sublime, although such aspects are barely visible in the dictionary entries.[94]

As we have seen, *wit* was often used interchangeably with *ingenuity* in its noun and adjective forms and, like *cunning*, derived from an Anglo-Saxon word for "knowledge": *witan*.[95] In its earliest appearances (e.g. Elyot 1538), *wit* is offered as a direct translation of *ingenium*, thus bearing all the many and varied meanings of that Latin word.[96] Baret 1574 emphasizes the *ingenium*-inflected meaning of the term (Lewis's *wit-ingenium*) by defining the headword as "nature, inclination." Yet as early as Huloet 1552 *wit* appears as a headword not just as *ingenium*, but also as a synonym for *sensus*: "the fyve naturall senses." For the adjective *wyttye*, Huloet offers a string of Latin terms: *acutulus, argutus, callidus, catus, cantus, conspicuus, ingeniosus, nasutus, præditus ingenio, sagax, situs, solers,* and *tenax*, while equating it also with the adjective "actyve. Industrius" and the headword *inventive*: "subtill, or wyttye. Technicus." Even taking into consideration the generic disposition toward *copia* of sixteenth-century lexica, this is a strikingly capacious list. It indicates that by the middle of the sixteenth century, *wit* traveled across semantic territory ranging from the natural faculties to abilities gained by experience, and could already include the particular capacity to invent.[97] *Wit*'s association with invention was especially pronounced in rhetoric, in part because of the significance of *ingenium* in that tradition.[98] We should note, however, given Huloet's connection of *witty* to *technicus*, that Rider 1589 offers *ingenium* in its objectified sense as a synonym for "a witty invention." Thus, by the end of the century (if not earlier), *wit* seems to have been associated not only with mental and linguistic qualities, but with the kinds of artisanal inventiveness denoted by *cunning*.

Toward the end of the sixteenth century, the lexica begin to reflect *wit*'s explicit relationship with "facetious speech" and what would later be termed *witticisms*, exemplified by John Lyly's hugely influential and much imitated *Euphues: The Anatomy of Wit* (1578).[99] For example, Rider 1589 offers *facetiae* and *scitamentum* as synonyms for "pleasant and witty sayings," of the kind collected and published in the many anthologies bearing *wit* in their titles: *Wits Commonwealth* (1597), *Palladis Tamia, Wits Treasury* (1598), *Wits Labyrinth* (1648), *The English Treasury of Wit and Language* (1655), etc.[100] The "sharpness" or concision of this kind of *wit* is captured especially in the polyglot dictionaries. For the French *accort*, Hollyband 1593 has "wittie, wise, circumspect, craftie, and subtile," while Florio 1598 includes *wittily* among the synonyms for the Italian *accortamente* and *argutamente*. In English, this "sharp" aspect is captured neatly in Coote 1596, who gives *wittie* as a single synonym for *acute*. It is well known that in late Elizabethan and Jacobean England, extremely short-form verse — especially epigrams, of the kind found in the *Greek Anthology* — was the most popular vehicle for displaying this kind of wit.[101] Hence Cockeram 1623 includes *atticke* (along with *ingenious* and *pregnant*) in the entry for the headword *witty*, while offering for "witty sayings short and pithy" none other than *aphorisms*.[102] Implicit in Cockeram's use of *atticke* is the relationship between *wit* and seasoning, as in the "attic salt" of sharp sayings, or the entry in Thomas 1587 for *persalse*: "very salt, verie wittily." This association of *wit* with *salt*, though seemingly throwaway, undergirds the highly important connection between *wit* and *taste*, which not only contributed toward the rise of the latter, in the eighteenth century, as a foundational aspect of aesthetics, but is central to the association of *wit* with *judgment*. For an early example of these connections in England we may turn again to Thomas 1587 in the entry for the Latin *săpio*: "to feele, to have savour, smell, taste, or a smack: also to be wise, to have a good wit, to know, to have knowledge, to be advised, to have a right minde or opinion, to understand and perceive well: also to resemble." We may thus better appreciate why later seventeenth- and eighteenth-century debates over what qualified as "good" poetry hinged on the nature of *wit*, including the repudiation of those short, "attic" forms that had previously qualified as *wit*'s quintessence.

Defining such forms, Cockeram 1623 offers:

A short and witty Poeme, which under a fayned name doth covertly praise or dispraise either some one person or thing: Epigramme.

Witty Sayings short and pithy. Aphorismes, Apothegms.

Poeme. A short matter wittily made in verse.

While Cockeram's definition of what constitutes a poem rapidly fell out of fashion, in these entries we may nevertheless observe the stirrings of semantic change that would lead eventually to Johnson's definition of *wit* in its objectified sense as "a man of fancy": a creative type, endowed especially with poetic ability, who might also be considered "a man of genius." Indeed, these poetic aspects of *wit* were in part responsible for the waning of its force as a term for technical inventiveness and its rise to prominence as a marker of literary capability. Hence the violent debates over the status of *wit* in later seventeenth-century criticism that responded, antagonistically, to the forced novelty and preciousness of the epigrammatic or metaphysical mode, while struggling to identify "true" (or at least critically sound) *wit*.[103] Much has been written about these debates in the work of Dryden, Rochester, Pope, Addison, and others such that we need not rehearse them here.[104] It is worth noting, however, that Dryden made an important distinction between "wit writing" (i.e. *wit* as *fancy* or *imagination*) and "wit written" (i.e. the products of *wit* on the page).[105]

Adopting Kenny's terminology for early modern *curiosity*, we might call Dryden's distinction as between "subjective" and "objective" senses of *wit*.[106] As we have seen, both senses are discernible in the early lexica, but they were not clearly distinguished in definitions of headwords until the mid-eighteenth century. The earliest instance seems to be Martin 1749, who offers for *wit* (in an entry that Johnson evidently drew upon): "1. the faculties of the rational soul. 2. genius, fancy, or understanding. 3. wisdom, judgment. 4. cunning, subtilty. 5. genius, or aptness for any thing. 6. witty or ingenious things in a discourse." A multitude of headwords from earlier lexica conform to Martin's sixth sense. We have already encountered some of the Latin terms, to which we may add loan words such as *laconic* and *drollery*.

The former appears in Florio 1598 as the Italian *lacònico*: "briefe, laconicall, short, witty, or compendious in speaking," although Blount 1656 explains its Latin root in an unusually discursive entry: "*Laconic* (laconicus): that speaks briefly or pithily. So used, because the Lacedaemonians or people of Laconia were wont to speak briefly and wittily." In the same lexicon, Blount identifies *drollery* as a French import: "Drolerie (French): is with us taken for a kinde of facetious way of speaking or wriating, full of merry knavish wit."[107] Thus, alongside its continuing sense as *understanding* or *fancy*, wit was increasingly associated with humour, such that Hogarth 1689 could define a *wag* thus: "from the Anglo Saxon. Waegan, to play the wag; or the Italian Vago, acute, witty." Sometimes, this humourous kind of *wit* could be of a base, sarcastic, or immoral nature. For example, Wilson, in his *Christian Dictionarie* (1612), defines *jesting* as "Pleasant and witty words, being offensive, and hindering edification," while the canting terms *peevish* and *arch rogue* are defined respectively as "witty, subtill" (Ray 1674) and "witty" (B. E. 1699).

If the dictionary definitions capture fairly accurately the ambivalence of *wit* ca. 1700 as a (potentially uncivil) sharp-tongued sarcasm, they nevertheless fail to identify other important aspects of the word.[108] The lexica give no hint of the extent to which *wit* could be gendered; they do not mark the distinction between learned and vulgar *wit*; and they are silent about the political implications of the term.[109] They did, however, reflect fairly swiftly the impact of philosophical debates about what *wit* was. Already by the turn of the seventeenth century Locke's views on the subject were being incorporated into the dictionary definitions, as in Phillips-Kersey 1706: "*Wit*: one of the Faculties of the Rational Soul, Fancy, Understanding; Genius, or aptness for any thing. Cunning. But Mr. Lock distinguishes it from Judgement and defines it, To be a quick and ready gathering of Idea's, and putting those together with great Ease and Variety, in which can be found any Resemblance or Agreeableness, so as to make up pleasant Pictures and delightful images in the fancy." However, Locke's distinction (which drew, as we have seen, on Hobbes) was not always adopted by lexicographers. Indeed, in an intriguing development that surely derives from the traditional association of *wit* with *ingenium* and the mind, toward the end of our period *wit* is identified not just with the rational soul, but with the very organ of thought: the brain. For example, Defoe 1735 contains the following entries:

To wit: to know.

Wit: one of the Faculties of the Rational Soul, Genius, Fancy, Aptness for any thing. Cunningness.

Brain: the general Organ of Sense, also Wit, Judgment.

It is not clear whether Defoe used *brain* in the anatomical sense, but the term was evidently sufficiently ambiguous for Bailey 1737 to specify that it could stand for *wit* only in a metaphorical sense: "Brain (metaphorically) is used for wit and judgment." Thus, even at the end of our period the old confusion between the concrete and abstract senses of *wit*-as-*ingenium* persisted, such that in his seventh sense of *wit* (in the plural) as "sound mind," Johnson 1755 thought it apt to quote Shakespeare's *Othello*: "Are his *wits* safe? is he not light of brain?"

As a headword, the definitions of *wit* are as follows:

Wit, n.
1. The proper nature of a thing (under *ingenium* as headword).
Elyot 1538 >; Baret 1574 (as "nature, inclination"). Not in the OED.
2. The five natural senses or a faculty of perception.
Huloet 1552 >. OED, I: "Denoting a faculty" (1000 >).
3. Understanding.
Hogarth 1689 > (but implicit in earlier definitions of *ingenium* as *wit*). OED I.2.a.: "The faculty of thinking and reasoning in general; mental capacity, understanding, intellect, reason" (589 >).
4. Ingenuity.
Hogarth 1689 > (but implicit in earlier definitions of *ingenium* as *wit*). OED II.5.b.: "Practical talent or cleverness; constructive or mechanical ability; ingenuity, skill" (see also period definitions of *ingenuity*, above) (c. 1325 >).
5. Sagacity, wisdom.
Kersey 1702 >. OED II.6.a.: "Wisdom, good judgement, discretion, prudence" (?c.1200 >).

6. Fancy.

Phillips-Kersey 1706 >; Johnson 1755 as "imagination"). Not in the OED, but related to I.3.a.: "Any one of certain particular faculties of perception" (?c.1225 >) and II. 7: "Quickness of intellect or liveliness of fancy, with capacity of apt expression; talent for saying brilliant or sparkling things, esp. in an amusing way" (1578 >).

7. Genius.

Phillips-Kersey 1706 >. OED, II.5.a.: "Good or great mental capacity; intellectual ability; genius, talent, cleverness; mental quickness or sharpness, acumen" (1297 >).

8. Cunning.

Phillips-Kersey 1706 >. OED, II.5.b.: "Practical talent or cleverness; constructive or mechanical ability; ingenuity, skill" (c.1325 >).

9. [In the plural] Judgment.

Kersey 1713 >. OED, II.6.a.: "Wisdom, good judgement, discretion, prudence" (?c.1200 >).

10. Witty or ingenious things in discourse.

Martin 1749 >. OED, II.8.a.: "That quality of speech or writing which consists in the apt association of thought and expression, calculated to surprise and delight by its unexpectedness; later always with reference to the utterance of brilliant or sparkling things in an amusing way" (1542 >).

11. A man of wit.

Martin 1749 >. OED, II.10: "A person of lively fancy, who has the faculty of saying smart or brilliant things, now always so as to amuse; a witty person" (1692 >).

12. A *virtuoso*.

Martin 1749 >. OED, II.9: "A person of great mental ability; a learned, clever, or intellectual person; a man of talent or intellect; a genius" (1508 >).

13. A man of fancy.

Johnson 1755 >. As per 10.

14. A man of genius.

Johnson 1755 >. As per 11.

15. Sanity.

Johnson 1755 (but in definitions from Baret, 1574 >). OED, I.4: "'right mind,' 'reason,' 'senses,' 'sanity'" (1000 >).

16. Stratagem.

Johnson 1755 > (see also Baret 1574, under "stratagem"). OED, II.6c: "A prudent measure or proceeding; an ingenious plan or device.

Wit, v.

1. To know.

Rare; appears only in Coles 1676. OED, 1. "To know."

Witty, adj.

1. Active.

Huloet 1552 >. Not in the OED.

2. Discrete.

Batman 1582 >. OED, 1.b.: "Having good judgement or discernment; wise, sagacious, discreet, prudent, sensible" (1027–34 >).

3. Ingenious.

Cockeram 1623 >. OED, 2.a.: "Having (good) intellectual ability; intelligent, clever, ingenious; skilful, expert, capable" (1100 >).

4. Attic.

Cockeram 1623 >. Not in the OED.

5. Pregnant.

Cockeram 1623 >. Not in the OED, but presumably related to 7.a.: "Possessing wit; capable of or given to saying (or writing) brilliant or sparkling things, esp. in an amusing way; smartly jocose or facetious" (1616 >).

6. Understanding.

Hogarth 1689 >. OED, 3.a.: "Endowed with reason, rational."

7. Full of imagination.

Johnson 1755 >. Not in the OED, but connected to 2.a. and 7.a.

8. Sarcastic.

Johnson 1755 >. OED, 7.b.: "Sharply critical, censorious, sarcastic" (1616 >).

Witticism, n.

1. A mean attempt at wit.

Johnson 1755 > (noting its derivation from *witty*). OED: "A piece of wit; a witty saying or remark; a smart joke. In earlier use often *contemptuous* ('a mean attempt at wit' (Johnson)), or applied esp. to a joke made at another's expense, a jeer, a witty sarcasm" (1677 >).

Cunning

Like *wit*, the word *cunning* in English is unmediated by Latin, deriving directly from the Anglo-Saxon *cunnung* (knowledge; experience). This meaning—knowledge, learning, or ability—is prominent and stable throughout our period, right up to Johnson 1755, in which the first sense of *cunning* offered is "Skilful; knowing; well instructed; learned." Johnson's definition shows that while *cunning* is enmeshed in the semantics of mental capacity and talent we have been tracing, it denotes especially "acquired" ability or skill. This distinguishes it somewhat from *genius*, *ingenuity*, and *wit*, which tend to indicate "natural" or "innate" ability or predisposition. Moreover, *cunning* was associated more frequently than were those terms with manual or technical arts, often denoting the practiced dexterity of an artisan or the inventive problem-solving of a mechanician. It is due in part to this association with craft that *cunning* frequently carried negative connotations, far more so than the other words in the ingenuity family of terms, even *wit*. *Cunning* was often used to describe the wiliness of artisans, who, owing to their traditional designation as (socially and intellectually) illiberal, were the natural purveyors of ruses and impostures. The "cunning" artisan was therefore *crafty* in two ways: skillful and deceitful, both senses emphasizing the "artifice" of his trade and his character. Cowell 1607 illustrates the point well in the entry for *deceyte (deceptio. fraus, dolus*): "a subtile wilie shift or devise, hauing noe other name. Hereunto may be drawn all maner of craft, subtiltie, guile, fraude, wilynes, slightnes, cunning, covin, collusion, practise, and offence, used to deceive another man by any meanes, which hath none other proper or particular name, but offence." With this in mind, it is not surprising that *cunning* was first deployed in the lexica to describe alchemy, an art traditionally associated with gulling and cozening: the anonymous (but sometimes attributed to Galfridus Anglicus) *Ortus vocabulorum* (1500) offers for *alchimia* "a cunnyng to turne won metal to another."

The definitions of the term as headword are as follows:

Cunning, n.
> 1. Ability or skill in a specific art.
> Anon. 1500 >. OED, 4: "A branch of knowledge or of skilled work; a science or art, a craft. In early times often = occult art, magic. *Obs.*"
> 2. Knowledge, learning.
> Huloet 1552 >. OED, 1: "Knowledge; learning, erudition. *Obs.*"
> 3. Ability, skill, expertise.
> Baret 1574 >. OED, 3a: "Knowledge how to do a thing; ability, skill, expertness, dexterity, cleverness."
> 4. Skilful deceit, artifice.
> Baret 1574 >. OED, 5a: "Now usually in bad sense: Skill employed in a secret or underhand manner, or for purposes of deceit; skilful deceit, craft, artifice. Cf, CRAFT, *n.* 4)."

Cunning, adj.
> As per *Cunning*, n. 1–4.
> 1. Of a thing: skillfully wrought, exquisite.
> Baret 1574 >; OED, 2b. "*transf.* Showing skill or expertness; skilfully contrived or executed; skilful, ingenious." And 5b: "Of things: Showing or characterized by craftiness; crafty."
> 2. Possessing magical knowledge or ability.
> Bullokar 1616 (under headword *magician*); OED 3. "3. *spec.* Possessing magical knowledge or skill: in cunning man, cunning woman, a fortune-teller, conjurer, 'wise man,' 'wise woman,' wizard or witch."

These definitions show that, as with all of our keywords, *cunning* was prodigiously variegated. For example, in Thomas 1587 it appears as a synonym for Latin headwords ranging from the overwhelmingly positive *bonus* (chief senses *good*, *honest*) to the rather more ambiguous *artificium* ("cunning, workmanship, sometime a thing cunningly wrought: art, subtiltie, deceite, guile"), including *ars*, *artifex*, *catus*, *doctus, elaboratio, expertus, humanus, intelligens, literatura, peritia, sagax, scientia, scitus, solertia,* and *technicus*. Given this range of meanings, it is tempting to claim that *cunning* is a sensible translation of the Greek *métis*, reaching across the semantic field mapped by Detienne and Vernant in *Cunning Intelligence in Greek Culture and Society*.[110] Yet Minsheu, in his *Ductor in linguas* (1617), eschewed *métis* in favour of *peira* (experience) as a synonym for the headword *cunning*. His Latin equivalents—*peritus* and *experientia*—are standard for *cunning* in the early lexica, although they are sometimes accompanied by *solers/solertia*. The entries for the latter terms in Elyot 1538 make plain the proximity of *cunning* to *wit* and, through their connotations of subtlety and sharpness, to *ingenuity*:

> *Solers*, wyttty, cunnyng.
>
> *Solertia*, sharpenes or quyckenes of wytte, craftynesse or subtiltie in practysynge, or wyttynesse.

For Thomas 1587 *solertia* is "wittines, craftines, skill, cunning, subtilty in practising good or ill." The last part of this definition underscores the flexibility of *cunning*, its applicability in a range of situations and to various ends. For Bullokar 1616 it defines such differing individuals as the *magician* ("A cunning man: a Sorcerer, a great learrned Clark, a Conjurer") and the *Machivilian* (A politicke states man: a cunning Polititian, such as Machivell was"). In the latter capacity, the association of *cunning* with wisdom, guile, and worldly problem solving explains why Cawdrey 1617 uses it to define *politique* ("craftie or cunning"), and (as per *métis*) its regular association with sophistry.[111] For example, Blount 1656 defines a Sophist as "He that professeth Philosophy for lucre or vain-glory; a deceiver, under an eloquent or crafty speaking; a cunning or cavilling disputer, who will make a false matter seem true." As a derogatory term, B. E. uses *cunning* somewhat similarly in his definition of *Gypsies*:

> [A] Counterfeit Brood of wandering Rogues and Wenches, herding together, and Living promiscuously, or in common, under Hedges and in Barns, Disguising themselves with Blacking their Faces and Bodies,

and wearing an Antick Dress, as well as Devising a particular Cant, Strolling up and down, and under colour of Fortune-telling, Palmestry, Physiognomy, and Cure of Diseases; impose allwaies upon the unthinking Vulgar, and often Steal from them, whatever is not too Hot for their Fingers, or too Heavy to carry off. A Cunning Gypsy, a sharp, sly Baggage, a Witty Wench. As Tann'd as a Gypsy, of a Gypsy-hue or colour.

Thus, "cunning folk" are the folkloric healers of rural communities, the conjurors and practitioners of popular magic, as in Kersey 1702: "A Wise-man, or Cunning-man; a Wizzard, or conjurer."[112]

In the eighteenth century especially, lexicographers dwelt on *cunning*'s capacity to denote the kind of manual or verbal dexterity that, in its deceptiveness, can both offend and amaze. It was thus associated with "facetious" forms of *wit*, such that Defoe 1735 used it to define such words as *arch* ("cunning, waggish"), *chicane* ("trickish, waggish"), and *a slight* ("a cunning Trick, Dexterity, a Disesteem"). Johnson sums this sense up neatly by defining the adjectival form of *cunning* as "Artfully deceitful; sly; designing; trickish; full of fetches and stratagems; subtle; crafty; subdolous." Such definitions explain why *cunning* could just as often be a term of abuse as of praise. For instance, B. E. 1619 defines the canting phrase *arch-whore* simply as *cunning*, perhaps drawing on its etymological connection to biblical "knowing," hence the word *cunt* and Kersey's definition of *to leer* as "to cast a cunning, or wishly look" (Kersey 1702).

Conclusion

The dictionary sources tend to skim lightly over the strongly gendered aspects of the ingenuity family of terms in English, as in other tongues. They thus conceal one of the more egregious aspects of that family's history: the identification of genius as an almost exclusively male preserve. We have caught glimpses of the lexicographical sources of this assumption in the "generative" aspects of *genius* and the connection of intelligence with masculine *virtue* (power or force).[113] Indeed, it was a commonplace that while human beings of both sexes possessed wit, only men were afforded the higher order form of wit that would become Romantic genius. As Walter Charleton put it in a paper on the distinction between human and animal brains, presented to The Royal Society in 1664: "It is commonly affirmed, that men have more brain than Women; but the great stupidity of some Men and the great ingeny of some women considered I am not unapt to refer this disparity to the smalnesse of their Bulk and stature, Women being for the most part lesser than Men."[114] A few decades later, Charleton's Royal Society colleague Thomas Willis commented on the case of a "most noble Lady, for above twenty years sick with almost a continual Headach," who "was skilled in the Liberal Arts, and in all sorts of Literature, beyond the condition of her sex; and as if it were thought too much by Nature, for her to enjoy so great endowments, without some detriment, she was extreamly punished with this Disease."[115]

The casting of mentally well-endowed women as curiosities and prodigies—that is, as aberrations—persisted well into the eighteenth and nineteenth centuries, despite occasional defenses of female equality (some penned by women themselves).[116] The *logodaedali* (without exception, men) were complicit in this state of affairs. John Kersey addressed his *New English Dictionary* (1702) to "all persons not perfectly Masters of the English Tongue," offering assistance to "young scholars, Tradesmen, Artificers and others, and particularly, the more ingenious Practitioners of the Female Sex." While this might seem an attempt to welcome women into the world of learning, he made clear elsewhere that his dictionary was intended to help women spell correctly, rather than to understand "hard" words.[117] Equally, Johnson might claim that the dictionary-maker was a simple "slave of science, the pioneer of literature, doomed only to remove rubbish and clear obstructions from the paths of Learning and Genius," but he nevertheless considered *verba* an inadequate, feminine stand-in for the sturdy, masculine *res*: "I am not yet so lost in lexicography as to forget that *words are the daughters of earth, and that things are the sons of heaven*. Language is only the instrument of science, and words are but the signs of ideas: I wish, however, that the instrument might be less apt to decay, and that signs might be permanent, like the things which they denote" (Johnson 1755).

Conclusion

When we began to write this book, we anticipated dictionary entries parsing intellectual abilities, focused on the cognitive aspects of ingenuity. What we found were lists of attributes of temperament and character, in which intelligence was only one among many qualities of an inborn nature. Early modern ingenuity was not Romantic genius. It defined each person, not just exceptional individuals. Through ingenuity, early moderns sought to understand differences between peoples, places, and things.

One of the main motivations for writing dictionaries was to serve one's linguistic community and usually one's compatriots. This motivation was often articulated in terms of the inborn (*gigno*) powers of a language or people (*gens*): tropes of ingenuity served the lexicographer's individual ambitions, and helped him map them onto the community he imagined.[1] The growing body of lexica identified and reinforced distinctions between languages, working out what lexicographers saw as the distinctive qualities innate to each.[2] They could draw on shared learned accounts of the *ingenium*—philosophical, rhetorical, medical, and so on—but equally each could take its own path.

In Italian and Spanish, *ingegno* and *ingenio* retained close ties to Latin *ingenium*, ensuring their clear definition as devices and machines, as a creative faculty, and even as the whole person. Neither language made *spiritus* or its cognates do the work of ingenuity. By contrast, already in the early sixteenth century French had poured all the meanings associated with *ingenium* (individual identity, courtly display, and even artistic style) into the lexis of *esprit*. Throughout the period, German and Dutch both explored alternative vocabularies to translate these terms, such as *Gemüt* for individual feeling, or *Geest* for artistic expression. English, a language barely known outside the British Isles before the eighteenth century, invested Anglo-Saxon *wit* with a wealth of Romance meanings of ingenuity. The list could go on; in the foregoing chapters we have aimed to clarify some of these distinct journeys and their different chronologies.

Whatever the national differences, this has also been a story of striking convergences, which justify seeing these very distinct word histories as traces of a pan-European culture of ingenuity. Certain terms inflated, deflated, and converged semantically owing to Europe's shifting political circumstances. In the early part of our period, the imperial legacies of Rome enforced the resemblances shared between Latin, Spanish, and Italian. Later on, the expanding cultural hegemony of France can be traced through Dutch, English, and finally German dictionaries, as *esprit* informed *spirit* or *geest*. The Low Countries in particular show the influences of their larger neighbors: Dutch *geest* was closely modeled on *esprit*, yet flexible enough to track *ingenio* closely in Dutch-Spanish dictionaries. Though German-speaking territories lacked a unified political center, German remained the language most resistant to imports. Nonetheless, even German definitions of *Scharfsinnigkeit* edged ever closer to *ingegno*, pulled in by the gravitational force of Italian and Spanish *conceptismo*. As English military and financial clout grew in the late seventeenth and early eighteenth centuries, so too did the influence of its language on the lexis of ingenuity. For the first time, *wit* registered as a term of art in other European discourses, and *genius* began to take on a non-classical meaning as exceptional ability. The transformation of ingenuity into genius was not immediate, and happened in different languages at different rates. While the process was complex, some of the many changes included: the eclipse of teachability by the immediacy of inspiration; the rise of sensibility at the expense of intellection; a growing emphasis on the isolation of the genius figure, which ironically inflated their fame in the public sphere; the commodification of outstanding talent; and a tighter correlation between genius and race, which developed sinister overtones in the rise of totalitarian ideologies.

Our lexicographical evidence has provided an outline map of the culture of ingenuity in Europe that needs to be filled in and expanded. "Siempre la lengua fue compañera del imperio" (language was always the companion of sovereignty), Nebrija famously stated in the prologue to his *Grammatica* of 1492, the first systematic grammar of Castilian—indeed of any Romance language. Dedicated to Queen Isabella I of Castile, and published in the same year as Columbus's "discovery" of America, this grammar, together with Nebrija's Latin-Spanish and Spanish-Latin dictionaries of 1492 and ca. 1495, illustrates the Europe-wide early modern concern with language as one of the building blocks of national identity and culture. Despite the longstanding debate over what *imperio* in Nebrija's statement may have meant (its period sense is likely to have been closer to "sovereignty" than "empire"), this declaration nonetheless suggests the extent to which language served political expansion and colonization. The lexicographical materials discussed in this book—and our focus on ingenuity and its word history—reveal this instrumentalization of language within the specific geographical domain of Europe. The argument could be amplified by examining the considerably large corpus of manuals, grammars, and dictionaries produced in the wake of fifteenth- and sixteenth-century European encounters with "new" territories and "new" peoples across the globe. The case of America is particularly germane. First, because of the logistical challenge of dealing with many cultures and languages across a vast territory. Second, for its importance as one of the key foci of early modern European debates on ingenuity, whether in relation to the anthropological status of their inhabitants or in terms of the reception and appreciation of their novel cultural products (e.g. by Dürer), among other issues.³

• • •

This global lexicographical exploration of the non-European features of early modern ingenuity remains largely uncharted territory, which we hope others might map.⁴ Our lexicographical investigation of ingenuity led us out of dictionaries and beyond our keywords in several ways. Lexica themselves gestured toward other sources: literature made canonical through repeated citations, and visual sources, such as the technical drawings required by increasingly complex verbal descriptions of mechanical engines.⁵ Beyond words and images, dictionaries are instances of lexicographical ingenuity and industry in their own right, which gesture toward the material culture of ingenuity. After all, the *logodaedalus* was a cunning wordsmith, his very name conflating the crafts of words and things.

Other sources make abundantly apparent what a purely lexicographical approach to early modern cultures of ingenuity might miss: works of poetics, artistic theory, natural philosophy, and medicine, not to mention the many artefacts that testify to ingenuity in their making.⁶ Girolamo Cardano's *De subtilitate* is a particularly illuminating example, since it touches on many of the domains of ingenuity, and provoked considerable dispute. First published in 1550, this book proved a spectacular success throughout Europe. It restructured in twenty-one chapters the subject matter of the scholastic natural-philosophical and mixed-mathematical encyclopaedia around the notion of subtlety, defined in the first chapter in terms of material fineness and intellectual complexity. Air and dust were subtle; so were knotty problems in mathematics and mechanics.⁷ Protean ingenuity in its many guises suffused *De subtilitate*. Yet the term *ingenium* only features once in the *index rerum* of the 1550 edition: 'ingenio acuendo" (for sharpening one's wit). This index entry corresponds to a passage concerned with the cognitive virtues of specific food and drink: chicken brains strengthen the memory, whereas lemon balm sharpens one's *ingenium*.⁸ This is one of the scarce—if telling—instances of the word *ingenium* encountered by the reader of Cardano's text. Like the remark on lemon balm, some of these instances echo medical views on ingenuity such as the impact of the expectant mother's diet on her child's brain and related intellectual abilities.[9] Other instances denote the various artful performances of one's nature and wit; these include machines (such as Juanelo Turriano's wonders), or a singer's melodic range.[10] Chapter fifteen, dedicated to "subtleties of an unknown kind or useless subtleties," lists games—interlocking rings and the "game of brigands," a Roman precursor of chess—the Lullian art, various forms of anagrammatic wordplay used in mnemonics and biblical hermeneutics, and finally arguments found in Proclus's commentary on Euclid's *Elements*. Cardano concludes the chapter by stating that "such subtleties are intended as pure displays of wit (*ingenium*) and were hardly invented to be useful."[11] Ingenuity is also assessed in the "Wonders and the way to generate belief through representation" of chapter eighteen: thus party tricks like fire- and sword-swallowing demonstrate more boldness than *ingenium*.[12] Whence a paradox: although the whole book is, in fact, concerned with ingenuity—with its natural-philosophical and medical explanations, with its natural-historical, mathematical, mechanical, social, and cultural expressions—the term *ingenium* itself barely featured. The cognitive operations made possible by spirits as fine cerebral matter, complex machines, and perspectival illusions in paintings were customarily qualified as ingenious by Cardano's contemporaries. Yet the striking paucity of *ingenium* the word in *De subtilitate* suggests that Cardano consciously refused to use the term in order to promote his own notion of *subtilitas*.[13]

Cardano's most exacting critic, Julius Caesar Scaliger, was quick to notice the absence of the term *ingenium* in his *Exotericarum exercitationum libri quintus decimus* (1557). By amending Cardano's definition of subtlety, he folds *subtilitas* back onto *ingenium* as it is described in the rhetorical tradition. In doing so, he denies subtlety the explanatory force with which Cardano had tried to endow it. He reminds the reader of the etymology of subtlety: the *sub-tela*, the warp of cloth. The term was first used by Cicero in rhetoric to denote the humid and bloodless temperament of the orator and its results. These consisted first in "tenuity of speech" (*tenuitas dicendi*) constitutive of a "plain style" (*stylus subtilis*) that shuns copious ornament; such style was best represented by the speeches of the Attic orator Lysias. It also stood for complex, knotty, and sophistical arguments (*argutia*). For Scaliger, tenuity and complexity were *both* integral to subtlety—yet such subtlety was primarily the product of the *ingenium* in rhetoric.[14]

• • •

By narrowing Cardano's subtlety to the rhetorical sense of *ingenium*, Scaliger reminds the word historian that the culture of ingenuity could take center stage even in a book that purposefully avoided its lexis. However, while his criticism has the merit of relocating explicitly Cardano's endeavor within the early modern culture of ingenuity, it fails to acknowledge the ways in which such a culture could take the form of what John Donne labeled "the New Science." In coining the notion of subtlety, Cardano attempted to describe those other forms of ingenuity concerned with inventions and novelties in natural philosophy, natural history, mixed mathematics, and mechanics—those seeds of the New Science. Alongside the poet, the courtier, and the engineer, the *novatores* were also wits: this wider history of early modern ingenuity remains to be written.

Notes

Introduction

1. Raymond Williams, *Keywords: A Vocabulary of Culture and Society* (Oxford: Oxford University Press, 1983), 143. "Genius" is one of twenty-one terms Williams introduced or modified since the publication of the first edition in 1976.

2. There is a vast literature on Romantic- and post-Romantic genius. Significant studies include James Engell, *The Creative Imagination: Enlightenment to Romanticism* (Cambridge, MA: Harvard University Press, 1981); Jochen Schmidt, *Die Geschichte des Genie-Gedankens in der deutschen Literatur, Philosophie und Politik, 1750–1945* (Darmstadt: Wissenschaftliche Buchgesellschaft, 1985); Darrin M. McMahon, *Divine Fury: A History of Genius* (New York: Basic Books, 2013); Ann Jefferson, *Genius in France: An Idea and Its Uses* (Princeton: Princeton University Press, 2014).

3. See Terence Cave, *Pre-histories*, 2 vols. (Geneva: Droz, 1999–2001). Richard Scholar, *The Je-Ne-Sais-Quoi in Early Modern Europe: Encounters with a Certain Something* (Oxford: Oxford University Press, 2005) uses a similar approach (see esp. 6). On Cave's methods, see Neil Kenny and Wes Williams, "Introduction," in Terence Cave, *Retrospectives: Essays in Literature, Poetics and Cultural History*, ed. Neil Kenny and Wes Williams (Oxford: Legenda, 2009).

4. Richard Scholar, "The New Philologists," in *Renaissance Keywords*, ed. Ita Mac Carthy (Oxford: Legenda, 2013), esp. 2. We consider our work to be situated also in the longer tradition of modern "word history," which may be traced at least to Lucien Febvre. See e.g. the essays "*Frontière*: the Word and the Concept" (1928) and "*Civilisation*: Evolution of a Word and a Group of Ideas" (1930), in Lucien Febvre, *A New Kind of History, from the Writings of Lucien Febvre*, ed. Peter Burke, trans. K. Folca (London: Routledge & Kegan Paul, 1973).

5. Neil Kenny, *The Uses of Curiosity in Early Modern France and Germany* (Oxford: Oxford University Press, 2004), 3. Examples of a retrojective approach to genius include Penelope Murray, "Introduction," in *Genius: The History of An Idea*, ed. Penelope Murray (Oxford: Basil Blackwell, 1989); Noel L. Brann, *The Debate Over the Origin of Genius During the Italian Renaissance: The Theories of Supernatural Frenzy and Natural Melancholy in Accord and in Conflict on the Threshold of the Scientific Revolution* (Leiden: Brill, 2002). On the dangers of analepsis in the history of ideas see Scholar, *The Je-Ne-Sais-Quoi*, 11–12.

6. See Neil Kenny, *Curiosity in Early Modern Europe: Word Histories* (Wiesbaden: Harrassowitz, 1998), esp. 24; Kenny, *Uses of Curiosity*, esp. 10–13. Kenny's work (and thus, ultimately, ours) is indebted to the approach developed by Skinner and others that

applied a modified version of J. L. Austin's speech act theory to intellectual history. His studies provide an excellent summary of the differing approaches to language and concepts in the versions of *Begriffsgeschichte* developed by Blumenberg and Koselleck.

7. Kenny, *Curiosity*, 25–6.

8. Kant 2000, 186.

9. The rhetorical account of ingenuity remains vital throughout our period, across all languages, right up to the eighteenth century, when it is given a prominent role in Vico's new philosophical system. The literature on ingenuity and rhetoric in various authors is large. See e.g. Emilio Hidalgo-Serna, "'Ingenium' and Rhetoric in the Work of Vives," *Philosophy & Rhetoric* 16, no. 4 (1983): 228–41; Mercedes Blanco, *Les Rhétoriques de la Pointe: Baltasar Gracián et le Conceptisme en Europe* (Paris: Champion, 1992); Stefano Gensini, "Ingenium/ingegno fra Huarte, Persio e Vico: le basi naturali dell'inventività umana," in *Ingenium propria hominis natura*, ed. Stefano Gensini and Arturo Martone, (Naples: Liguori, 2002), 29–70; David L. Marshall, *Vico and the Transformation of Rhetoric in Early Modern Europe* (Cambridge: Cambridge University Press, 2010).

10. While early modern lexica in manuscript have, for the most part, been excluded from this study, see John Considine, *Dictionaries in Early Modern Europe: Lexicography and the Making of Heritage*. (Cambridge: Cambridge University Press, 2008), e.g. 38, 93, 102, 166–171, 191, 198, 231; John Considine, *Academy Dictionaries 1600–1800* (Cambridge: Cambridge University Press, 2014), 9–10, 93ff, for a discussion of these sources and their significance.

11. See e.g. Franco Moretti, *Distant Reading* (New York: Verso, 2013). An alternative approach to the large-scale structural analysis of concepts is offered by Peter de Bolla, *The Architecture of Concepts: The Historical Formation of Human Rights* (New York: Fordham University Press, 2013).

12. See Kenny, *Curiosity*, 18; Scholar, *The Je-Ne-Sais-Quoi*, 45ff.

13. See, for example, the entries for *genius* and *ingenium* in *Dictionary of Untranslatables: A Philosophical Lexicon*, ed. Barbara Cassin (Princeton: Princeton University Press, 2014).

14. Here, we join Scholar in departing somewhat from Kenny's approach. See Kenny, *Curiosity*, 21–32; Scholar, *The Je-Ne-Sais-Quoi*, 56–7.

15. Our selection of keywords has also been informed by the earlier lexicographical investigations in: Edgar Zilsel, *Die Entstehung des Geniebegriffes: ein Beitrag zur Ideengeschichte der Antike und des Frühkapitalismus* (Tübingen: J. B. C. Mohr, 1926); Wido Hempel, "Zur Geschichte von *spiritus, mens* und *ingenium* in den romanischen Sprachen," *Romanistisches Jahrbuch* 16 (1965): 21–33; Jean Lecointe, *L'idéal et la différence: la perception de la personnalité littéraire à la Renaissance* (Librairie Droz, 1993), 219–25; Hubert Sommer, *Génie: zur Bedeutungsgeschichte des Wortes von der Renaissance zur Aufklärung* (Frankfurt am Main: Peter Lang, 1999); Cristina Vallini, "Genius/ingenium: derive semantiche," in *Ingenium propria hominis natura*, ed. Stefano Gensini and Arturo Martone (Naples: Liguori, 2002); Elisabetta Graziosi, *Questioni di*

lessico: l'ingegno, le passioni, il linguaggio (Modena: Mucci Editore, 2004). There is a small but growing literature on early modern ingenuity in context. Significant studies include Blanco, *Les Rhétoriques de la Pointe*; Jürgen Klein, "Genius, Ingenium, Imagination: Aesthetic Theories of Production from the Renaissance to Romanticism," in *The Romantic Imagination: Literature and Art in England and Germany*, ed. Frederick Burwick and Jürgen Klein (Amsterdam: Rodopi, 1996); Riccardo Campi, "*Ingenio* ed *esprit* tra Gracián e Bouhours: una questione di metodo," *Studi di estetica* 25, no. 16 (1997): 185–209; Brann, *Origin of Genius*; Stefano Gensini and Arturo Martone, eds., *Ingenium propria hominis natura* (Naples: Liguori, 2002); Aurora Egido, "Estudio preliminar," in *Agudeza y arte de ingenio (Huesca, Juan Nogués, 1648) Edición facsímil*, by Baltasar Gracián (Zaragoza: Institución Fernando el Católico, 2007); Rhodri Lewis, "Francis Bacon and Ingenuity," *Renaissance Quarterly* 67, no. 1 (2014): 113–63; Ann Jefferson and Jean-Alexandre Perras, eds., *Thinking Genius, Using Genius / Penser le génie à travers ses usages*, special issue of *L'Esprit Créateur* 55, no. 2 (2015); Jean-Alexandre Perras, *L'Exception exemplaire: Inventions et usages du génie (XVIe-XVIIIe siècle)* (Paris: Classiques Garnier, 2016).

16. We use the term "culture" as per the OED sense 7a: "Chiefly as a count noun. The distinctive ideas, customs, social behaviour, products, or way of life of a particular nation, society, people, or period. Hence: a society or group characterized by such customs, etc." See also Douglas Bruster, *Shakespeare and the Question of Culture: Early Modern Literature and the Cultural Turn* (New York: Palgrave Macmillan, 2003), 211–16. On the absence of an "ineliminable core" in concepts, see Kenny's comments on Wittgenstein's rope metaphor in Kenny, *Uses of Curiosity*, 10.

17. See e.g. Stephen Perkinson, "Engin and Artifice: Describing Agency at the Court of France, ca. 1400," *Gesta* 41, no. 1 (2002): 51–67; Jonathan Morton, "Ingenious Genius: Invention, Creation, Reproduction in the High Middle Ages," *L'Esprit Créateur* 55, no. 2 (2015): 4–19; Marcel Detienne and Jean-Pierre Vernant, *Cunning Intelligence in Greek Culture and Society*, trans. Janet Lloyd (Chicago: University of Chicago Press, 1991). Studies of genius in antiquity include Jane C. Nitzsche, *The Genius Figure in Antiquity and the Middle Ages* (New York: Columbia University Press, 1975); Hille Kunkel, *Der römische Genius* (Heidelberg: Kerl, 1974); and the early sections of McMahon, *Divine Fury*.

18. Detienne and Vernant, *Cunning Intelligence* (first, French edition 1974), 3.

19. Detienne and Vernant, *Cunning Intelligence*, 3.

20. On which see Raphaële Garrod with Alexander Marr, eds., *Cartesian Ingenuity: Embodied Cognition in the Philosophy of René Descartes* (forthcoming).

21. C. F. Goodey, *A History of Intelligence and Intellectual "Disability": The Shaping of Psychology in Early Modern Europe* (Farnham: Ashgate, 2011), 58. See also Rodrigo Cacho Casal, *La esfera del ingenio. Las silvas de Quevedo y la tradición europea* (Madrid: Biblioteca Nueva, 2012), which offers the metaphor of a "sphere" of *ingenio* for the intellectual-cum-literary culture of early modern Spain.

22. These fields each have a large literature. Landmarks on melancholy include Raymond Klibansky, Erwin Panofsky, and Fritz Saxl, *Saturn and Melancholy: Studies in the History of Natural Philosophy, Religion and Art* (London: Thomas Nelson and Sons, 1964); Teresa Scott Soufas, *Melancholy and the Secular Mind in Spanish Golden Age Literature* (Columbia: University of Missouri Press, 1990); Winfried Schleiner, *Melancholy, Genius, and Utopia in the Renaissance* (Wiesbaden: Harrassowitz, 1991); Brann, *Origin of Genius*; Angus Gowland, *The Worlds of Renaissance Melancholy: Robert Burton in Context* (Cambridge: Cambridge University Press, 2006); Felice Gambin, *Azabache. El debate sobre la melancolía en la España de los Siglos de Oro* (Madrid: Biblioteca Nueva, 2008). On inspiration, see e.g. Liane Nebes, *Der "furor poeticus" im italienischen Renaissance-Platonismus: Studien zu Kommentar und Literaturtheorie bei Ficino, Landino und Patrizi* (Marburg: Tectum-Verlag, 2001); Lecointe, *L'idéal et la différence*, 219–374; Patricia A. Emison, *Creating the "Divine" Artist: From Dante to Michelangelo* (Leiden: Brill, 2004); Jackie Pigeaud, *Melancholia: Le malaise de l'individu* (Paris: Payot, 2008); Sarah Eron, *Inspiration in the Age of Enlightenment* (Newark: University of Delaware Press, 2014). On the sublime, see e.g. Nicholas Cronk, *The Classical Sublime: French Neoclassicism and the Language of Literature* (Charlottesville, VA: Rookwood, 2002); Caroline A. van Eck et al., eds., *Translations of the Sublime: The Early Modern Reception and Dissemination of Longinus' Peri Hupsous in Rhetoric, the Visual Arts, Architecture and the Theatre* (Leiden: Brill, 2012).

23. For the case of artistic temperament in our period, see Rudolf Wittkower and Margot Wittkower, *Born under Saturn: The Character and Conduct of Artists: A Documented History from Antiquity to the French Revolution* (New York: Norton, 1969).

24. Vives 1538, 77–83.

25. Goodey, *History of Intelligence*, 52, 54.

26. Sallust, *Bel. Cat.* 8: "ingenium nemo sine corpore exercebat." This aspect of ingenuity intersects in subtle but important ways with emerging notions of grace and *sprezzatura*, notably in Castiglione 1528. See e.g. Ita Mac Carthy, "Grace and the 'Reach of Art' in Castiglione and Raphael," *Word and Image* 25, no. 1 (2009): 33–45.

27. Tesauro 1654, 551. Quoted in Peter Schwenger, "Crawshaw's Perspectivist Metaphor," *Comparative Literature* 28, no. 1 (1976): 65–74, at 67. Similar arguments are found in the works of Baltasar Gracián.

28. See Horst Bredekamp, *Der Künstler als Verbrecher: Ein Element der frühmodernen Rechts- und Staatstheorie* (Munich: Carl Friedrich von Siemens Stiftung, 2008).

29. On representations of craftsmen, see James S. Amelang, *The Flight of Icarus: Artisan Autobiography in Early Modern Europe* (Stanford: Stanford University Press, 1998).

30. Calepino 1502, [a1v]: "Non enim tam instruendorum aliorum quam mei exercendi ingenii gratia id efficere aggressus sum."

31. The expression features for the first time in a 1635 discourse in front of the newly founded French Académie, where it denotes "cette energie, et ne sais quel esprit" (this energy, and that certain something of wit), i.e., a linguistic form of the *Je-ne-sais-quoi*,

a paradoxical unknown which defined singularity and identity. See Jürgen Trabant, "Du génie aux gènes des langues," in *Et le génie des langues?*, ed. Henri Meschonnic (Saint Denis: Presses Universitaires de Vincennes, 2000), 80.

32. In the rhetorical tradition, judgment is traditionally involved in disposition rather than invention, that is, in the ordering of the copia generated by invention.

33. The word *logodaedalus*, coined by Socrates and popularized by Erasmus (Adagia 1262), was widely found in these lexica. *Lododaedali* were famously critiqued by Kant in the *Metaphysics of Morals* (6" 206), for which see Jean-Luc Nancy, *The Discourse of the Syncope: Logodaedalus*, trans. Saul Anton (Stanford: Stanford University Press, 2008).

34. Perotti 1496.

35. See chapter 1.

36. On the epigram as a witty genre, see chapter 6. On its status as "minor genre," stemming from the salon culture of wit, see chapter 4. On the European fortunes of the genre in the early modern period, see Susanna de Beer, K. A. E. Enenkel and David Rijser, eds., *The Neo-Latin Epigram: A Learned and Witty Genre* (Leuven: Leuven University Press, 2009); Hoyt Hopewell Hudson, *The Epigram in the English Renaissance* (Princeton: Princeton University Press, 1947); Tatjana Schäffer, *The Early Seventeenth-Century Epigram in England, Germany and Spain: A Comparative Study* (Oxford: Peter Lang, 2004); Pierre Laurens, *L'abeille dans l'ambre: célébration de l'épigramme de l'époque alexandrine à la fin de la Renaissance* (Paris: Belles Lettres, 1989).

37. Cockeram offers *apothegms* as well as *aphorisms*, suggesting that his source may well have been Cotgrave 1611, for whom *apothegme* is "a short and pithy sentence." Earlier Latin-English dictionaries attribute these qualities to *sententiae*. See e.g. the entries for *sententiose/us* in Thomas 1587.

38. A trajectory similar to *genius > aphorism* could be found for *genius > epigram*; see e.g. Bullokar 1616: "Epigramme. It properly signifieth a superscription or writing set upon any thing; now it is commonly taken for a short *wittie* poeme, which under a fained name, doth covertly praise or tax some particular person or thing." Italics added.

39. Estienne, 1531, I: *3v. We surely should read "arte" for "Marte," a delightfully apropos typographic error.

40. Estienne, 1531, *2v. "Quibus cum laborem nostrum haudquaquam disciplicere animadverterem, iique summopere me ad perseverandum hortarentur, eos iam sponte currenti stimulos addiderunt, ut non solum rem domesticam, sed et curam corporis in totum ferme negligerem."

41. Calepino 1502, s.v. *Industria*. Estienne 1531 glosses this definition by adding that "Etiam pro ingenii sive artis exercitatione accipitur . . ." (It denotes either a natural or a methodical (artful) practice . . .). As so often, in this entry Estienne excerpts from Calepino.

42. On humanist Latin as a hegemonic, pan-European language of intellectual and cultural reform, see Ann Moss, *Renaissance Truth and the Latin Language Turn* (Oxford: Oxford University Press, 2003); Françoise Waquet, *Latin, or the Empire of the Sign:*

From the Sixteenth to the Twentieth Century, trans. John Howe (New York and London: Verso, 2001). For a different history of polyglot sociocultural exchange, see Eric R. Dursteler, "Speaking in Tongues: Language and Communication in the Early Modern Mediterranean," *Past & Present* 217, no. 1 (2012): 47–77. For a nuanced account of the relationships between Latin and vernaculars, and their stakes, see Karen Newman and Jane Tylus, eds., *Early Modern Cultures of Translation*, (Philadelphia: University of Pennsylvania Press, 2015) and José María Pérez Fernández, "Translation, Sermo Communis, and the Book Trade," in *Translation and the Book Trade in Early Modern Europe*, José María Pérez Fernández and Edward Wilson-Lee (Cambridge: Cambridge University Press, 2014), 40–60; Sietske Fransen and Niall Hodson, eds., *Translating Early Modern Science* (Leiden: Brill, 2017).

43. Covarrubias 1611, "Al letor" (preface to the reader).

44. Nebrija 1492, sig. A4r-v: "Nulla fuit arbor inter antiquos nobilior platano, sed an extet hodie apud aliquas gentes, ego non ausim affirmare: certe in Hispania nusquam esse audio. Fuisse autem antiquis temporibus, vel ex eo manifestum est, quod C. Caesar in memoria nominis sui, propria manu Cordubae consevit."

45. Preface to Johnson 1755.

46. Johnson 1755: "If an academy should be established for the cultivation of our stile, which I, who can never wish to see dependance multiplied, hope the spirit of *English* liberty will hinder or destroy, let them . . . endeavour . . . to stop the licence of translatours, whose idleness and ignorance . . . will reduce us to babble a dialect of *France*."

47. Goropius 1569, 604–611. Goropius was deemed ingenious by Baltasar Sebastian Navarro de Arroyta in the prefatory letter that opens Covarrubias' *Tesoro* (Covarrubias 1611, [5]v).

48. Howell 1660, "To the Philologer," 2v.

49. The notion, if not the expression itself, features in Richelet's preface to his 1680 dictionary: "certaines perfections tres-avantageuses qui ne se trouvent point dans les autres langues" (some truly advantageous perfections, not to be found in other languages) (Richelet 1680, 1:5v). See also Vaugelas 1647 about the improper use of the article in front of proper nouns ("L'Aristote, Le Petrone") as "cest tres-mal parler, et contre le genie de notre langue" (Vaugelas 1647, 253).

50. Port Royal equated grammar and thought in its 1681 *Grammaire générale et raisonnée contenant les fondemens de l'art de parler, expliqués d'une manière claire et naturelle*, thus integrating grammar into logic, the scholastic art of good thinking. This view still holds to this day; its grammaticality is still what characterizes French according to Albert Dauzat, *Le génie de la langue française* (Paris: Payot, 1943) and in the much more recent account of Gilles Siouffi, *Le génie de la langue française: étude sur les structures imaginaires de la description linguistique à l'âge classique* (Paris: Champion, 2010), 28–9.

51. Académie françoise 1694 [4]v: " . . . & surtout cette Construction directe, qui sans s'esloigner de l'ordre naturel des pensées, ne laisse pas de rencontrer toutes les delicatesses que l'art est capable d'y apporter."

1 Latin

1. See Cicero, *De orat.*, 2.217–90.
2. Goodey, *A History of Intelligence*, 51.
3. André du Laurens, *Historia anatomica humani corporis* (Paris: M. Orry, 1600), 535: "Idem in *Arte parva*, ingenium ad tenuem & crassam cerebri substantiam refert. Ingenium autem appellat ἀγχίνοια, id est, solertiam, quae definitur 'εὐστοχία, id est, inveniendi et conjiciendi promptitudo. Eodem lib. εὐμαθία, id est, discendi facilitas, substantiam cerebri mollem et humidam ostendit" (Similarly, in the *Small arts*, [Galen] ascribes *ingenium* to the matter of the brain, whether it is fine or rough. Indeed, he calls *ingenium* ἄγνοιά, that is, quick-wittedness, which is defined as εὐστοχία, that is, speed in finding and combining. In the same book, εὐμαθία, that is, the ability to learn with ease, reveals that the matter of the brain is soft and wet).
4. See pages 47–50.
5. See for example Cicero, *De orat.* 1.4.15, which thematizes the role of *ingenium* in making exceptional orators—especially among the Roman people: "ingenia vero . . . nostrorum hominum multum ceteris hominibus omnium gentium praestiterunt" (but the *ingenia* of our people far surpass the other people of all nations). *Ingenium* as natural ability is also found in Quintilian's *Institutio oratoria*, as the quality that cannot be taught (2.8, 7.10) or imitated (10.2).
6. This was a deeply embodied set of terms too, exemplified in the humoral illness of the *ingenium* famous as melancholy, based on pseudo-Aristotle, *Problemata* §30.1 (953a10–955a40). The passage is translated at length in Klibansky, Panofsky, and Saxl, *Saturn and Melancholy*, 18–29.
7. One arena in which this focus is particularly pronounced is Jesuit pedagogy. See Raphaële Garrod, "*Subtilis—Inutilis*: The Jesuit Pedagogy of Ingenuity at La Flèche in the Seventeenth Century," in *Teaching Philosophy in the Seventeenth Century: Image and Text*, ed. Susanna Berger and Daniel Garber (forthcoming).
8. Cicero, *De orat.* 2.35.147–50; Quintilian, *Inst. orat.* 2.12–13 and 10.2 (where we find the famous contrast of *ars* and *ingenium*).

    This capacity for invention, or composition of a discourse, is the primary meaning given by Alain Pons, "Ingenium," in *Dictionary of Untranslatables: A Philosophical Lexicon*, ed. Barbara Cassin, Steven Rendall, and Emily S. Apter (Princeton: Princeton University Press, 2014).
9. Petrarch reframed literary production in the same terms in his letters: see Francesco Petrarca, "De inventione et ingenio," in *Epistolae de rebus familiaribus et variae*, ed. Giuseppe Fracassetti (Florence: Le Monnier, 1859), I.7, 57–64. Morhof begins the second book of his *Polyhistor* (Lübeck: Böckmannus, 1688) with the "method" of the polyhistor and comments on the *ingenia* of various authors.
10. Michael Baxandall, *Giotto and the Orators: Humanist Observers of Painting in Italy and the Discovery of Pictorial Composition* (Oxford: Oxford University Press,

1986), 15–17; Hélène Vérin, *La Gloire des ingénieurs. L'Intelligence technique du XVIe au XVIIIe siècle* (Paris: Albin Michel, 1993); Edgar Zilsel, *Die Entstehung des Geniebegriffes: ein Beitrag zur Ideengeschichte der Antike und des Frühkapitalismus* (Tübingen: J. B. C. Mohr, 1926).

11. See e.g. the Italian *inganno* as fraud (chapter 2), and *engaño* in Spanish (chapter 3), and *engin* in French (chapter 4). See Simon Schaffer and Steven Shapin, *Leviathan and the Air Pump: Hobbes, Boyle, and the Experimental Life* (Princeton: Princeton University Press, 1985), 129–131; Robert A. Greene, "Whichcote, Wilkins, 'Ingenuity,' and the Reasonableness of Christianity," *Journal of the History of Ideas* 42, no. 2 (1981): 227–52; Jim Bennett, "Instruments and Ingenuity," in *Robert Hooke: Tercentennial Studies*, ed. Michael Cooper and Michael Hunter (Aldershot, UK: Ashgate, 2006).

12. Isidore's influence is difficult to overstate, as the most common reference work—the source, for example, of the acquaintance of most medieval scholars with Quintilian's rhetoric. As for Balbus's *Catholicon*, it is a five-part work, of which the fifth is a massive lexicon based on Papias, Hugutio and others—by far the most important medieval dictionary, with well over 200 manuscripts known and over 25 editions printed before 1521. On medieval lexicography, see Hans Sauer, "Glosses, Glossaries, and Dictionaries in the Medieval Period," in *The Oxford History of English Lexicography*, ed. Anthony Paul Cowie, vol. 1 (Oxford: The Clarendon Press, 2009); Olga Weijers, "Lexicography in the Middle Ages," *Viator* 20 (1989): 139–53; and Olga Weijers, *Dictionnaires et répertoires au moyen âge: une étude du vocabulaire* (Turnhout: Brepols, 1991).

13. Reuchlin 1478. Ludwig Geiger, *Johann Reuchlin, sein Leben und seine Werke* (Leipzig: Duncker & Humblot, 1871), 68ff, gives an overview of Reuchlin as lexicographer.

14. Marianne Pade, "Niccolò Perotti's *Cornu Copiae*: Commentary on Martial and Encyclopedia," in *On Renaissance Commentaries*, ed. Marianne Pade, Noctes Neolatinae (Hildesheim: Olms, 2005); Marianne Pade, "Niccolò Perotti's *Cornu Copiae*: The Commentary as a Repository of Knowledge," in *Neo-Latin Commentaries and the Management of Knowledge in the Late Middle Ages and the Early Modern Period (1400 -1700)*, ed. Karl A. E. Enenkel and Henk Nellen (Leuven: Leuven University Press, 2013); Martine Furno, *Le* Cornu copiae *de Niccolò Perotti: culture et méthode d'un humaniste qui aimait les mots* (Geneva: Droz, 1995). On the epigram, see above, Introduction, n.36.

15. On the organization of Perotti, see Ann Blair, *Too Much to Know: Managing Scholarly Information before the Modern Age* (New Haven: Yale University Press, 2010), 352–355.

16. Albert Labarre, *Bibliographie du* Dictionarium *d'Ambrogio Calepino (1502–1779)* (Baden-Baden: Koerner, 1975).

17. Percy Stafford Allen et al., eds., *Opus epistolarum Des. Erasmi Roterdami* (Oxford: Clarendon, 1906-), ix: 177 (esp. 2446).

18. The reason for Estienne's influence is not repeated print runs—the *Dictionarium seu latinae linguae thesaurus* was printed only a few times (ed. princeps 1531, 1536, 1543, 1734–1735, and again in 1740–1743). But the book became the starting point for many other dictionaries, including Latin-French dictionaries from Estienne's own *Dictionarium latinogallicum* and *Dictionnaire francois latin* to the various versions of the *Dictionariolum puerorum* (first published 1542), which in turn inspired Latin-vernacular dictionaries for other languages. From the middle of the sixteenth century, even editions of Calepino were augmented using material from Estienne. Following Estienne's example, the most ambitious dictionaries of Latin and Greek usually took the name *Thesaurus*. See Considine, *Dictionaries in Early Modern Europe*, 46, where he observes that the most influential dictionaries across the European languages were "written in the tradition of 'Estienne and others.'"

19. Allen et al. *Opus epistolarum Erasmi*, iv:575–80 (no. 1233) at 579–80: "Video qua in re plurimum adiumenti possis adferre Graecanicis studiis, nimirum si copiosissimo Lexico nobis non tantum recenseas vocabula, verumetiam idiomata et Graeci sermonis tropos." Cit. Considine, *Dictionaries*, 36.

20. Estienne 1531, *2r.

21. The 1536 edition was presented as a continuation of the first; the third edition of 1543 (labeled *editio secunda* on the title page) abandoned French translations and added many more examples, growing from 940 single-column folios in 1531 to a total of 1584 folio leaves (often bound in three volumes) in 1543. This expansion is explored by Considine, *Dictionaries*, 38–55.

22. See chapters 4 and 5 on French and German.

23. Considine, *Dictionaries*, 43, suggests that Estienne stayed close to Calepino partly because he had begun with Calepino, and partly because he needed to sell his new dictionary in a market dominated by Calepino.

24. Estienne, *Sententiae et proverbia ex omnibus Plauti & Terentii comoediis* (Paris: Estienne, 1530). For these editions, see Fred Schreiber, *The Estiennes* (New York: E.K. Schreiber, 1982), 54 (item 43).

25. See nos. 68–70 in Labarre, *Bibliographie du* Dictionarium *d'Ambrogio Calepino*.

26. The 1590 edition of Calepino included Hebrew, Hungarian, and Polish, besides the languages we discuss in this book.

27. First published in 1562 and 1573, respectively.

28. This line became a standard example of the use of *genius* in Estienne's *Thesaurus*; it was not originally an adage, but the description of a character hard done by in Terence, *Phormio* 1.43–45.

29. Goropius 1569, 881. "Melius igitur et prudentius dixisset, suum defraudans defectum, is qui dixit 'Quod ille unciatim vix de demenso suo suum defraudans Genium comparsit miser.' Nam qui unciatim comparcit, is Genium suum minime defraudat, sed promovet ad genus vitae beatius, quod in parimonia consistere paulo prius ostendi. Immo

vero facit, ut Genio nihil unquam desit, etsi naturae convenienter vivere, et semper suavissime comedere et bibere Genio gratum est; nihil ei potest accidere optatius, quam eum, quicum agit, parce sibi alimenta administrare, quo nunquam desit condimentum cibi et potus delicatissimum."

30. Emblematics was an ingenious genre *par excellence*: an emblem was wit made into image and text for the early moderns. The first French emblem book by Guillaume de la Perrière (written in 1535, printed in 1540) makes this point in its very title, the *Theatre des bons engins*, where *engins* denotes emblems as the products of one's *ingenium*. In the seventeenth century, the invention and interpretation of emblems was integral to Jesuit pedagogy, geared toward sharpening one's wit: Nicolas Caussin lists the emblem among witty inventions: see Caussin, 1623, fol. [9r]. On emblematics in the Jesuit programme of studies, see Jouvancy, 1725, 94–100.

31. See the entry "Genius" in Yves Bonnefoy, ed., *Dictionnaire des mythologies et des religions des sociétés traditionnelles et des mondes antiques* (Paris: Flammarion, 1981); E. C. Knowlton, "The Allegorical Figure Genius," *Classical Philology* 15. 4 (1920): 380–84.

32. The grammarian Censorinus (3rd century AD) wrote the *De die natali* (*The Birthday Book*) as a birthday present—it tackles, among other, topics relating to birth: that of the genius, the astrological problem of ascertaining the time of birth. This interest in astronomy and chronology partly account for the early modern success in print of Censorinus. See Censorinus; *The Birthday Book*, trans. H. Parker (Chicago: University of Chicago Press: 2007).

33. Calepino 1502.

34. Terence, *Andr.* Act I. sc.5, 54–55: "Quod ego per hanc te deteram oro et genium tuum, per tuam fidem, perque huius solitudinem, te obtestor" (By this right hand I do entreat you, and by your good Genius, by your own fidelity and her bereft condition).

35. The recurring sources are Persius and Terence: "indulge genio" (Persius, *Sat.* 5.151); "Suum defraudans genium" (Terence, *Phorm.* Act.1, sc.1, 10). See also infra 30.

36. *Genius* as individual nature or inner dynamism features in Nebrija 1492; Kahl 1600; Roboredo 1621 and Pereira 1653. *Genius* as defining quality of a person or thing can be found in Reuchlin 1478; Calepino 1590; Roboredo 1621; and Holyoake 1676.

37. We use "naturalism" here to mean the appeal to natural rather than transcendent explanatory principles. For example, the naturalization of Pagan *genii* into the elements in sixteenth-century dictionaries parallels the rationalizing natural-philosophical interpretation of Roman gods and their myths in Renaissance mythography. See Natale Conti, *Mythologiae sive explicationum fabularum libri X* (Venice: 1567).

38. Aristotle, *Metaphysica*: 1014a16–7; Genius as the God of nature exercising generative power—"qui vim [habet] gignendarum rerum" (which has the power to generate things)—features in Reuchlin 1478; Fernández de Palencia 1490; Nebrija 1492; Perotti 1525; Dasypodius 1536; Roboredo 1621; Holyoake 1676; Hofmann 1677; Calepino 1654; Calepino 1728.

39. The parallel etymology of *genius* from *gerere* also accounts for their naturalization as astrological influences: the stars, the sun, and the moon can determine one's action. See I.a.5, p. 28.

40. Genius as the god of hospitality and pleasure can be found in Calepino 1502; Estienne 1531; Dasypodius 1536; Estienne 1538; Frisius 1541; Estienne 1571; Calepino 1590; Roboredo 1621; and Holyoake 1676.

41. See e.g. the case of Spanish, chapter 3.

42. Santra is the mysterious author of a lost *De antiquitate verborum* mentioned by later grammarians. Maurus Servius Honoratus (4th-5th century A.D) was a grammarian, polymath, and author of a commentary on Virgil: Maurus Servius Honoratus, *In tria Virgilii expositio* (Florence: Bernardo Cennini, 1471).

43. This comedic understanding of the agency of natural drives and of their ability to define identities in sixteenth-century lexica should be contrasted, in our view, with the comedy of characters of the following century, where passions have replaced genial inclinations and assume another model of singular characterization and comedic agency. However, it is also worth noting that, since antiquity, Terence is a *favorite* source for the grammatical and lexicographical traditions, which are more concerned with linguistic propriety and style than with the anthropological bases of classical comedy: see Giulia Torello-Hill and Andrew J. Turner, eds., *Terence Between Late Antiquity and the Age of Printing* (Leiden: Brill, 2015).

44. Kahl 1600: "Budaeus lib.I de contemptu rerum fortuitarum scribit, genium pro vi naturae atque pro ingenio usurpari ab Antistitibus eruditionis nostrae aetatis," under entry *Genius*. Yet the entry for δαίμων in Budé's *Lexicon Graeco-Latinum* seu *Thesaurus linguae graecae* (1530) reads "deus, genius, larva, ingenium, sapiens" (God, tutelary spirit, spirit of the deceased, one's given nature, man of knowledge).

45. Kahl 1600: "non dissimulet e vetustis quoque Theologis extitisse, qui binos singulis quibusque, iam inde a natali die, adesse angelos affirmarent, bonum alterum, alterum malum: illum tuendae vitae propugnandaeque salutis, hunc divexandi causa additum" (he does not hide that it came from the old theologians as well, who asserted that two angels attend to every individual, from the very day of their birth, a good one and a bad one; one to protect life and defend health; the other in order to generate trouble).

46. The example of Caligula imposing that oaths be sworn by his own genius features in Hofmann 1677; the *genii* as *lares* and *penates* feature in Estienne's gloss on Censorinus (Estienne 1531, repeated in Estienne 1543, and in Calepino 1590 and 1718). *Genius* denoting good and bad angels also features in Nebrija 1492; Estienne 1531; Dasypodius 1536; Estienne 1538; Frisius 1541; Kiliaan 1562; Hadrianus 1567; Etienne 1571; Kahl 1600; Hadrianus 1620; and Hofmann 1677.

47. See for example Martin Mulsow, "Antiquarianism and Idolatry: The *Historia* of Religions in the Seventeenth Century," in *Historia: Empiricism and Erudition in Early Modern Europe*, ed. Gianna Pomata and Nancy G. Siraisi (Cambridge MA: MIT Press,

2005); Euan Cameron, *Enchanted Europe: Superstition, Reason, and Religion, 1250–1750* (Oxford: Oxford University Press, 2010); Anthony Ossa-Richardson, *The Devil's Tabernacle: The Pagan Oracles in Early Modern Thought* (Princeton, NJ: Princeton University Press, 2013).

48. Calepino 1502, 1590 and Estienne 1531: "Genius veterum sententia dicitur uniuscuiusque anima rationalis. Id eoque esse singulos singulorum" (The views of the ancients state that genius denotes the individual rational soul, hence the reason why individuals differ from one another). Cf. Hofmann 1677: "iuxta profanam veterum opinionem, quanquam Varronem affirmare dicat, genium esse uniuscuiusque rationalem animum" (According to the profane ancient opinion, upheld by Varro, which states that genius is the individual rational soul); and Calepino 1718: "Genium animum uniuscuiusque esse tradunt" (It is said that Genius is the individual soul).

49. Augustine, *Civ. dei* VII.13.

50. Hence, perhaps, the insistence in Calepino 1718: "Mens expers corporis" (the mind, separate from the body).

51. Grace features in Estienne 1543 and 1571; Holyoake 1676; Calepino 1718. This notion of *genius* as grace in the Latin lexica remains distinct from its theological counterpart.

52. Martial, VI.61, quoted in Estienne 1543 and Calepino 1590, 1718.

53. Perotti 1513, col. 330, line 38ff.

54. Ibid.

55. *Ingenuus*, an especially rich term, yields a noun (*ingenuitas*) and an adverb (*ingenue*) in Perotti. One acts *ingenue* whose comportment is "as befits a free man, without fear, without anything to do with servility" (ingenue adverbium, hoc est libere, unde ingenue loqui dicimus eum, qui ita loquitur, ut liberum hominem decet, nihil dimidum, nihil servile habens).

56. E.g. Isidore of Seville, *Etymologiarum sive Originum libri XX*, ed. W.M. Lindsay (Oxford: Oxford University Press, 1911), at I.3, on *ingenium* as the natural origin of discourse, and I.3, on artless arguments.

57. Isidore of Seville, *Etymologiae*, XIX.20: "Sed hoc poetice fingitur; non enim Minerva istarum artium princeps est, sed quia sapientia in capite esse dicitur hominis, et Minerva de capite Iovis nata fingitur, hoc est ingenium; ideoque sensus sapientis, qui invenit omnia, in capite est. Ideo et dea artium Minerva dicitur quia nihil excellentius est ingenio, quo reguntur universa."

58. Balbus 1506: "notat ingenium esse intrinsecam vim anime et naturalem . . . unde et dicitur ingenium quasi intus genitum, scilicet a natura." The *ingenium* as inner power of the soul also features in Kahl 1600; Micraelius 1653; and Volckmar 1675.

59. Reuchlin 1478: "est interior vis animi quo sepe invenimus que ab aliis non didicimus." Reuchlin takes this definition from Balbus, who in turn took it from Papias. It also features in Perotti 1525.

60. Balbus 1506.

61. Calepino 1502, repeated in Estienne 1531: "a gignendo proprie natura dicitur cuique ingenita" Calepino likely adapted this from Perotti 1525: "a gignendo ingenium, quod proprie significat naturam, cuique ingenitam." This very stable definition also features in Dasypodius 1536; Kiel 1562; Estienne 1571; Kahl 1600; Megiser 1603; Fontecha 1606; Roboredo 1621; and Holyoake 1676.

62. This metonymic definition can be found in Nebrija 1492; Estienne 1531, 1543 and 1571, which include many new examples from Quintilian and Ovid; Frisius 1541; Kiel 1562; Calepino 1590; Kahl 1600; Fontecha 1606; Glocenius 1613; Alsted 1626; Volckmar 1675; and Holyoake 1676.

63. Sallust, *Historiarum fragm.*, III.15: "Castrisque collatis pugna tamen ingenio loci prohibebatur" (Once the forts came together, the character of the terrain nevertheless kept a battle from starting). The *ingenium loci* can be found in Estienne 1541; Frisius 1541; Estienne 1571; Calepino 1590; and Holyoake 1675. It features as an improper use of the term *ingenium* in Glocenius 1613.

64. Terence, *Andr.* 1, 1.: "Nam que cum ingeniis conflictatur eiusmodi, id est cum hominibus eiusdem natura."

65. Calepino 1502, repeated in Estienne 1531: "Est etiam ingenium vis quaedam naturalis nobis insita suis viribus praevalens ad inveniendum quod ratione iudicari possit." This definition or its variants also feature in Dasypodius 1536; Calepino 1590; Kahl 1600; Glocenius 1613; Micraelius 1653; Volckmar 1675; and Chauvin 1692.

66. Calepino 1502: "Una ingenii appellatur docilitas et memoria, easque virtutes que habent ingeniosi vocant." This slightly rephrases Cicero, *De finibus*, 5.13.36: " . . . docilitas, memoria; quae fere omnia appellantur uno ingenii nomine, easque virtutes qui habent, ingeniosi vocantur." The identification of *ingenium* with the cognitive power to learn and to memorize also features in Estienne 1531; Dasypodius 1536; Calepino 1590; Kahl 1600; Glocenius 1613; Micraelius 1653; Volckmar 1675; Chauvin 1692.

67. Plautus, *Bacch.* 5.42. Here we follow the tellingly loose translation of Estienne himself: "Immò ingenium avidi haud pernoram hospitis. *Ie ne cognoissoie point sa nature.*"

68. Plautus, *Milit.* 10.44: "Nam qui ipse amavit, amantis aegre ingenium inspicit."

69. Plautus, *Bacch.* 21.11: "Scio fecisse: eo ingenio est."

70. Terence, *Adelph.* 1.1.45–6: "Dum id rescitum iri credit, tantisper cavet: si sperat fore clàm, rursum ad ingenium redit."

71. Not all verbs, of course, imply development; e.g. one can also follow or have an *ingenium*.

72. By the edition of 1740, these lists take up more than two full columns for *ingenium* alone. Such lists also grow in the later editions of Estienne's *Latinogallicum dictionarium*.

73. Petrarch 2003, 4, 196, 244. For Petrarch and *ingenio*, see chapter 2.

74. Petrarca 1859, I.7, 57–64.

75. See this in relation to Castiglione in chapter 2.

76. "Vel ingenue vitam ducens, *Noblement vivant.*"

77. Terence, *Mil.*, 10.135: "Itidem divos dispertisse vitam humanam æquum fuit, qui lepide ingenuatus esset, vitam longinguam darent."

78. Kiel 1562.

79. The debates over the expansion of nobility across Europe at this time were at least in part responsible for this lexicographical concern. See introduction.

80. Kahl 1600.

81. The first, third, and fourth definitions are lifted verbatim from the expanded 1554 edition of the *Lexicon iuris civilis* by his predecessor, Jakob Spiegel. The first edition is Spiegel 1538.

82. Kahl 1600: "Ingenium ubi intenderis, valet" (When you meant a trick, he yet fares well), quotation from Sallust, *De coniur. Cat.* 51. This legal sense of *ingenium* can also be found in Du Cange 1678.

83. Kahl cites *Digest* 35.1.109, "Scaevola 20 Dig. A testatore rogatus, ut acceptis centum nummis restitueret hereditatem titiae coheredi suae, adita hereditate decessit: similiter et titia, antequam daret centum: quaesitum est, an heres titiae offerendo centum fidei commisso partem hereditatis consequi possit. respondit heredem condicioni parere non posse. claudius. magno ingenio de iure aperto respondit, cum potest dubitari an in proposito condicio esset." (Scaevola, 20 Dig. An heir, requested by the testator to accept one hundred pieces of money and to make over the estate to his co-heiress, Titia, died after taking the inheritance; so did Titia, before giving the hundred. The issue was whether Titita's heir, by offering the hundred of the *fideicommossum*, could acquire the portion of the inheritance; the reply was that the heir could not fulfill the condition. CLAUDIUS: From the perspective of the law in general he replied with great ingeniuity, as one can indeed doubt whether or not there was a condition in the original contract). Scaevola (d. 82 BC) wrote a long treatise on civil law that was compiled into the Justinian *Digest*.

84. See also Johnson 1755 on wit, in chapter 6.

85. The example is taken from a text on the old relationship between cunning and trade, the section *De commercio et mercato* in *Codex Just.*, 4.63.2: "Non solum aurum barbaris minime praebeatur, sed etiam si apud eos inventum fuerit, subtili auferatur ingenio. Si ulterius aurum pro mancipiis vel quibuscumque speciebus ad barbaricum fuerit translatum a mercatoribus, non iam damnis, sed suppliciis subiugentur, et si id iudex repertum non vindicat, tegere ut conscius criminosa festinat."

86. No similar meaning is found in the *Thesaurus linguae Latinae* (Leipzig and Berlin: Teubner & De Gruyter, 1894), vol. VII, pars 1, fasc. IX; for the medieval origins, cf. R.E. Latham, ed., *Dictionary of Medieval Latin from British Sources*, 17 vols. (London: Oxford University Press for the British Academy, 1975–2013).

87. Spelman 1626: "Pro *spectaculo*. Mat. Par. in An. *Ornata est civitas tota vexillis, coronis et pallis, cereis et lampadibus, et quibusdam prodigiosis ingeniis.*"

88. This meaning also features in Holyoake 1676.

89. Glocenius 1613: "varietas ingenii pendet tum ex temperamento corporis, tum

ex varia dispositione mentis, & constitutione organorum, & facultatum administratum, ut φανταστικης (imaginatrix)"; Chauvin 1692: "ingenii vires intendi a corpore."

90. E.g. Alsted 1626.

91. Aristotle, *Post. An.* 1.34 ; Aristotle, *Nic. Eth.* 6.9.

92. Glocenius 1613 had already criticized this same typology by limiting the *ingenium* to docility and invention only: good judgment and memory are consequences of a good *ingenium*, but not its proper expressions as such.

93. See Aristotle, *Problem.* 953a10–955b41. On Huarte de San Juan's *Examen de los ingenios* (1575), see chapter 3, The *Examen* was translated twice into Latin: once as the *De cultura ingeniorum* by the Jesuit Antonio Possevino, who integrated it into his pedagogical programme, the *Bibliotheca selecta* (1603). The second Latin translation is that of the German humanist Joachim Caesar, the *Scrutinium ingeniorum, pro iis qui excellere cupiunt* (1622).

94. See chapter 2.

95. The Galenic and Aristotelian typology of *ingenia* also features in Alsted 1626, which uses it to distinguish between degrees and modes of learning ability, in accordance with Antonio Possevino's pedagogical re-gearing of Huarte's typology in his own *Cultura ingeniorum* (1603). Aristotle's reference to Heracles's melancholy in Problem 30.1 is an early instance of the relationship between heroism and ingenuity (Aristotle, *Problem.* 953a14). Alongside the valorization of ingenuity, the early modern period witnessed the growth of the hero cult. This joint history is instantiated in the works of Baltasar Gracián, who wrote on both ingenuity and heroism with his *Agudeza y arte de ingenio* (first edition 1642) and *El héroe* (first edition 1637). On the Renaissance rebirth of the hero cult, see Randolph Starn, "Reinventing Heroes in Renaissance Italy," *The Journal of Interdisciplinary History* 17, no. 1 (1986): 67–84.

96. In fact, Chauvin ascribes to Alexander of Aphrodisias the identification of *ingenium* with the agent intellect. Although this identification does not in fact feature in Alexander's commentary on Aristotle's *De anima*, by mentioning it Chauvin highlights the philosophical valence of *ingenium* as an individuating principle alongside the agent intellect at the end of the seventeenth century.

97. The key philosophical issue at stake here is the immortality of the soul, on which see e.g. Eckhard Kessler, "The Intellective Soul," in *The Cambridge History of Renaissance Philosophy*, ed. Charles B. Schmitt et al. (Cambridge: Cambridge University Press, 1988).

2 Italian

1. Persio 1576, 12. On Persio and his sources, see Persio 1999. For a brief account of the (largely unstudied) content of the *Trattato*, see Gensini, "Ingenium/ingegno." See also Brann, *The Debate*, 309–11, who mistranslates Persio's *ingegno* as *genius*.

2. Persio 1576, 12. The correct quotation is Cicero, *De finibus bonorum et malorum*, V.13 ("prioris generis est docilitas . . ."). The misquotation appears in e.g. Calepino 1558

and is quoted also in Huarte 1594. Cicero's definition also appears in the definition of *ingegno* in Minerbi's Italian translation of Calepino (Calepino 1553).

3. Persio 1576, 13 (quoting Boccaccio, *Decameron*, IV.I.3): "... né mai lo prendiamo per natura, o radissime volte, come fu preso da' migliori della nostra lingua, *Tancredi prencipe di Salerno fu signore assai humano, et di benigno ingegno*" (... nor do we ever take it for nature, or at least very rarely, as it has been taken by the best in our language, *Tancred, prince of Salerno, was a very humane man, and of a benign nature*).

4. Ibid., quoting Terence, *Andr.*, 93; Paterculus, *Hist. Rom.*, II.36; Petrarch, *Canzoniere*, CCCLX.20.

5. On *spiritus* see Marta Fattori and Massimo Bianchi, eds., *Spiritus. IVo Colloquio Internazionale del Lessico Intellettuale Europeo* (Rome: Edizioni dell'Ateneo, 1984).

6. See Robert Klein, "Spirito Pelegrino," in idem., *Form and Meaning: Writings on the Renaissance and Modern Art*, trans. Madeline Jay and Leon Wiseltier (Princeton: Princeton University Press, 1979); Richard Spear, *The "Divine" Guido: Religion, Sex, Money, and Art in the World of Guido Reni* (New Haven and London: Yale University Press, 1997); Michael Cole, "The Demonic Arts and the Origin of the Medium," *The Art Bulletin* 84, no. 4 (2002): 621–40; Idem., *Cellini and the Principles of Sculpture* (Cambridge: Cambridge University Press, 2004); Christine Göttler and Wolfgang Neuber, eds., *Spirits Unseen: The Representation of Subtle Bodies in Early Modern European Culture* (Leiden: Brill, 2008); Henry Keazor, "'Spirito abile' ed 'elevatissimo ingegno.' Giovan Pietro Bellori e Carlo Cesare Malvasia," *Mitteilungen des Kunsthistorischen Institutes in Florenz* 52, no. 1 (2008): 73–82; David Young Kim, *The Traveling Artist in the Italian Renaissance* (New Haven and London: Yale University Press, 2014).

7. As *ingegno*, see the entries for *spirito* in Marinelli 1565; Crusca 1612; as *genio*, see the entry for *spirito* in Alunno 1539b.

8. See chapter 1.

9. On this aspect of *genio*, see e.g. Klibansky, Panofsky, and Saxl, *Saturn and Melancholy*; Brann, *The Debate*.

10. There is a large literature on this topic. Key studies that touch upon the role of *ingegno* (and, to a lesser extent, *genio*) in the discourse on artistic character and artistic freedom include Erwin Panofsky, "Artist, Scientist, Genius: Notes on the Renaissance *Dämmerung*," in idem., *The Renaissance: Six Essays* (New York: Harper & Row, 1962); Baxandall, *Giotto and the Orators*; idem, "A Dialogue on Art from the Court of Leonello d'Este," *Journal of the Warburg and Courtauld Institutes* 26 (1963): 304–26; Martin Kemp, "From 'Mimesis' to 'Fantasia': The Quattrocento Vocabulary of Creation, Inspiration and Genius in the Visual Arts," *Viator* 8 (1977): 347–98; idem, "The 'Super-Artist' as Genius: The Sixteenth-Century View," in *Genius: The History of an Idea*; David Summers, *Michelangelo and the Language of Art* (Princeton: Princeton University Press, 1981); idem., *The Judgment of Sense: Renaissance Naturalism and the Rise of Aesthetics* (Cambridge: Cambridge University Press, 1987); Patricia Emison, *Creating the "Divine" Artist from Dante to Michelangelo* (Leiden: Brill, 2004); Pari Riahi, *Ars et ingenium: The*

Embodiment of Imagination in Francesco di Giorgio Martini's Drawings (Abingdon: Routledge, 2015). On the emergence of artisans as liberal artists more broadly, see e.g. Rensselaer W. Lee, *Ut pictura poesis: The Humanistic Theory of Painting* (New York and London: W. W. Norton, 1967); Anthony Grafton and Lisa Jardine, eds., *From Humanism to the Humanities: Education and the Liberal Arts in Fifteenth- and Sixteenth-Century Europe* (Cambridge, MA: Harvard University Press, 1986); Martin Warnke, *The Court Artist: On the Ancestry of the Modern Artist* (Cambridge: Cambridge University Press, 1993); Pamela O. Long, *Openness, Secrecy, Authorship: Technical Arts and the Culture of Knowledge from Antiquity to the Renaissance* (Baltimore and London: The Johns Hopkins University Press, 2001).

11. See e.g. Summers, *Michelangelo*; Danilo Aguzzi-Barbagli, "*Ingegno, acutezza*, and *meraviglia* in the Sixteenth Century Great Commentaries to Aristotle's *Poetics*," *Petrarch to Pirandello: Studies in Italian Literature in Honour of Beatrice Corrigan*, ed. Julius A. Molinaro (Toronto: University of Toronto Press, 1973); Martin Kemp, "'Equal excellences': Lomazzo and the Explanation of Individual Style in the Visual Arts," *Renaissance Studies* 1, no. 1 (1987): 175–200; Robert Williams, *Art, Theory, and Culture in Sixteenth-Century Italy. From Techne to Metatechne* (Cambridge: Cambridge University Press, 1997); Patricia Emison, "Grazia," *Renaissance Studies* 5, no 4 (1997): 427–60; Philip Sohm, *Style in the Art Theory of Early Modern Italy* (Cambridge: Cambridge University Press, 2001); Stefano Gensini, "L'ingegno e le metafore: alle radici della creatività linguistica fra Cinque e Seicento," *Studi di estetica* 25, no. 16 (1997): 135–62; Ita Mac Carthy, "Grace and the 'Reach of Art' in Castiglione and Raphael," *Word and Image* 25, no. 1 (2009): 33–45.

12. The role of *ingegno* in these authors has been insufficiently explored. See e.g. for Alberti, Emison, *Creating the "Divine" Artist*; for Castiglione, Fiorella di Stefano, "Ingegno et Sprezzatura du Cortegiano dans la traduction de l'Abbé Duhamel (1690)," in *Traduire en français à l'âge classique. Génie national. Génie des langues*, ed. Y. M. Tran-Gervat, (Paris: Presses Sorbonne Nouvelle, 2013), 139–51; for Michelangelo, Irving Lavin, "David's Sling and Michelangelo's Bow: A Sign of Freedom," in idem., *Past-Present: Essays on Historicism in Art from Donatello to Picasso* (Berkeley: University of California Press, 1993), 29–61; Summers, *Michelangelo*; Emison, *Creating the "Divine" Artist*; for Vasari, Andrew Steptoe, *Artistic Temperament in the Italian Renaissance: A Study of Giorgio Vasari's Lives*, in *Genius and the Mind: Studies of Creativity and Temperament*, ed. Andrew Steptoe (Oxford: Oxford University Press, 1998).

13. For *ingegno* in Tesauro, see Ezio Raimondi, "Ingegno e metafora nella poetica del Tesauro," *Verri* 2, no. 2 (1958): 53–75; Denise Aricò, "Prudenza e ingegno nella *Filosofia morale* di Emanuele Tesauro," *Studi Secenteschi* 42 (2001): 187–208. For Vico, see Stefano Gensini, "Ingenium/ingegno fra Huarte, Persio e Vico: le basi naturali dell'inventività umana," in Gensini and Martone, *Ingenium*; Marshall, *Vico*.

14. See Werner L. Gundersheimer, "Bartolommeo Goggio: A Feminist in Renaissance Ferrara," *Renaissance Quarterly* 33, no. 2 (1980): 175–200; Frederika H. Jakobs,

"Woman's Capacity to Create: The Unusual Case of Sofonisba Anguissola," *Renaissance Quarterly* 47, no. 1 (1994): 74–101; Philip Sohm, "Gendered Style in Italian Art Criticism from Michelangelo to Malvasia," *Renaissance Quarterly* 48, no. 4 (1995): 759–808.

15. See John R. Hale, *Renaissance Fortification: Art or Engineering?* (London: Thames & Hudson, 1977); Simon Pepper and Nicholas Adams, *Firearms and Fortifications: Military Architecture and Siege Warfare in Sixteenth-Century Siena* (Chicago: University of Chicago Press, 1986); Alexander Marr, *Between Raphael and Galileo*.

16. For the semantics of the language of early modern engineering in context, see Hélène Vérin, *La gloire des ingénieurs*.

17. See Brian Richardson, "The Concept of a *lingua comune* in Renaissance Italy," in *Languages of Italy: Histories and Dictionaries*, ed. Arturo Tosi and Anna Laura Lepschy (Ravenna: Longo, 2007); Paola Gambarota, *Irresistible Signs: The Genius of Language and Italian National Identity* (Toronto: University of Toronto Press, 2011). Given this context it is surprising that early modern Italian lexicographers did not define *genio* as the nature of a given language.

18. For a useful summary, see Claudio Marrazzini, "Questione della lingua," in *Enciclopedia dell'Italiano* (2011) (Treccani online: http://www.treccani.it/enciclopedia/questione-della-lingua_(Enciclopedia-dell%27Italiano)/). Important overviews include Robert A. Hall, *The Italian 'Questione della Lingua': An Interpetative Essay* (Chapel Hill: University of North Carolina Press, 1942); Maurizio Vitale, *La questione della lingua*, 2nd edn (Palermo: Palumbo, 1970). See also the essays in Tosi and Lepschy, *Languages of Italy*.

19. Robert Hastings, "Questione della lingua," in *The Oxford Companion to Italian Literature*, ed. Peter Hainsworth and David Robey (Oxford: Oxford University Press, 2002). Retrieved 8 March 2018, http://www.oxfordreference.com/view/10.1093/acref/9780198183327.001.0001/acref9780198183327-e-2629.

20. See Riccardo Drusi, *La lingua "cortigiana romana." Note su un aspetto della questione cinquecentesca della lingua* (Venice: Il Cardo, 1997); Claudio Giovanardi, *La teoria cortigiana e il dibattito linguistico nel primo Cinquecento* (Rome: Bulzoni, 1998); Caterina Mongiat Farina, *Questione della lingua: L'ideologia del dibattita sul italiano nel Cinquecento* (Ravenna: Longo, 2014).

21. There is a large literature on the Accademia della Crusca and its dictionaries. An excellent summary is provided in John Considine, *Academy Dictionaries 1600–1800* (Cambridge: Cambridge University Press, 2014), chapter 2.

22. Ibid., 17.

23. Considine, *Academy Dictionaries*, 19.

24. In this respect, Crusca 1612 may be compared to the copious confusion of predecessors such as Alunno 1543, Marinelli 1565, and Pergamino 1602.

25. See Andrea Bocchi, "I Florio contro la Crusca," in *La nascita del vocabolario. Convegno di studio per i quattrocento anni del Vocabolario della Crusca*, ed. Antonio Daniele and Laura Nascimben (Padua: Esedra, 2014); Desmond Connor, *A History of*

Italian and English Bilingual Dictionaries (Florence: L. S. Olschki, 1990). Florio 1611 was even longer than the *Vocabolario*: 70,000 entries compared to 24,595 in Crusca 1612.

26. Little work has been done of the history of the word *genio* in the early modern period, but see Teresa Graviana, "Breve storia della parola 'genio,'" *Lingua nostra* 28, no. 2 (1967): 37–43; idem., "Genio/ingegno," special issue of *Studi di estetica* 25 no. 16 (1997): 135–62; Valentina Conticelli, "Prometeo, Natura e il Genio sulla volta dello Stanzino di Francesco I: fonti letterarie, iconografiche e alchemiche," *Mitteilungen des Kunsthistorischen Institutes in Florenz* 46, no. 2/3 (2002): 321–56.

27. "Quel santo precettor, quell'alma guida, [/] Genio appellato, il qual come ministro [/] Della ragion lo sproni al bene oprare, [/] E dall'opere ingiuste il tiri, e frene" (That spiritual preceptor, which guides the soul, is called Genius, which as a minister of reason prompts it to do good, and draws it away from wrongdoing, and reins it in).

28. On the shift from genius as natural inclination to genius as "will" (*voluntas*) see chapter 1.

29. See chapter 1.

30. Emphasis added. It is not clear whether the festive sense of *geniale* lies behind this entry. See chapters 1 and 6.

31. "Andare. Andare a genio. Andare all'animo, andare a cuore, andare a genio, andare a sanguem, ec. Aver genio, indursi di buona voglia, far volentieri. Latino *placere, arridere*. Andare, come la biscia all'incanto, a che che sia, vale Farlo malvolentieri, e contra genio. Figuratam. Far che sia con allegrezza, di buona voglia, di genio. Latino *libenter agere*. Andare a stomaco. Lo stesso, che Andare a genio, a sangue: Confarsi. Latino *arridere, placere*." Crusca 1691.

32. See also Florio's definition of the term in chapter 6.

33. See chapter 1.

34. On *ars* see e.g. Paul O. Kristeller, "The Modern System of the Arts," in idem., *Renaissance Thought and the Arts: Collected Essays* (Princeton: Princeton University Press, 1990).

35. See e.g. Kemp, "The 'Super-Artist,'" 49: "all art theorists during the fifteenth and sixteenth centuries [insisted] that rational learning, studious application are absolutely essential if inborn talent is to reach fruition."

36. Summers, *The Judgment of Sense*, 99.

37. See Baxandall, *Giotto*; Kemp, "From 'Mimesis' to 'Fantasia'; idem., "The 'Super-Artist'; Summers, *Michelangelo*. See also Emison, *Creating the "Divine" Artist*, 325–33 (for a critique of Baxandall), 335–36 (for a curious reading of Summers). Notably, Ghiberti uses *ingegno* as a term of praise for artists in his *Commentarii*. See Jim Harris, "Lorenzo Ghiberti and the Language of Praise," *Sculpture Journal* 26, no 1 (2017): 107–18.

38. Baxandall, *Giotto*, 74. The same passage is cited by Baldinucci in his entry for *stile* (Baldinucci 1681). Notably, in his translation of Calepino, Minerbi quotes Boccaccio (using the classical topos) to distinguish *arte* from *ingegno*: *Ingegno*[,] in vece d'astutia.

Boccaccio, Con arte, & ingegno. Vedi astutia" (Calepino 1553). See also Castiglione's comments on Boccaccio, quoted below.

39. The word *ingegno* appears in the definitions (but not as headword) slightly earlier, in Anon. 1435–60: "Ingeniolum . . . poco o pizolo inzegno."

40. *Vocabolario italiano-latino: edizione del primo lessico del volgare: secolo XV*, ed. Federico Pelle (Florence: L. S. Olschki, 2001). It is notable that Tranchedini includes *perspicax* as a synonym. There is a minor tradition in which *ingegno* is associated with the eyes and perspicacity. For example, Scoppa 1511/15 gives for *ingegno*: "s.v. acies, ei. visola, lucziola, pupilla, pronella delo occhio, *ponitur pro oculo*, occhio; *lata acies*: ponta, taglio de ogni strumento ferreo; exercito parato ad combattere, *et* quando combatte, la schera, squatrone, compagnia, *ordinanza militum; hic acies certare solebat*: scaramuza, pugna, combattere; *acie dicitur ab acumine telorum*: la suttilita, perspicacita de ingegno." Calderino Mirani 1586 defines *ingegnoso* as "accorto," which is connected to the "acute" senses of the word, for which see below.

41. In this wide-ranging and suggestive list, the appearance of *gratia* and *diligentia* reflect the connection of these terms to *ingenium* in certain ancient writings (e.g. Pliny), which informed the reception of the words/concepts by Italian humanists. See e.g. Emison, *Creating the "Divine" Artist*, esp. chapter 1. Notably, *industria* is defined as "diligenza ingegnosa" in Crusca 1612.

42. The definition of *ingegno* as *ars/arte* is regularly repeated up to Crusca 1612 (from which it is absent). See Marinelli 1565; Ruscelli 1588; Florio 1598; Pergamino 1602; Florio, 1611 (repeated in Torriano, 1659); For *ingegno* as *industria*, see Ulloa 1553 ("[Ingegno] Maña. per industria et ingenio"); Bevilacqua 1567; Venuti 1576; Calderino Mirani 1586; Ruscelli 1588; Pergamino 1602. Calderino Mirani 1586 [1730] defines *ingegnarsi* as "studeo, operamdo." While the definition of *ingegno* as "sense" is not common, *ingegno* occurs regularly in entries for *senno* as a headword.

43. The variant spelling of *ingegno* as *inzegno* is attested in several dialects, including Tuscan.

44. "[I]ngenium industria alitu." Cicero, *Pro caelio*, 19.45. This presumably informed Venuti 1561: "ingegno, discorso naturale] hoc Ingenium . . . Haec industria . . . Cicero."

45. Tranchedini may also have had in mind the association of *genio* with study, attested later in Luna (1536). See below. Matteo Palmieri, *Libro della vita civile* (1529), quoted in Emison, *Creating the "Divine" Artist*, 74.

46. See chapter 1. However, on the relationship between humanism and machinery, see Jessica Wolfe, *Humanism, Machinery and Renaissance Literature* (Cambridge: Cambridge University Press, 2004).

47. See Summers, *Judgment*, 136–7; Kemp, "From 'Mimesis' to 'Fantasia,'" 389.

48. Vitruvius, *De architectura*, I.1.3. Alberti's contemporary, the architect-engineer Francesco di Giorgio Martini, treated *ingegno* in a similar fashion in his *Trattati*. See Riahi, *Ars et ingenium*, where *ingegno* is mistakenly translated as "creativity" rather than "talent" (57). While this notion is to some extent implicit in Francesco's writings, this

meaning does not come to the fore in the Italian vernacular until significantly later. On Vitruvius and *ingenium/ingegno*, see also Emison, *Creating the "Divine" Artist*, 20

49. David Summers, "Pandora's Crown: On Wonder and Imitation in Western Art," in *Wonders, Marvels, and Monsters in Early Modern Culture*, ed. Peter G. Platt (Newark: University of Delaware Press, 1999), 67. Brunelleschi was thus considered a modern who had equaled or even surpassed the "divine" *ingenium* of Archimedes (as Cicero put it): one of the great engineers of antiquity, whose fortunes were also in the process of revival during the Renaissance. See W. Roy Laird, "Archimedes amongst the Humanists," *Isis* 82, no. 4 (1991): 629–38; Emison, *Creating the "Divine" Artist*, 7, 19, 73–5, 240, 288; Matteo Burioni, *Die Renaissance der Architekten. Profession und Souveränität des Baukünstlers in Giorgio Vasaris Viten* (Berlin: Gebr. Mann Verlag, 2008), 109–14.

50. Patent of Federico da Montefeltro, Count of Urbino, on behalf of Luciano Laurana, 10 June 1468. Published in David S. Chambers, ed., *Patrons and Artists of the Italian Renaissance* (London: Macmillan, 1970), 164–66. See also Marr, *Between Raphael and Galileo*, 30ff.

51. This sense is presumably why the Sicilian *ingegnu* is translated as "argumentum" in Scobar 1519, although it is presumably connected also to the *argu*-stemmed Latin words (e.g. *argutus*) with which *ingegno* was routinely associated.

52. See chapter 1.

53. See, for example, the debate concerning the sources of *arguzie* in Castiglione 1528, II. 43. See also chapter 3.

54. These are not by any means the only examples of the word in the works of the *Tre Corone*. Liburnio 1526, for example, offered different quotations: "[Dante] Qual. Tutto (quel che sia sia) il mio ingegno. Paradiso 22 [113–14]." "[Petrarch] Ingegno. M'accorsi di vostr'ingegno, per fur voi certo. Sonnetto [*Rime*, "Quelle pietose rime in ch'io m'accorsi]." The index of Alunno 1539a offers dozens of entries in Petrarch for *ingegni*, *'ngegno nome, ingegno, mio ingegno, ogni ingegno, qual ingegno*, and *ingegnose*; see also Alunno 1539b. Likewise, Alunno 1543 offers multiple instances of *ingegno* in Boccaccio.

55. On medieval and Renaissance commentaries on Dante, see Deborah Parker, *Commentary and Ideology: Dante in the Renaissance* (Durham, NC: Duke University Press, 1993); Simon Gilson, *Reading Dante in Renaissance Italy: Florence, Venice and the Divine Poet* (Cambridge: Cambridge University Press, 2018). On *ingegno* in Dante, see Mario Trovato, "The Semantic Value of Ingegno and Dante's Ulysses in the Light of the Metalogicon," *Modern Philology* 84, no. 3 (1987): 258–66; Paul Arvisu Dumol, *The Metaphysics of Reading Underlying Dante's Commedia: The Ingegno* (New York: Peter Lang, 1998). The latter, which proposes *ingegno* as the "key" to the *Commedia*, should be treated with caution

56. Buti 1858, 60. For the importance of Papias and his definition see also chapter 3.

57. "Val diligentemente, cercar con ingegno" (Calepino 1553); "Scrutor, prevalersi con l'accortezza, investigare, addestrarsi ad alcuna cosa" (Sansovino 1568).

58. Landino 1487, unpaginated [under "Canto secondo della prima cantica"].

59. On this much studied phenomenon, see now Emison, *Creating the "Divine" Artist*; Gilson, *Reading Dante*. A particularly suggestive account, drawn from seventeenth-century artistic theory, is Olivier Bonfait, "'Ingegno divino' o 'beauté du génie': Bellori, Félibien e il 'super-artist' nel Seicento," in *Begrifflichkeit, Konzepte, Definitionen. Schreiben über Kunst und ihre Medien in Giovanni Pietro Belloris Viten und der Kunstliteratur der Fühen Neuzeit*, ed. Elisabeth Oy-Marra, Marieke von Bernstorff, and Henry Keazor (Wiesbaden: Harrassowitz, 2014).

60. Ripa 1603, 178–79. I translate "numeri" as "gods" based on "nùme" (god; divinity).

61. On the *furor poeticus*, see Jean Lecointe, *L'idéal et la différence: la perception de la personnalité littéraire à la Renaissance* (Geneva: Libraire Droz, 1993); Liane Nebes, *Der furor poeticus im italienischen Renaissanceplatonismus. Studien zu Kommentar und Literaturtheorie bei Ficino, Landino und Patrizi* (Marburg: Tectum-Verlag, 2001); Brann, *The Debate*.

62. The association of Dante with the poetic inspiration implied by the phrase "alto ingegno" was contested. Leonardo Bruni, for example, claimed Dante as a liberal artist who worked through study and diligence rather than through *ingegno* and *furore*. See Emison, *Creating the "Divine" Artist*, 79.

63. In the same dictionary, the diminutive *ingegnevole* is defined as "voce antiqua ... ghiribizzo, astuta invenzione." See Maria Pia Mannini, "Chimere, capricci, ghiribizzi e altre cose. Esempi periferici di grottesca del tardo Cinquecento," *Annali / Fondazione di Studi di Storia dell'Arte Roberto Longhi* 1 (1984): 71–86; Sabine Eiche, ed., *I Gheribizzi di Muzio Oddi* (Urbino: Accademia Raffaello, 2005). *Ghiribizzo* is very close to *arzigogolo*, given a synonym for *ingegno* in Ménage 1669 and defined in Crusca 1623 as "Invenzione sottile, e fantastica. Latino Inventum."

64. On the early modern *capriccio*, see Francesco Paolo Campione, *La regola del capriccio. Alle origini di un'idea estetica* (Palermo: Centro internazionale studi di estetica, 2011); Roland Kanz, *Die Kunst des Capriccio. Kreativer Eigensinn in Renaissance und Barock* (Munich: Deutscher Kunstverlag, 2002).

65. Aquinas, *Physicorum Aristoteles Commentaria*, 2.13, quoted in Kemp, "From 'Mimesis' to 'Fantasia,'" 355.

66. See Felicia M. Else, "Ammanati's Shield of Achilles: Making a Virtue out of Necessity," *Source: Notes in the History of Art* 28, no. 1 (2008): 30–38.

67. Kemp, "From 'Mimesis' to 'Fantasia,'" 354.

68. See e.g. Marinelli 1565: "Ingegno. Vale senno, intelletto, avedimento, valore, & virtù." Kemp insightfully connects *virtù* to *ingegno* in "The 'Super Artist' as Genius," 34. On Marino, Marinism and the marvelous, see e.g. James V. Mirollo, *The Poet of the Marvelous, Giambattista Marino* (New York: Columbia University Press, 1963); Christine Ott, "Terribile meraviglia: animismo artistico, empatia ed ekplexis nella poesia di Giovan Battista Marino." *MLN* 130, no. 1 (2015): 63–85. For the wider aesthetic terrain of *meraviglia*, see Summers, "Pandora's Crown."

69. We may note that among his several synonyms for *ingegno*, Florio provides "fantasy, imagination" (Florio 1598). On the vocabulary and semantics of fantasy, see e.g. Kemp, "From 'Mimesis' to 'Fantasia.'"

70. On Dante and Florence see Simon Gilson, *Dante and Renaissance Florence* (Cambridge: Cambridge University Press, 2005).

71. On which see Bruni, noted above, n. 62.

72. *Sprezzatura* first occurs in the definition of *naturalezza* in Crusca 1691, before becoming a headword in Crusca 1729–38. For *sprezzatura*'s connection to *ingegno*, see Emison, *Creating the "Divine" Artist*, 48.

73. Castiglione 1556, 6v. There is a growing literature on Boccaccio and *ingenium*. See, most recently, Michaela Paasche Grudin and Robert Grudin, *Boccaccio's Decameron and the Ciceronian Renaissance* (New York: Palgrave, 2012). On the *questione della lingua* in the *Cortegiano*, see Uberto Motta, "La 'questione della lingua' nel primo libro del Cortegiano: dalla seconda alla terza redazione," *Aevum* 72, no. 3 (1998): 693–732.

74. Games and game-players were considered to be especially *ingenious*. See Alison Levy, ed., *Playthings in Early Modernity: Party Games, Word Games, Mind Games* (Kalamazoo: Medieval Institute Publications, 2017).

75. Intriguingly, it merited a place in Emanuele Tesauro's *Vocabulario Italiano* when the more important term in his poetics—*ingegno*—did not. See Tesauro 2008.

76. See chapter 6.

77. Ménage cited Charles de Bovelles in defence of his derivation of *inganno* from *ingenium*: "Inganno. Da *ingenium*. Carlo Bovillo nelle Origini delle Voci Francesi a 59. Enigner, id est fallere; ab ingenio. Nella nostra lingua non se ne può dubitare." *Inganno* likely derives from the post-classical Latin *engannum*, which may itself derive from *ingenium*. See also chapter 3 for the Spanish *engaño*.

78. While *ingegno* occurs in the definitions for *sottile* and *sottigliezza* as headwords, the reverse is the case only in Italian-French lexica, e.g. Canal 1603: "Ingegno. Finesse, subtilité."

79. This sense is emphasized in Crusca 1612, which quotes Bartolomeo di San Concordio: "Inganno è una insidiosa malizia, quando alcuno si fa involar la cosa, che gli è stata data in serbanza."

80. Notably, the verb *ingegnare* is defined in Crusca 1612 as "gabbar con doppiezza. Latino decipere, fraudare" (to swindle dishonestly, deceive, defraud).

81. See chapter 6.

82. See Paul Barolsky, *Infinite Jest. Wit and Humor in Italian Renaissance Art* (Columbia: University of Missouri Press, 1978); Paul Barolsky and Andrew Ladis, "The Pleasurable Deceits of Bronzino's So-called London 'Allegory,'" *Source: Notes in the History of Art* 10, no. 3 (1991): 32–36.

83. *Ingegnoso* is defined as *vezzoso* ("very beautiful") in Crusca 1691.

84. See e.g. Summers, "Pandora's Crown"; R. J. W. Evans and Alexander Marr, eds., *Curiosity and Wonder from the Renaissance to the Enlightenment* (Aldershot: Ashgate,

2006); Alexander Marr, "Understanding Automata in the Late Renaissance," *Journal de la Renaissance* 2 (2004): 205–22; Sharon Gregory and Sally Anne Hickson, eds., *Inganno—The Art of Deception* (Aldershot: Ashgate, 2012).

85. The relationship is certainly present in non-lexicographical sources, however. Georgia Clarke has noted the relationship between *artificio* (as "beautiful work of art"), *lavoro* and *ingegno* in "'La più bella e meglio lavorata opera': Beauty and Good Design in Italian Renaissance Architecture," in *Concepts of Beauty in Renaissance Art*, ed. Francis Ames-Lewis and Mary Rogers (Aldershot: Ashgate, 1998), 111.

86. Florio 1598 captures this sense well with "a toole, a devise, an artifice, an invention, an implement."

87. On which see Marr, *Between Raphael and Galileo*, chapter 5.

88. There is ample evidence that artefacts associated with the arts of war were considered especially ingenious and highly prized by Italian elite. Moreover, artists and art theorists traded on the fact that Vulcan forged the fabled arms of Achilles. See e.g. Summers, "Pandora's Crown"; Else, "Ammanati's Shield of Achilles." On the early modern mind-hand relationship, see Lissa L. Roberts, Simon Schaffer, and Peter Dear, eds., *The Mindful Hand: Inquiry and Invention from the Late Renaissance to Early Industrialisation* (Chicago: University of Chicago Press, 2008). In connection to *ingegno*, see Hana Gründler, "'Gloriarsi della mano e dell'ingegno': Hand, Geist und pädagogischer Eros bei Vasari und Bellori," in *Begrifflichkeit, Konzepte, Definitionen*, ed. Oy-Marra et al., 77–103.

89. It seems likely that in amplifying Florio's definition, Torriano drew on the entries in Covarrubias's dictionary. See chapter 3.

3 Spanish

1. Cited from Thomas Shelton's 1612 English translation; Cervantes 1612, 5. Note the addition of the expressions "artificiall manner" and "invention." The original Spanish version reads as follows: "Y lo primero que hizo fue limpiar unas armas que habían sido de sus bisabuelos, que, tomadas de orín y llenas de moho, luengos siglos había que estaban puestas y olvidadas en un rincón. Limpiólas y aderezólas lo mejor que pudo; pero vio que tenían una gran falta, y era que no tenían celada de encaje, sino morrión simple; mas a esto suplió su industria, porque de cartones hizo un modo de media celada que, encajada con el morrión, hacían una apariencia de celada entera. Es verdad que, para probar si era fuerte y podía estar al riesgo de una cuchillada sacó su espada y le dio dos golpes, y con el primero y en un punto deshizo lo que había hecho en una semana; y no dejó de parecerle mal la facilidad con que la había hecho pedazos, y, por asegurarse deste peligro, la tornó a hacer de nuevo, poniéndole unas barras de hierro por de dentro, de tal manera, que él quedó satisfecho de su fortaleza y, sin querer hacer nueva experiencia della, la diputó y tuvo por celada finísima de encaje." Cervantes 1998, 41.

2. For a discussion of this passage, with bibliography, see Cervantes 1998, 41;

Verónica Vivanco, "El Quijote y la industria," *Destiempos. Revista de Curiosidad Cultural* 4, no. 20 (2009): 26–40.

3. Cervantes 1687, 4 (emphasis our own). Note the interesting allusions to invention, workmanship and the biblical smith Tubal-Cain. In John Stevens's 1700 translation the word *industria* in this passage is translated as "ingenuity." On Philips's singular translation, and other early modern English translations of Cervantes's novel see e.g. Carmelo Cunchillos Jaime, "Traducciones inglesas del 'Quijote' (1612–1800)," in *De clásicos y traducciones. Versiones inglesas de clásicos españoles (ss. XVI-XVII)*, ed. Julio-César Santoyo and Isabel Verdaguer (Barcelona: Promociones y Publicaciones Universitarias, 1987); J. A. G. Ardila, ed., *The Cervantean heritage. Reception and influence of Cervantes in Britain* (Oxford: Legenda, 2009).

4. The literature on Cervantes's characterization of Don Quijote as *ingenioso* is large. Some of the early important works include: Rafael Salillas, *Un gran inspirador de Cervantes: El Dr. Juan Huarte y su "Examen de ingenios"* (Madrid: Eduardo Arias, 1905); Mauricio de Iriarte, *El doctor Huarte de San Juan y su Examen de ingenios. Contribución a la historia de la psicología diferencial* (Madrid: CSIC, 1948); Harald Weinrich, *Das Ingenium Don Quijotes. Ein Beitrag zur literarischen Charakterkunde* (Munich: Aschendorf, 1956); Otis H. Green, "El Ingenioso Hidalgo," *Hispanic Review* 25, no. 3 (1957): 175–93; Francisco Maldonado de Guevara, "Del 'Ingenium' de Cervantes al de Gracián," *Revista de estudios políticos*, no. 100 (1958): 147–66. For further information, with bibliography, see e.g. Nobuaki Ushijima, "Sobre los títulos del *Quijote*. La función del ingenio," in *Actas del Tercer Coloquio Internacional de la Asociación de Cervantistas* (Barcelona: Anthropos, 1993), 325–29; Cervantes 1998; Guillermo Serés, "Don Quijote, ingenioso," in *Los rostros de Don Quijote. IV Centenario de la publicación de su primera parte*, ed. Aurora Egido (Zaragoza: IberCaja, 2004); Fernando Bouza and Francisco Rico, "Digo que yo he compuesto un libro intitulado El ingenioso hidalgo de la mancha," *Cervantes: Bulletin of the Cervantes Society of America* 29, no. 1 (2009): 13–30; Aurora Egido, "La fuerza del ingenio y las lecciones cervantinas," *Boletín de la Real Academia Española*, t. 96, cuaderno 314 (2016): 771–794. On the challenge of translating the adjective *ingenioso* in the title, see Germán Colón, *Las primeras traducciones europeas del Quijote* (Barcelona: Universitat Autònoma de Barcelona, 2005), 41–45. Interestingly, at least three of the authors in charge of translating *Don Quijote* into other European vernaculars were also involved in the production of dictionaries: Cesar Oudin, Lorenzo Franciosini and the above-mentioned John Stevens.

5. Huarte de San Juan 1575 and 1594. In addition to the references listed in note 4, on Huarte de San Juan's *Examen* see e.g. C. M. Hutchings, "The *Examen de Ingenios* and the Doctrine of Original Genius," *Hispania* 19, no. 2 (1936): 273–82; Esteban Torre, *Ideas lingüísticas y literarias del Doctor Huarte de San Juan* (Seville: Universidad de Sevilla, 1977); Guillermo Serés, "Introducción," in *Examen de ingenios para las ciencias*, by Juan Huarte de San Juan, ed. Guillermo Serés (Madrid: Cátedra, 1989); Guillermo Serés, "El

ingenio de Huarte y el de Gracián. Fundamentos teóricos," *Ínsula*, no. 655–656 (2001): 51–53; Véronique Duché-Gavet, ed., *Juan Huarte au XXIe siècle* (Anglet: Atlantica, 2003); Gambin, *Azabache*; Rocío Sumillera, "Introduction," in *The examination of men's wits*, by Juan Huarte de San Juan, trans. Richard Carew (London: Modern Humanities Research Association, 2014), 1–67.

6. See, in particular, his treatise *Arte de ingenio, tratado de la agudeza* (1642) and its expanded version *Agudeza y arte de ingenio* (1648). The literature on Gracián and his work on *ingenio* is extensive. See, among other works, Blanco, *Rhétorique de la Pointe*; Emilio Hidalgo-Serna, *El pensamiento ingenioso en Baltasar Gracián* (Barcelona: Anthropos, 1993); Emilio Blanco, "Introducción," in *Arte de ingenio, tratado de la agudeza*, by Baltasar Gracián, ed. Emilio Blanco (Madrid: Cátedra, 1998); Guillermo Serés, "El ingenio en Gracián: de la invención a la elocución," in *Baltasar Gracián IV Centenario (1601–2001). Actas I Congreso Internacional "Baltasar Gracián: pensamiento y erudición,"* ed. Aurora Egido, Fermín Gil Encabo, and José Enrique Laplana Gil (Huesca-Zaragoza: Instituto de Estudios Altoaragoneses; Institución Fernando el Católico, 2003); Jorge Manuel Ayala, "Introducción," in *Agudeza y arte de ingenio*, by Baltasar Gracián, ed. Ceferino Peralta, Jorge Manuel Ayala, and José María Andreu (Zaragoza: Prensas Universitarias de Zaragoza, 2004); Aurora Egido, "Estudio preliminar," in *Arte de ingenio, tratado de la agudeza. Edición facsímil (Madrid, Juan Sánchez, 1642)*, by Baltasar Gracián (Zaragoza: Institución Fernando el Católico, 2005), VII–CXLVIII; Aurora Egido, "Estudio preliminar," in *Agudeza y arte de ingenio (Huesca, Juan Nogués, 1648) Edición facsímil*, by Baltasar Gracián (Zaragoza: Institución Fernando el Católico, 2007), XI–CLXXI; Egido Aurora, *Bodas de arte e ingenio. Estudios sobre Baltasar Gracián* (Barcelona: Acantilado, 2014).

7. "Si frecuento a los españoles, es porque la agudeza prevalece en ellos . . ." (I often use examples by Spanish authors, because *agudeza* prevails in them.) Gracián 1969, I, 46.

8. See, among others, Otis H. Green, *The Literary Mind of Medieval and Renaissance Spain* (Lexington: University Press of Kentucky, 1970); Terence May, *Wit of the Golden Age. Articles on Spanish Literature* (Kassel: Reichenberger, 1986); Maxime Chevalier, *Quevedo y su tiempo. La agudeza verbal* (Barcelona: Crítica, 1992); Michael J. Woods, *Gracián Meets Góngora: The Theory and Practice of Wit* (Warminster: Aris & Phillips, 1995); Arturo Zárate Ruiz, *Gracián, Wit, and the Baroque Age* (New York: Peter Lang, 1996); Jeremy Robbins, *Arts of Perception. The Epistemological Mentality of the Spanish Baroque, 1580–1720* (Abingdon: Routledge, 2007); Cacho Casal, *La esfera del ingenio*; Mercedes Blanco, *Góngora o la invención de una lengua* (León: Universidad de León, 2012); Ignacio Arellano, *El ingenio de Lope de Vega. Escolios a las Rimas Humanas y divinas del Licenciado Tomé de Burguillos* (New York: Instituto de Estudios Auriseculares, 2012); Juan Carlos Cruz Suárez, *Ojos con mucha noche. Ingenio, poesía y pensamiento en el Barroco español* (Bern: Peter Lang, 2014); Christine Orobitg, "Del '*Examen de ingenios*' de Huarte a la ficción cervantina, o cómo se forja una revolución

literaria," *Criticón*, no. 120 (2014): 23–39. See also the references in previous and subsequent notes. We thank Jim Amelang for generously sharing his bibliography and notes on early modern *ingenio*.

9. *Ingenio, engenio, engeno, engenno, engeño, engeño, engeynno, engueño, ingenho, ingeño, yngenio*. For further details see the entry for *ingenio* in Nieto Jiménez and Alvar Ezquerra 2007 and the online databases *Corpus Diacrónico del Español* (CORDE) and *Nuevo Diccionario Histórico del Español*, Real Academia Española.

10. See e.g. Emilio Hidalgo-Serna, "'Ingenium' and Rhetoric in the Work of Vives," *Philosophy and Rhetoric* 16, no. 4 (1983): 228–41; Jorge M. Ayala, "El 'ingenio' en Huarte de San Juan y otros escritores españoles," in *Actas del VI Seminario de Historia de la Filosofía Española e Iberoamericana*, ed. Antonio Heredia Soriano (Salamanca: Universidad de Salamanca, 1990); Scott Soufas, *Melancholy and the Secular Mind*; Christine Orobitg, *L'humeur noire. Mélancolie, écriture et pensée en Espagne au XVIe et au XVIIe siècle* (Bethesda: International Scholars Press, 1997); Cristina Müller, *Ingenio y melancolía. Una lectura de Huarte de San Juan* (Madrid: Biblioteca Nueva, 2002); Jon Arrizabalaga, "Huarte en la medicina de su tiempo," in *Huarte au XXIe siècle*, ed. Duché-Gavet; Gambin, *Azabache*; Roger Bartra, *Melancholy and Culture. Essays on the Diseases of the Soul in Golden Age Spain* (Cardiff: University of Wales Press, 2008); José Pardo Tomás, "Ancora su Michel de Montaigne e Huarte de San Juan. Ricezione dei lettori e comunità interpretative tagli *Essais* e l'*Examen de ingenios para las sciencias*," in *Michel de Montaigne e il termalismo*, ed. Anna Bettoni, Massimo Rinaldi, and Maurizio Rippa Bonati (Florence: Leo S. Olschki Editore, 2010); José Luis Peset, *Las melancolías de Sancho. Humores y pasiones entre Huarte y Pinel* (Madrid: Asociación Española de Neuropsiquiatría, 2010); Elena Carrera, "Madness and Melancholy in Sixteenth- and Seventeenth-Century Spain: New Evidence, New Approaches," *Bulletin of Spanish Studies* 87, no. 8 (2010): 1–15; Andrés Vélez Posada, "Ingenia: puissances d'engendrement. Philosophie naturelle et pensée géographique à la Renaissance" (PhD diss., Paris, EHESS, 2013).

11. Antonio García Berrio has referred to this as an "exaltation of *ingenium*" in sixteenth-century Spanish literary theory. See Antonio García Berrio, *Formación de la teoria literaria. Volumen 1* (Madrid: Cupsa, 1977), 256–62, and Antonio García Berrio, *Formación de la teoria literaria. Volumen 2* (Murcia: Universidad de Murcia, 1980). The relevance of *ingenio*—and the influence of Huarte de San Juan's *Examen*, among other works—is particularly noteworthy in the case of treatises like Alonso López Pinciano's *Philosophía antigua poética* (1596) and Luis Alfonso de Carvallo's *El cisne de Apolo* (1602). See e.g. Joaquín Roses Lozano, "Sobre el ingenio y la inspiración en la edad de Góngora," *Criticón*, no. 49 (1990): 31–49; Marina Mestre Zaragoza, "La 'Philosophía antigua poética' de Alonso López Pinciano, un nuevo estatus para la prosa de ficción," *Criticón*, no. 120 (2014): 57–71; Rocío G. Sumillera, "From Inspiration to Imagination: The Physiology of Poetry in Early Modernity," *Parergon* 33, no. 3 (2017): 17–42.

12. See Manuel Silva Suárez, "Sobre Técnica e Ingeniería: en torno a un excursus lexicográfico," in *Técnica e ingeniería en España I. El Renacimiento. De la técnica imperial y la popular*, ed. Manuel Silva Suárez (Zaragoza: Real Academia de Ingeniería, Institución Fernando el Católico, Prensas Universitarias de Zaragoza, 2008); Cristina Martín Herrero, "El léxico de los ingenios y máquinas en el Renacimiento" (PhD diss., Universidad de Salamanca, 2013). For a discussion of the larger context see, among others, Nicolás García Tapia, *Ingeniería y arquitectura en el renacimiento español* (Valladolid: Universidad de Valladolid, 1990); Alicia Cámara, *Los ingenieros militares de la Monarquía Hispánica en los siglos XVII y XVIII* (Madrid: Centro de Estudios Europa Hispánica, 2005); Alicia Cámara and Bernardo Revuelta, eds. *Ingenieros del Renacimiento* (Madrid: Fundación Juanelo Turriano, 2014).

13. See chapter 1 for further discussion on *genius*.

14. On *ingenuus* and *ingenuitas* see chapter 1.

15. See e.g. Gaspar Gutiérrez de los Ríos, *Noticia general para la estimacion de las artes, y de la manera en que se conocen las liberales de las que son mecanicas* (1600), Juan de Butrón, *Discursos apologéticos, en que se defiende la ingenuidad del arte de la pintura* (1626), Antonio Palomino, *El museo pictórico y escala óptica* (1715–1724), and the contributions of literary figures such as Lope de Vega and Calderón de la Barca. On these debates, see Julián Gállego, *El pintor de artesano a artista* (Granada: Universidad de Granada, 1976); María Elena Manrique Ara, "De memoriales artísticos zaragozanos (I), Una defensa de la ingenuidad de la pintura presentada a Cortes de Aragón en 1677," *Artigrama: Revista del Departamento de Historia del Arte de la Universidad de Zaragoza*, no. 13 (1998): 277–94; José Manuel Cruz Valdovinos, "El fuero y el huevo. La liberalidad de la pintura: textos y pleitos," in *"Sacar de la sombra lumbre." La teoría de la pintura en el Siglo de Oro (1560–1724)*, ed. José Riello (Madrid: Abada, 2012); Antonio Urquízar, "La ingenuidad de la pintura y la teoría jurídica y social de los clásicos," in *Siete memoriales españoles en defensa del arte de la pintura*, ed. Antonio Sánchez Jiménez and Adrián J. Sáez (Madrid and Frankfurt: Iberoamericana Vervuert, 2018).

16. In this regard, see, for instance, the etymological discussion of the term *hidalgo* in Huarte de San Juan's *Examen de los ingenios*; Huarte de San Juan 1989, 551–558. The case of such celebrated Spanish painters as Diego Velázquez or Bartolomé Esteban Murillo is also particularly illustrative of these arguments connecting *ingenio*, liberality and nobility. For a recent discussion, see José Ramón Marcaida, "Examen de ingenios en la pintura de género de Murillo," in *Murillo ante su IV Centenario: Perspectivas historiográficas y culturales*, ed. Benito Navarrete Prieto (Seville: Instituto de la Cultura y las Artes de Sevilla, forthcoming, 2018), and Antonio Urquízar, "La profesión de pintor (principiante, aprovechado o perfecto) en la teoría artística de Francisco Pacheco," *Studi Ispanici* 43 (2018): 183–99. For a broader discussion of *ingenio* and the arts, see e.g. Jon R. Snyder, *La estética del Barroco* (Madrid: Antonio Machado Libros, 2014).

17. See chapters 1 and 6 for further discussion. In the Spanish context, the social and moral dimensions of *ingenio* and its relationship with *juicio* (judgment) are amply

discussed by early modern authors. For further information, especially in connection to Gracián, see references in the next note.

18. See e.g. Aurora Egido, "Introducción," in *El discreto*, by Baltasar Gracián, ed. Aurora Egido (Madrid: Alianza, 1997); José María Andreu Celma, *Gracián y el arte de vivir* (Zaragoza: Institución Fernando el Católico, 1998); Aurora Egido, *Las caras de la prudencia y Baltasar Gracián* (Madrid: Editorial Castalia, 2000); Aurora Egido, *Humanidades y dignidad del hombre en Baltasar Gracián* (Salamanca: Universidad de Salamanca, 2001).

19. Among other references, the German artist Albrecht Dürer famously praised the "subtle *ingenia*" of native Americans (Dürer 1958, 102). See Alessandra Russo, "An Artistic Humanity. New Positions on Art and Freedom in the Context of Iberian Expansion, 1500–1600," *RES* 65/66 (2015): 352–63, and Davide Domenici, "Missionary Gift Records of Mexican Objects in Early Modern Italy," in *The New World in Early Modern Italy, 1492–1750*, ed. Elizabeth Horodowich and Lia Markey (Cambridge: Cambridge University Press, 2017); with thanks to Peter Mason for the latter reference. See also Benito Jerónimo Feijoo's reflections on the ingenuity of different countries and places, including America, in his *Teatro crítico universal* (1726–1740) and *Cartas eruditas y curiosas* (1742–1760); on which see e.g. José Manuel Rodríguez Pardo, *El alma de los brutos en el entorno del Padre Feijoo* (Oviedo: Pentalfa Ediciones, 2008).

20. Gracián 1642 and 1648; Blanco, *Les Rhétoriques de la Pointe*; Antonio Pérez Lasheras, "Arte de ingenio y agudeza y arte de ingenio," in *Baltasar Gracián. Estado de la cuestión y nuevas perspectivas*, ed. Aurora Egido and María Carmen Martín (Zaragoza: Institución Fernando el Católico, 2001); Carmen Codoñer, "Ingenio y agudeza. Reflexiones léxicas," in Egido, Gil Encabo, and Laplana Gil, *Gracián Actas I*, 203–33; Elena Cantarino and Emilio Blanco, eds., *Diccionario de conceptos de Baltasar Gracián* (Madrid: Cátedra, 2005). As is well known, Gracián's work on *agudeza* is part of a corpus of early modern treatises devoted to this subject, including Maciej Kazimierz Sarbiewski's *De acuto et arguto* (ca.1627), Matteo Peregrini's *Delle acutezze* (1639) and *I fonti dell'ingegno ridotti ad arte* (1650), and Emanuele Tesauro's *Il Cannocchiale aristotelico* (1654).

21. Blanco, *Les Rhétoriques de la Pointe*. Blanco is writing in a tradition of scholarship on *concettismo* that is traceable at least to the works of Benedetto Croce and Joel Spingarn; Alexander A. Parker, "'Concept' and 'Conceit': An Aspect of Comparative Literary History," *The Modern Language Review* 77, no. 4 (1982): xxi–xxxv. See also chapters 2 and 6.

22. Among other accounts, see, with further bibliography, Ignacio Ahumada, ed., *Cinco siglos de lexicografía del español* (Jaén: Universidad de Jaén, 2000); Manuel Alvar Ezquerra, "Dictionaries of Spanish in their historical context," in *Lexicography. Reference Works across Time, Space and Languages*, ed. Reinhard R. K. Hartmann (London: Routledge, 2003); Antonia María Medina Guerra, ed., *Lexicografía española* (Barcelona: Ariel, 2003).

23. The identification of lexicographical sources and the compilation of materials owes much to the useful entries in Lidio Nieto Jiménez and Manuel Alvar Ezquerra, eds., *Nuevo tesoro lexicográfico del español (s. XIV-1726)*, 11 vols (Madrid: Arco-Libros, 2007). However, not all the lexica used there have been considered here, particularly non-printed sources. Furthermore, the numerous early modern publications devoted to the grammar and vocabulary of extra-European languages have not been considered in this survey. The support of the online lexicographical resources provided by the Real Academia Española must be acknowledged, as well as the information in Samuel Gili Gaya, *Tesoro lexicográfico, 1492–1726* (Madrid: CSIC, 1947), and Joan Corominas and José A. Pascual, *Diccionario crítico etimológico castellano e hispánico*, 6 vols. (Madrid: Gredos, 1991).

24. On the lexicographical context in this period, see Barbara Freifrau von Gemmingen, "Los inicios de la lexicografía española," in Medina Guerra, *Lexicografía española*, 151–74. On Fernández de Palencia's dictionary see Antonia María Medina Guerra, "Modernidad del *Universal vocabulario* de Alfonso Fernández de Palencia," *Estudios de Lingüística* 7 (1991), 45–60.

25. Nebrija 1492 and Nebrija ca. 1495.

26. Not in vain one of Nebrija's most often quoted statements—featured in the prologue to his important grammar of the Castilian language, the *Grammatica* of 1492—is "*siempre la lengua fue compañera del imperio*," which John Considine translates as "language has always gone hand in hand with sovereignty;" John Considine, *Academy Dictionaries 1600–1800* (Cambridge: Cambridge University Press, 2014), 111. On Nebrija see, among others, Manuel Alvar, "Nebrija, Lexicógrafo," in Ahumada, *Cinco siglos de lexicografía*, 179–201; Miguel Ángel Esparza Torres and Hans-Josef Niederehe, *Bibliografía nebrisense. Las obras completas del humanista Antonio de Nebrija desde 1481 hasta nuestros días* (Amsterdam, Philadelphia: John Benjamins Publishing Company, 1999).

27. As Medina Guerra shows, the lack of examples in Nebrija's dictionaries (and, hence, its obscurity) was criticized in the early modern period; Medina Guerra, "Modernidad del *Universal vocabulario*," 60.

28. "Engañador con palabras. logodedelus.i." Nebrija ca. 1495, f. XLVIIIr; "logodaedalus," in later editions. See Jorge Ledo and Harm den Boer, eds., *Moria de Erasmo Roterodamo. A Critical Edition of the Early Modern Spanish Translation of Erasmus's Encomium Moriae* (Leiden: Brill, 2014), pp. 141–42.

29. In relation to this lexicographical context, it is worth noting the Spanish editorial efforts that in the sixteenth century saw the publication of such polyglot works as the *Biblia complutense* and the *Antwerp Polyglot Bible* (or *Biblia regia*), associated, respectively, with the figures of the cardinal Francisco Jiménez de Cisneros and the polymath Benito Arias Montano. See e.g. Theodor Dunkelgrün, "The Multiplicity of Scripture: The Confluence of Textual Traditions in the Making of the Antwerp Polyglot (1568–1573)" (PhD diss., University of Chicago, 2012). See also, for instance, Juan de Valdés's *Diálogo*

de la lengua (1535), which includes an often-quoted discussion of the difference between *ingenio* and *juicio*; Valdés 1969, 165–66.

30. Manuel Alvar Ezquerra, "El largo camino hasta el diccionario monolingüe," *Voz y Letras* 5, no. 1 (1994): 47–66. On the *Tesoro* see, among others, Dolores Azorín Fernández, "Sebastián de Covarrubias y el nacimiento de la lexicografía española monolingüe," in Ahumada, *Cinco siglos de lexicografía*, 3–34; José Ramón Carriazo and María Jesús Mancho, "Los comienzos de la lexicografía monolingüe," in Medina Guerra, *Lexicografía española*, 205–34; Sebastián de Covarrubias, *Tesoro de la lengua castellana o española. Edición integral e ilustrada*, ed. Ignacio Arellano and Rafael Zafra (Madrid; Frankfurt am Main: Iberoamericana; Vervuert, 2006).

31. Azorín Fernández, "Sebastián de Covarrubias," 17, citing Manuel Seco, *Estudios de lexicografía española* (Madrid: Paraninfo, 1987), 100–101. In his address to the reader (*Al Letor*), Covarrubias claims that he uses this term "por conformarme con los que han hecho diccionarios copiosos y llamándolos *Tesoros*, me atrevo a usar deste término por título de mi obra."

32. Etymologies, according to Quevedo, "are for the most part the product of ingenuity rather than a testimony of truth" ("las mas vezes son obra del injenio, i no testimonios de la berdad"). Francisco de Quevedo, *España defendida* (1609), quoted in Enrique Martínez Bogo, *Retórica y agudeza en la prosa satírico-burlesca de Quevedo* (Santiago de Compostela: Universidad de Santiago de Compostela, 2010), 247, note 801. Quevedo also notoriously referred to the *Tesoro* as "a large work, of untidy erudition," where "there is more paper than reason" ("También se ha dicho tesoro de la lengua española, donde el papel es más que la razón: obra grande, y de erudicion desaliñada"). Francisco de Quevedo, *Cuento de cuentos*, cited in Seco, *Estudios*, 107.

33. Covarrubias 1674. The so-called *Suplemento*, a complementary appendix that Covarrubias compiled as he wrote his dictionary, remained in manuscript form. For a modern edition of this text, see the modern edition of the *Tesoro* published by Arellano and Zafra.

34. See Fernando Lázaro Carreter, *Crónica del Diccionario de Autoridades (1713–1740)* (Madrid: Real Academia Española, 1972), 18–20. For the lexicographical context of the *Diccionario de autoridades* (abbreviated to *Autoridades* in the notes) see Pedro Álvarez de Miranda, "La lexicografía académica de los siglos XVIII y XIX," in Ahumada, *Cinco siglos de lexicografía*, 35–61; Stefan Ruhstaller, "Las obras lexicográficas de la Academia," in Medina Guerra, *Lexicografía española*, 235–61.

35. "Su libro [Covarrubias's *Tesoro*] ha merecido la estimación de proprios y Extrangeros; pero como es facil al ingenio añadir y limar lo mismo que se halla inventado: los Franceses, Italianos, Ingleses y Portugueses han enriquecido sus Patrias, e Idiomas con perfectissimos Diccionarios, y nosotros hemos vivido con la gloria de ser los primeros, y con el sonrojo de no ser los mejores." *Autoridades*, "Historia de la Real Academia Española," XI.

36. *Autoridades*, Prólogo, I.

37. Considine, *Academy dictionaries*, 113; Álvarez de Miranda, "Lexicografía académica," 37.

38. "... nunca podrán servir, para autorizar la voz, los Diccionarios, Vocabularios, ó Indices, á excepcion de los que á este efecto ha admitido hasta aqui la Academia, como son Covarrubias, Nebrixa, el Padre Alcalá, etc. Pues la evacuación de aquellos solo ha de servir, para excitar á la memoria las voces, que se hayan olvidado." Real Academia Española 1743, Section IV ("Para la autoridad"). On the *autoridades* in *Autoridades* in general, see Stefan Ruhstaller, "Las autoridades del *Diccionario de autoridades*," in *Tendencias en la investigación lexicográfica del español: El diccionario como objeto de estudio lingüístico y didáctico*, ed. Stefan Ruhstaller and Josefina Prado Aragonés (Huelva: Universidad de Huelva, 2000); Margarita Freixas, "Las autoridades en el primer diccionario de la Real Academia española" (PhD diss., Universidad Autónoma de Barcelona, 2003).

39. Lázaro Carreter, *Crónica*, 20.

40. *Autoridades*, "Historia de la Real Academia Española," XIII.

41. On the nationalistic drive in the work of Spanish lexicographers, from Nebrija and Covarrubias to the compilers of the *Diccionario de autoridades*, see Aurora Egido, "De las academias a la Academia," in *The Fairest Flower. The Emergence of Linguistic National Consciousness in Renaissance Europe* (Florence: Presso L'Accademia, 1985), 87.

42. "*Geniolus.* se toma por ingenioso ombre que resplandece en ingenio. Geniolus. id est ingeniosus qui pollet ingenio." Fernández de Palencia 1490.

43. On this dictionary, see Antonia María Medina Guerra, "Rodrigo Fernández de Santaella, *Vocabularium Ecclesiasticum*," *Analecta Malacitana* 13, no. 2 (1990): 329–42.

44. On this lexicographical context see, among others, Isabel Acero Durántez, "La lexicografía plurilingüe del español," in Medina Guerra, *Lexicografía española*, 175–204.

45. Nebrija 1545.

46. See, for example, Bravo 1599: "Abilidad. *Ingenium, ii. Acumen ingenii. Vide ingenio.*"

47. Huarte de San Juan 1575. In the first chapter of the 1594 expurgated and expanded edition of the *Examen*, Huarte de San Juan indicates that the subjet of the book is the *ingenio* and *habilidad* of human beings; Huarte de San Juan 1594, 21v.

48. Anonymous 1533. See Lidio Nieto Jiménez, "Repertorios lexicográficos españoles menores en el siglo XVI," in Ahumada, *Cinco siglos de lexicografía*, 203–24.

49. Anonymous 1556 (headwords: *abelheyt, loosheyt* and *scherp*).

50. Anonymous 1565.

51. Calepino 1590.

52. Junius 1606, chapters 8 and 72 respectively.

53. See Robert Verdonk, "La importancia del 'Recueil' de Hornkens para la lexicografía bilingüe del Siglo de Oro," *Boletín de la Real Academia Española* 70 (1990): 69–109. See also Vittori 1609.

54. Minsheu, 1599. Note the range of meanings, from instruments to stage contraptions. By contrast, the entry for *ingenio* is simply "wit."

55. Herrera 1580. On this work see, with further bibliography, Inoria Pepe and José María Reyes, "Introducción," in *Anotaciones a la poesía de Garcilaso*, by Fernando de Herrera, ed. Inoria Pepe and José María Reyes (Madrid: Cátedra, 2001).

56. *Ingenio*: "es aquella fuerça i potencia natural, i aprehension facil i nativa en nosotros, por la cual somos dispuestos a las operaciones peregrinas i a la noticia sutil de las cosas altas. Procede del buen temperamento del animo i del cuerpo. Sinifica propriamente aquella virtud del animo i natural abilidad, nacida con nosotros mesmos, i no adquirida con arte o industria. Llaman los Griegos i Latinos ingenio a la naturaleza de cualquiera cosa." (*Ingenio*: is a natural force and power, a native ability to understand with ease, which disposes us towards rare operations and the subtle grasp of elevated matters. It derives from the good temperament of the body and the soul. Its proper meaning is the inborn power of the soul and a natural ability, which cannot be acquired through art and industry. The Greeks and Latins define 'ingenio' as the nature of any thing.) Herrera 1580, 581.

57. Another interesting case study would be an exploration of how the language of ingenuity—in particular the language of *sprezzatura*—is articulated in Juan Boscán's 1534 Spanish translation of Baltasar Castiglione's *Il libro del cortegiano*. See Margherita Morreale, *Castiglione y Boscán: el ideal cortesano en el Renacimiento español*, 2 vols., Anejos del *Boletín de la Real Academia Española* (Madrid: S. Aguirre Torre, 1959). In relation to this, it is worth noting the relevance of such terms as *despejo* (amply discussed by Gracián), and *desenvoltura* (of interest for authors like Francis Bacon and Feijoo).

58. "*Latine ingenium, a gignendo, proprie natura dicitur cuique ingenita, indoles*;" Covarrubias 1611. See chapter 1 for similar expressions in the definitions of *ingenium*.

59. See Huarte de San Juan's discussion of the definitions of *ingenio*, including its etymology (*ingenero*), in the first chapter of the expurgated version of the *Examen*; Huarte de San Juan 1594, chapter 1. Though *ingenio* as mental capacity was the main sense in earlier accounts like Vives's, the influence of Huarte de San Juan's *Examen* and other treatises dealing with ingenuity would largely explain this shift. See, for example, the definition of *ingenio* provided by Alonso de Freylas in his discourse on melancholy and foretelling *Si los Melanchólicos pueden saber lo que está por venir con la fuerça de su ingenio o soñando* (1606): "Es ingenio una fuerça, o potencia natural de entender, con la qual entendemos, conocemos, hallamos, y juzgamos las cosas dificultosas, por muy secretas y ocultas que sean; sin que nadie nos las muestre, o enseñe, y resumiendo su naturaleza en breve digo; ser una fuerça natural, de entender lo dificultoso con presteza: a ésta llamó Nonio Marcello, natural sabiduría, porque con ella sin maestro se halla lo que se busca, lo muy dificultoso, entricado, y oscuro; se entiende con facilidad lo que está confuso, se explica con claridad y distinción. Con ella se conoce la verdad, y falsedad de las cosas, las consequencias, propiedades, y fines dellas y de sus contrarios." (*Ingenio* is a force or natural power, by which we understand, learn, find out and judge the most

difficult subjects, no matter how secret and hidden they are, with no one showing or teaching them to us; in a nutshell, it is a natural force that allows us to understand difficult matters with promptness. Nonio Marcello called it 'natural wisdom,' for without a master it allows us to find what is difficult and hidden, understand with ease what is confusing, and explain it with clarity and distinction. With it [*ingenio*] the truth and falseness of things can be known, as well as the consequences, properties, and aims of things and their contraries.) Freylas 1606, 2–3; Orobitg, "Del '*Examen de ingenios*,'" 23–39.

60. The literature on Turriano is large. See, among others, José Antonio García Diego, *Juanelo Turriano, Charles V's Clockmaker. The Man and His Legend* (Wadhurst: Antiquarian Horological Society, 1986); Alfredo Aracil, *Juego y artificio. Autómatas y otras ficciones en la cultura del Renacimiento a la Ilustración* (Madrid: Cátedra, 1998); Nicolás García Tapia and Jesús Carrillo, *Tecnología e imperio: Turriano, Lastanosa, Herrera, Ayanz* (Madrid: Nivola, 2012); Cristiano Zanetti, *Janello Torriani and the Spanish Empire* (Leiden: Brill, 2017).

61. The expression "*ingenio de azúcar*" (or simply "*ingenio*") would retain its currency for centuries, even in other languages. Covarrubias's third example, *seda de ingenio*, refers to a method and contraption for silk spinning.

62. Bovelles 1533, 59; see Blanco, *Rhétoriques de la Pointe*, 26. See chapter 2 on *inganno*.

63. These references to dexterity, ingenuity, and liberality are key to understanding early modern debates on the status of the art of painting in Spain; as indicated earlier, Velázquez is a case in point.

64. See also the definition of *engeño* in Francisco del Rosal's (unpublished) *Origen y etimología de todos los vocablos originales de la lengua castellana*: "la traza y artificio, de ingenio, que es la industria y natural" (Rosal ca. 1601).

65. See, for example, how Nebrija equates *arte engaño* and *artimaña*–both translated as *ars*, *dolus* and *techna*–in Nebrija ca. 1495. See also, for instance, the entries for *ingegno* in the 1604 edition of Cristóbal de las Casas's *Vocabulario de las dos lenguas toscana y castellana*, where one features just one word: *maña*.

66. See, for instance, Covarrubias's emblem on Daedalus' and Icarus's flight, whose motto, extracted from Ovid's *Ars Amatoria*, says: "*Ingenium mala saepe movent*" (adversities often stir one's wit). Note the use of *sagaz* and *maña* in the poem. Covarrubias 1610, Centuria 3, emblem n. 85, 285r.

67. Earlier, in 1599, Minsheu's expanded version of Percival's dictionary defined *maña* as: "wit, quicknesse of spirit, skill, deceit, cunning;" Minsheu 1599.

68. Blanco, *Rhétoriques de la Pointe*, 26.

69. Latin equivalents provided: *Ingenii vis, vel acumen*.

70. On *ingenuitas* and *ingenuidad* see notes 14 and 15 in this chapter.

71. See e.g. Minsheu 1599, Oudin 1607 and Franciosini 1620, among others.

72. "Otros dixeron, que Genio no era otra cosa que la sitrimea [sic], y commensuracion

de los elementos, la qual conserba los cuerpos humanos, y los de toda cosa viviente. Otros una virtud, e influencia de los planetas que nos inclinan a hazer esto, o aquello, y no solo constituyan genios a los hombres, pero tambien a las plantas, y a los edificios, y como dixo Marcial, a los libros. *Victurus Genium debet habere liber*." Covarrubias 1611. It is not clear whether the first meaning above connects with the sense of symmetry and grace that we encounter in other keywords.

73. Gracián 1997, 163–72.

74. Gracián 2003, 161. English translation cited from Gracián 2011, 3, where Jeremy Robbins translates *genio* and *ingenio* as "inclination" and "ingenuity" respectively. On Gracián's use of the notion of *genio* see, with further bibliography, Cantarino and Blanco, *Diccionario de conceptos*, 121–25.

75. On the important notion of *gusto*, particularly in relation to the work of Gracián, see, among others, Blanco, *Rhétorique de la Pointe*; Hidalgo-Serna, *El pensamiento ingenioso*, chapters 2 and 5 and Cantarino and Blanco, *Diccionario de conceptos*, 127–32.

76. Herrera 1580, 581.

77. On which see e.g. Brann, *The Debate Over the Origin of Genius*. For the early modern Spanish context see e.g. Pedro Azara, "Furor divino. Contribución a la historia de la teoría del arte" (PhD diss., Universitat Politècnica de Catalunya, 1986); Roses, "Sobre el ingenio y la inspiración."

78. *Espíritu*: "Vale assimismo génio, inclinación, hábito, y passión, que nos inclina a obrar y executar, con más propensión y afecto, unas cosas que otras." *Autoridades*.

79. Several decades later, in the 1812 edition of his *Filosofía de la eloquencia*, the Spanish polygraph Antonio de Capmany y de Montpalau would point to the influence, via translation, of the French term *génie*, leading to the "abuse of constantly using *genio* to refer to what our parents and grandparents constantly called *ingenio*." Capmany y de Montpalau 1812, 49–50.

80. On this notion of *genio* in the Spanish context see e.g. José Luis Peset, *Genio y desorden* (Valladolid: Cuatro, 1999).

81. *Industria*: "Destreza y habilidad en qualquier arte. Es voz puramente Latina *Industria*;" *Industria*: "Se toma tambien por ingenio y sutileza, maña u artificio. Latín. *Industria. Gnavitas. Solertia.*" *Autoridades*.

82. *Artificioso*: "Cosa de artificio, de ingenio y primor, o la que está hecha segun arte y sus reglas. Latín. *Artificiosus, a, um*;" *Artificioso*: "Vale tambien astuto, caviloso, sagaz, diestro, ingenioso, cauto, y de maña y disimulacion: y tambien se dice de las cosas hechas con cautela, simulacion y artificio. Latín. *Simulationis artificio eruditus. Fallax. Callidus, a, um.*" *Autoridades*.

83. *Talento*: "Tomamoslo por el ingenio y natural, porque fue lo significado por los talentos de la parabola del evangelio." Rosal ca. 1601.

84. Covarrubias 1674. A later example of this equivalence can be found in Requejo 1729, where "*talento, habilidad*" is defined as "*ingenium, facultas, atis*."

85. *Talento*: "Metaphoricamente se toma por los dotes de naturaleza: como ingenio, capacidad, prudencia, etc. que resplandecen en alguna persona, y por antonomásia se toma por el entendimiento. Latín. Ingenium. Animi dotes, vel facultas." *Autoridades*.

86. The English translation is cited from Huarte de San Juan 1698, 153. The text in Spanish reads as follows: "A los ingenios inventivos llaman en lengua toscana *caprichosos*, por semejanza que tienen con la cabra en el andar y pacer. Esta jamás huelga por lo llano; siempre es amiga de andar a sus solas por los riscos y alturas, y asomarse a grandes profundidades; por donde no sigue vereda ninguna ni quiere caminar con compaña." Huarte 1989, 344–45.

87. The term *capriccio* is recorded in the lexica too, as in the entry for *concepto* (*capriccio, concetto*) in Cristóbal de las Casas's *Vocabulario de las dos lenguas toscana y castellana*s (1570).

88. *Caprichoso*: "Se toma tambien por ingenioso y de idea: y assí del que concibe, halla, y hace alguna cosa con novedad y buen gusto, se dice que es caprichoso y de rara phantasía. Latín. *Ingeniosus. Solers.*" *Autoridades*. Note the use of the expression "*buen gusto*" (good taste), on which the literature is large; see e.g. Blanco, *Rhétorique de la Pointe*; Hidalgo-Serna, *El pensamiento ingenioso*, chapters 2 and 5; Cantarino and Blanco, *Diccionario de conceptos*, 127–32.

89. Fernando Marías, "El género de *Las Meninas*: los servicios de la familia," in *Otras Meninas*, ed. Fernando Marías (Madrid: Siruela, 1995). In Marías's account, Palomino's characterization of *Las Meninas* as a "*capricho nuevo*" and the use that Velázquez himself makes of this expression are associated with the notion of *capricho* as an "artist's new concept" presented by Velázquez's contemporary Vicente Carducho in his *Diálogos de la pintura* (1633). Interestingly, those artists that can come up with new concepts are compared by Carducho to the goat, in a passage in his *Diálogos* that is likely to have been inspired directly by Huarte de San Juan's abovementioned characterization of *ingenios inventivos/caprichosos* as goats. Carducho owned an edition of the *Examen de los ingenios* and explicitly refers to the treatise and its author in his *Diálogos*. On *capricho/capriccio* see Roland Kanz, *Die Kunst des Capriccio: Kreativer Eigensinn in Renaissance und Barock* (Munich: Deutscher Kunstverlag, 2002); Veronica Maria White, "Serio Ludere: Baroque Invenzione and the Development of the *Capriccio*" (PhD diss., Columbia University, 2009); Francesco Paolo Campione, *La regola del Capriccio. Alle origini di una idea estetica* (Palermo: Centro Internazionale Studi di Estetica, 2011). See also Aurora Egido, "Arte y literatura: lugares e imágenes de la memoria en el Siglo de Oro," in *El Siglo de Oro de la pintura española*, ed. Alfonso E. Pérez Sánchez et al. (Madrid: Mondadori, 1991).

90. *Mania*: "Significa tambien extravagancia, capricho, tema y ridiculez de genio: y assi se dice, Ha dado en la manía de que ha de hacer esto. Latín. *Insania. Exotica voluntas.*" *Autoridades*. In relation to this, see, for instance, Fernando Marías's account of El Greco in Fernando Marías, *El Greco. Biografía de un pintor extravagante* (Madrid: Editorial Nerea, 1997).

91. *Phantasia*: "Significa assimismo ficción, cuento, novela o pensamiento elevado y ingenioso: y assí se dice, las phantasías de los Poetas y de los Pintores. Latín. Excogitatio subtilis." *Autoridades*.

92. *Agudo*: "transfierese al alma, y dezimos agudo al que tiene ingenio sutil, y penetrante." Covarrubias 1611. Elsewhere in the dictionary, Covarrubias offers variants of this definition, qualifying as *agudos* those individuals endowed with a "good nature" (*buen natural*) and a "good understanding" (*buen entendimiento*); see entries on *bastón* and *tonto* respectively. Covarrubias rehearses this idea in the entry for *alesna* (a sharp-pointed tool, a piercer): "Al que es muy vivo y presto, decimos que es agudo como una alesna." This very example is recorded by Stevens 1706: "*Agudo como una alezna*, as sharp as an awl, a man of a sharp wit."

93. The list of Latin terms used as equivalent to "*agudeza de ingenio*" remains practically the same throughout our period.

94. For a discussion of *ingenio*, *agudeza* and velocity see Cacho Casal, *La esfera del ingenio*, 25–33.

95. "Tambien llamamos agudo al inquieto que anda de aqui para alli bullendo." Covarrubias 1611.

96. In this regard, it is interesting to note Feijoo's allusion—in one of his *Cartas eruditas y curiosas*—to the "*espiritosidad volátil*" (volatile spiritedness) of the sublime *ingenios*, and how this makes them more subject to distractions. "Carta XVIII. De la crítica" (On criticism), in Feijoo 1773, II, 239–56. See chapter 2 for further references to *spirito*.

97. *Agudo*: "Metaphoricamente vale ingenioso, pronto, perspicaz, y sutil: lo que compete y se dice no solo de los hombres, sino de sus operaciones. Latín. *Acri ingenio vir.*" *Autoridades*.

98. *Agudeza*: "Metaphoricamente: la sutileza, prontitud y facilidad de ingenio en pensar, decir o hacer alguna cosa. Latín. *Perspicacitas. Acumen.*" *Autoridades*.

99. *Hombre de capricho*: "Se llama el que tiene agudeza para formar idéas singulares, y con novedad, que tengan feliz éxito." *Espíritu*: "Significa tambien vivacidad, prontitúd y viveza en concebir, discurrir y obrar: y assí del que es ingenioso y descubre viveza en sus dichos y acciones, se dice que tiene o descubre espíritu. Latín. *Spiritus. Vivacitas.*" *Autoridades*.

100. *Agudo*: "Tambien por metaphora vale picante, ingenioso, y que pica en satyrico: como son los pasquines, y modos de decir, o hablar con alusion y equívoco. Latín. *Mordax. Lividus. Detractor.*" *Autoridades*.

101. *Viveza*: "Se toma tambien por el dicho agudo, pronto, è ingenioso." *Sal*: "Metaphoricamente vale tambien la agudeza, gracia, ò viveza en lo que se dice." *Autoridades*. In the early modern Spanish context, an important treatise on salt is Bernardino Gómez Miedes's *Commentarii de sale libri IV* (1572, expanded edition 1579); see Bernardino Gómez Miedes, *Comentarios sobre la sal*, ed. and trans. Sandra Inés Ramos Maldonado, 3 vols. (Madrid: Alcáñiz, 2003). In relation to this, see also Gracián's allusions to *sazonar*

(food seasoning) and *buen gusto*; Ayala, "Introducción," LXV. On salt see also the discussion in the chapter 6.

102. Before this entry in the *Diccionario*, a reference to *concepto* in the sense of (ingenuity-related) conceit in the Spanish lexicographical corpus can be found in Stevens 1706: "Conceto, a conceit, or conception, a piece of wit, a fine fancy." Pineda 1740, which expands Stevens 1706, adds the entry for *conceptista*: "a merry facetious Man, one who conceives witty Sayings;" and *conceptuar*: "to express elegantly whatsoever is conceiv'd in the Mind." Minsheu 1599 connects *agudeza* and *conceit* too, but the ingenuity-related inflection is less clearly articulated.

103. See *conceptuar* below. *Concepto*: "Se toma y dice muchas veces por senténcia, agudeza y discreción. Latín. *Praeclara sententia.*" *Autoridades*.

104. *Conceptuar*: "Discurrir con agudeza y primor, usando de palabras alusívas y discretas, que ocasionen atención particular a los oyentes. Es voz formada del nombre Concepto en el sentido de agudeza sentenciosa. Latín. *Praeclara animi sensa eleganter exprimere, enunciare.*" *Conceptuoso*: "Sentencioso, discreto, grave y lleno de agudezas y conceptos. Dicese del Orador, razonamiento, discurso u otra qualquiera obra del entendimiento, que es docta, ingeniosa y discreta. Latín. *Sententiis scatens, affluens, tis.*" *Autoridades*.

105. See, in particular, Blanco, *Rhétorique de la Pointe*.

106. Gracián is cited as Lorenzo Gracián, the pseudonym with which he signed the majority of his works. The quotations are taken from the *Agudeza y arte de ingenio*, Discurso 4 and Discurso 1 respectively.

107. See Egido, "Estudio," LXXI and CL-CLI; also, Blanco, *Rhétorique de la Pointe*.

108. Egido, "Estudio," CL-CLI.

109. "Hay palabras que no están porque todo el mundo sabe lo que significan." Cited, in the original Spanish, in José Antonio Pascual Rodríguez, "El comentario lexicográfico: tres largos paseos por el laberinto del diccionario," in Medina Guerra, *Lexicografía española*, 355.

110. Gracián 1969, I, 51: "[Agudeza] Déjase percibir, no definir."

4 French

1. Bayle 1740, 2:609, for the Critical Remark α under the entry "Gretserus."

2. The French critique, on linguistic and stylistic grounds, of other national expressions of ingenuity features in the *Entretiens d'Ariste et d'Eugène* (1671) written by the Jesuit Dominique Bouhours. In the *entretien* 'La langue francoise,' Bouhours mocks Spanish for its verbose obscurity, and Italian for its mannerist lack of seriousness: Dominique Bouhours, *Les Entretiens d'Ariste et d'Eugène* (Paris: Sébastien Mabre-Cramoisy, 1671), 40–46.

3. On the genius of the French language as a period category, see Introduction; see also Aisy 1685; Gilles Siouffi, *Le génie de la langue française: études sur les structures*

imaginaires de la description linguistique à l'âge classique (Paris: H. Champion, 2010); Wendy Ayres-Bennett and Magali Seijido. *Remarques et observations sur la langue française: histoire et évolution d'un genre* (Paris: Classiques Garnier, 2011); Marc Fumaroli, *Quand l'Europe parlait français* (Paris: Fallois, 2001); Jean-Alexandre Perras, "Ce que les Latins appelleroient *genius*," in *L'exception exemplaire: inventions et usages du génie (XVIe-XVIIIe siècle)* (Paris: Garnier, 2016).

4. The question of identity is constructed alongside both the moral semantic axis of ingenuity, and the presentational one described in the introduction: style, author, and critic envisage identity through modes of display, that is, from a presentational perspective.

5. See chapter 1.

6. On the *Grand dictionnaire françois-latin*, see chapter 1.

7. According to the ARTFL project, these late editions represent the best state of Bayle's *Dictionnaire*. See https://artfl-project.uchicago.edu/content/dictionnaire-de-bayle. None of the keywords feature as headwords in their own rights in Bayle's *Dictionnaire*.

8. *Académie* 1694.

9. See Girard 1718, 85–7, under the entry "RAISON BON-SENS ESPRIT GÉNIE CONCEPTION INTELLIGENCE ENTENDEMENT JUGEMENT."

10. Marc Fumaroli, *L'âge de l'éloquence. Rhétorique et res literaria en France de la Renaissance au seuil de l'âge classique* (Paris: Droz, 2002), 694. Bouhours, *Entretiens*, 152–53; Marc Fumaroli, "Le génie de la langue francaise," in *Trois institutions littéraires* (Paris: Gallimard, 1994).

11. Hélène Merlin-Kajman, *La Langue est-elle fasciste? Langue, pouvoir, enseignement* (Paris: Seuil, 2003), 169–78. For an extreme instance of this debate, see the 2004 Fumaroli/Cavaillé polemic: Marc Fumaroli, "La République des lettres, l'Université et la grammaire," *Nouvelle revue d'histoire littéraire de la France*, 2 (2004): 463–74 and Jean-Pierre Cavaillé, "Le paladin de la République des lettres contre l'épouvantail des sciences sociales," *Les Dossiers du Grihl*, 2 (2007), online 12 april 2007.

12. Furetière 1690, preface [6r].

13. Albert Dauzat, *Le génie de la langue française* (Paris: Payot, 1943). Bernard Cerquiglini has shown that the philological fabrication of a "pure" French language was driven by the political demands of defining national identity: Bernard Cerquiglini, *Une langue orpheline* (Paris: Editions de Minuit, 2007).

14. Gournay 1634, 267, 269, quoted and discussed in Gilles Siouffi, *Le Génie de la langue*, 301–3. Gilles Philippe historicizes this debate in "Un ramage subtil et faible" in *Le français, dernière des langues. Histoire d'un procès littéraire* (Paris: PUF, 2010), 21–59; while Henri Meschonnic tackles the myth of the 'génie de la langue' as clarity and distinctness from the perspective of poetics: Henri Meschonnic, *De la langue française: essai sur une clarté obscure* (Paris: Hachette, 1997).

15. "Vieux mot" in Richelet 1680, Furetière 1690, Corneille 1694, Académie 1694, Trévoux 1740.

16. Rabelais 1532, 103: "engin mieulx vault que force" (cunning is worth more than strength).

17. However, since Estienne 1536, the relationship between *ingenium* and *industria* is complex: industry is therefore not altogether a mistranslation. See Introduction and chapters 1 and 2.

18. "On dit proverbialement, mieux vaut engin que force; pour dire, que l'adresse & l'esprit, la douceur, la complaisance, font réüssir en des choses dont on ne viendroit pas à bout par la violence. M. Voiture les a joint ensemble, *force & engin, en ce cas j'emploirois.*" (One uses the proverb: cunning is worth more than strength in order to mean that craftiness, wit, kindness, and leniency succeed where violence would not. Mr Voiture has joined them together: *strength and cunning in this case I shall use.*)

19. See chapter 3.

20. Vérin, *La gloire*, 20–31 (on *engin*).

21. "un peu trop crûment dans la conversation, ou dans le stile comique, & satirique. C'est la partie qui fait les Empereurs & les Rois . . . la partie naturelle de l'homme." (A bit crude in conversation, or as a satirical and burlesque term. This is the part that makes Emperors and kings. . . . Man's natural part.) This—quite stable, if rarely emphasized—generative meaning lingers in other terms in the ingenuity lexical family: thus *nature* can denote reproductive organs and sperm (Nicot 1606, Richelet 1680, Furetière 1690, Académie 1694, Trévoux 1740), and *naturel* can denote the sexual drive (Trévoux 1740). Similar associations can be found in English: see chapter 6.

22. These four semantic clusters are: A. The theological meanings of esprit as *spiritus*: the Holy Ghost, God, angels, demons, the soul inasmuch as it is immaterial and separate from the body; B. The medical meanings of esprit as *spiritus*: the finest, most volatile particles of the human body involved in the generation of motion; C. The chemical/alchemical meanings of *esprits* as *spiritus*: mercury, or any concentrate resulting from distillation; D. The natural-philosophical meanings of esprit as *anima*: the principle of life and motion. The medical and chemical meanings dominate the entries dedicated to *Esprit* in the technical dictionaries: Furetière 1690, Corneille 1694, and Des Bruslons 1726.

23. Nicot 1606: "Esprit: Il vient du Latin *spiritus*. L'esprit et entendement que l'homme a de nature: *ingenium*."

24. This analytical tendency reached the excessive level of pedantic hair-splitting in the 'distinctionary' of Girard 1718. See note 14 above.

25. Académie 1694: Faculties of the rational soul (imagination and conception, imagination alone, conception alone, judgment)

26. "I am a thing that thinks: that is, a thing that doubts, affirms, denies, understands a few things, is ignorant of many things, is willing, is unwilling, and also which imagines and has sensory perceptions," René Descartes, "Third Meditation" in *Meditations on First Philosophy*, in *The Philosophical Writings of Descartes*, ed. and trans. John Cottingham, Anthony Kenny, Dugald Murdoch and Robert Stoothoff, 3 vols (Cambridge: Cambridge University Press: 1984–91), 2 (1984): 24. "But when it [the spiritual power through

which we know things] forms new ideas in the corporeal imagination, or concentrates on those already formed, the proper term for it is 'native intelligence' (*ingenium*)." René Descartes, rule 12 in *Rules for the Direction of the Mind* in the *Philosophical Writings of Descartes*, ed. Cottingham et al., 1 (1985): 42.

27. The relationship between memory—or rather, reminiscence—and *ingenium* features in the scholastic Latin reception of Aristotle's *De memoria et reminiscentia* (in the *Parva naturalia*), and in Descartes' technical definition of the *ingenium* in the *Regulae*: see Aristotle, *Parv Nat*, 449b4–453b11, and note above.

28. On thin and thick wits, see Saint Évremond as quoted in Trévoux 1740: "Cet esprit si fin et délicat s'est usé et épuisé en peu de tems" (This wit, such a fine and delicate one, soon wore out and dried out); "Un esprit judicieux est d'ordinaire pesant et ennuyeux" (a judicious mind is usually a heavy and boring one). Voltaire 1755: "C'est un mot générique qui a toûjours besoin d'un autre mot qui le détermine" (It is a generic word which always requires another word to qualify it).

29. In Greek and Latin, a character originally denotes a letter or sign before referring to the individual features of a person's physiognomy and temperament and finally, to the idiosyncrasies of one's style in rhetoric. The term features in that latter, stylistic sense in Voltaire 1755.

30. Richelet 1680.
31. Académie 1694.
32. Académie 1694.
33. Académie 1694; Trévoux 1740, quoting Le Chevalier d'Harcourt.
34. Académie 1694.
35. Trévoux 1740.
36. Trévoux 1740: "'Il a bien de *l'esprit* pour un Allemand,' disoit le Cardinal du Perron du Jésuite Grétser. Cette femme n'a pas assez d'*esprit* pour sçavoir qu'elle n'en a point" ("He's quite witty for a German," Cardinal du Perron used to say about the Jesuit Gretser. This woman is not witty enough to be aware that she has no wit.) This was excerpted from Bouhours and Saint Évremond respectively. Richelet 1680: "Il ne sort aucun livre de chez nous qui n'est l'esprit de la société" (we do not issue any book that does not bear the mark of the society's spirit), a paraphrase of "Il ne sort aucun Ouvrage de chez nous, qui n'ait l'esprit de la Société," from Blaise Pascal, "Neusviesme lettre" of the *Provinciales* (1656–1657): Blaise Pascal, *Provinciales, Pensées et opuscules divers*, ed. Philippe Sellier and Gérard Ferreyrolles (Paris: La Pochothèque, 2004), 406.

37. The conflation between the types of the wit and that of the philosophical libertine in Trévoux 1740 is telling in this respect. See Jean-Pierre Cavaillé, *Postures libertines. La culture des esprits forts* (Toulouse, Anacharsis, 2011).

38. Trévoux 1740, quoting Bouhours, *Les Entretiens*: "Le bon sens qui brille; . . . un juste tempérament, de la vivacité et du bon sens" (the shine of good sense . . . a well-tempered disposition, liveliness and good sense); "Un esprit qui a beaucoup de faux brillans, mais peu de vrai et de solide; un esprit de pointes qui a de l'afféterie" (A flashy

wit, yet nothing true or solid to it; a conceited wit, full of affectation); "Se dit aussi des effets et des inventions que produit cet esprit; des pensées ingénieuses épandües dans un livre, ou dans quelque Ouvrage que ce soit" (Can be said of the effects and inventions generated by such a wit, of ingenious thoughts displayed in a book, or in any other kind of work).

39. See Andry de Boisregard 1692, 346: "Ouvrages d'esprit. Ouvrages de l'esprit." Retrieved from the *Corpus numérique des remarques sur la langue française (XVIIe siècle)*, https://www.classiques-garnier.com/numerique-bases/index.

40. Trévoux 1740: "Esprit. . . . On ne sauroit avoir trop d'esprit dans une conversation enjouée. Le CH. de M. Les compositions ingénieuses des gens de lettres sont des ouvrages d'esprit" (Wit. . . . One cannot have too much wit in a lively conversation. Le Chevalier de Méré. . . . The ingenious compositions of men of letters are works of wit).

41. Trévoux 1740: "Un vrai bel esprit songe plus aux choses qu'aux mots" (a true wit ponders more about things than words), which contrasts with the quotation from Saint Évremond on the false *bel esprit*: "Un bel esprit, si j'en sçai bien juger? / Est un diseur de bagatelles" (A wit, if I am an able judge? / is a speaker of trifles).

42. Trévoux 1740: "Un véritable bel esprit. . . . ses expressions sont polies & *naturelles*. Il n'a rien de faux, ni de vain dans ses discours et dans ses manières" (A true wit . . . his expressions are polite and *natural*. There is nothing either false or vain in his speech and manners). Emphasis ours.

43. An early example is the "Querelle des Lettres," which pitted Guez de Balzac (1597–1654) and his supporters such as René Descartes against Jesuit polemicists and rhetoricians on the question of good style in the epistolary genre. See Mathilde Bombart, *Guez de Balzac et la querelle des lettres: écriture, polémique et critique dans la France du premier XVIIe siècle* (Paris: Champion, 2007). Fumaroli locates the origin of good classical French prose in conversation: Marc Fumaroli, *La diplomatie de l'esprit: de Montaigne à La Fontaine* (Paris: Gallimard, 2002). This view is contested by Meschonnic in *Le génie de la langue*.

44. Recent scholarship has identified dis/simulation and its rhetorical expression, irony, not only as a trademark of philosophical free-thinking and political subversion, but also as one of the structures of political control deployed by institutions of the state: see Jean-Pierre Cavaillé, *Dis/simulations: Jules-César Vanini, François La Mothe Le Vayer, Gabriel Naudé, Louis Machon et Torquato Accetto: religion, morale et politique au XVIIe siècle* (Paris: Honoré Champion, 2002); Fernand Hallyn, *Descartes: Dissimulation et ironie* (Geneva: Droz, 2006).

45. Trévoux 1740, quoting Bouhours: "Ces messieurs les baux [sic] esprits auroient beau faire valoir leurs madrigaux, leurs bouts-rimez et leurs impromptus, &c" (This bunch of wits could try as hard as they could to talk up their madrigals, their rhymed snippets and their improvisations, &c); Ogier de Gombauld 1658, *Aiiiv*. "our infamous mores and actions have brought me down to the epigram." Meschonnic contests the genealogy of literature out of conversation: Meschonnic, *De la langue française*.

46. See Pascal, *Pensées*, 1128, fr. 554 'Style' : "Quand on voit le style naturel on est tout étonné et ravi, car on s'attendait de voir un auteur et on trouve un homme" (When we encounters the natural style, we are stunned and ravished, for we expected to encounter an author, and we find a man). On the emergence of the writer as a social category in the seventeenth-century, see Alain Vialla, *Naissance de l'écrivain: sociologie de la littérature à l'âge classique* (Paris: Editions De Minuit, 1985).

47. Compare with chapter 1, 37 and 45.

48. Richelet 1680: "Avoir un esprit de vengeance. Voici quel est l'esprit de notre contrat" (To nurture a revengeful spirit. This is the spirit of our contract). The second example is repeated in Trévoux 1740. See chapter 1.

49. The preface of the *Dictionnaire de l'Académie* justifies the choice not to (self-) quote by appealing to the humility *topos*—some of the greatest orators in the French language have compiled the dictionary. Moreover, the dictionary records "la langue commune, telle qu'elle est dans le commerce ordinaire des honnestes gens" (the common language, as it is used in the ordinary conversation of honest men), and therefore includes "les phrases les plus receuës" (the most received sentences). Richelet is the first to cite Scarron's "Il mourra sans rendre l'esprit." (he will die without giving up the ghost/ squeezing any wit out). His quotations assumes a shared literary culture from his readers for these allusive puns and conceits to work: Pascal's "esprit de la société" from the anti-Jesuit satire of the *Provinciales*, or the following example for meaning A of *esprit*: "Elle a peur des esprits et ne couche jamais seule" (she is scared of ghosts and never sleeps alone), borrowed from Jean Ogier de Gombauld's "Marthe," epigramme 37, book 1. See Jean Ogier de Gombauld, *Les épigrammes* (Paris: Augustin Courbé, 1658), 18.

50. On comedy of characters as a disciplining genre, see Larry F. Norman, *The Public Mirror: Molière and the Social Commerce of Depiction* (Chicago: University of Chicago Press, 1999).

51. Here French *politesse* adopts the Italian culture of *sprezzatura*. See chapter 2. On aspects of this paradox relating to the relationship between absolutism and the emergence of the modern notion of literature, see Christian Jouhaud, *Les pouvoirs de la littérature: histoire d'un paradoxe* (Paris: Gallimard, 2000).

52. Six of these clusters do not relate to *nature/ naturel* as *ingenium*: A. Active force which maintains the universal order (*natura naturans, providentia, principia rerum, spiritus universalis*); B. The universe as an orderly collection of beings and creation (*mundi machina, universus, rerum creaturarum collectio*); the orderly nature of those beings (*genera, species*); C. Inner power and organisation in living beings which incline them towards what is necessary for their conservation (*conatus*); D. innate human faculty to distinguish good from evil (*instinctus, naturae indoles*), E. The natural state of man (theology) (*status naturae integrae et elevatae, status naturae lapsae et per Christum reparatae, humanitas, fragilitas, infirmitas naturalis*); F. Nature as the model of all arts. See Bernard Tocanne, *L'idée de nature en France dans la seconde moitié du XVIIe siècle: contribution à l'histoire de la pensée classique* (Paris: Klincksieck, 1978).

53. "C'est le *naturel* des lions d'estre cruels & farouches," "Neron estoit [etoit in Trévoux 1740] d'un *naturel* cruel & farouche."

54. Genie features as a synonym of *naturel* in Trévoux 1740.

55. Nicolas Perrot d'Ablancourt (1606–1664) academician, quoted in Richelet 1680: "Theophile avoit un beau naturel pour la poesie" (Theophile was naturally gifted in poetry).

56. Repeated in Trévoux 1740.

57. Ascribed to d'Ablancourt, quoted in Richelet 1680, Furetière 1690, Trévoux 1740.

58. Saint-Réal 1726, 242: "Premiere journée; de la difficulté d'avancer dans le monde, lors meme qu'on a de l'esprit." Quoted in Furetière 1690 and Trévoux 1740.

59. quoted in Furetière 1690 and Trévoux 1740.

60. This view remains widespread: see Paul Duro, "'The Surest Measure of Perfection': Approaches to Imitation in Seventeenth-century French Art and Theory," *Word & Image* 25, 4 (2009): 363–83. See also Christian Michel, "La peinture peut-elle être réduite en art?" in *Réduire en art: la technologie de la Renaissance aux Lumières*, ed. Pascal Dubourg-Glatigny and Hélène Vérin (Paris: Éditions de la maison des sciences de l'homme, 2008); Jean-Gerald Casteix, "Réduire la gravure en art et en principes: lecture et réception du *Traité des manières de graver a l'eau-forte* d'Abraham Bosse," in Glatigny-Vérin, *Réduire en art*.

61. Richelet 1680: from D'Ablancourt: "Théophile avoit un beau naturel pour la poësie & c'est dommage qu'il n'ait pas assez châtié ses vers" (Theophile was naturally gifted for poetry; it is a shame that he did not castigate his lines enough).

62. Furetière 1690: "se dit aussi en l'homme, de ce qui n'y est point fixe, ni general, mais qui change suivant son temperament, ou son education" (what is neither fixed nor general in man and fluctuates according to his education and talent).

63. On the overlap between moral qualities and social status implied by ingenuousness, see chapters 1 and 6.

64. Trévoux 1740.

65. Trévoux 1740 (quoting Saint Évremond): "le Poëte doit s'attacher à bien imiter la nature, à bien peindre la nature" (the poet should strive to imitate nature properly, to depict it well). From Longinus translated by Boileau: "On gâte le sublime si on l'abandonne à l'impétuosité d'une nature ignorante et téméraire" (sublimity is spoiled when left in the impetuous hands of a ruthless, ignorant nature). On *ars* and *ingenium*, see chapters 1 and 2.

66. Nicot 1606.

67. "Talent naturel . . . la disposition qu'on a à une chose plutôt qu'à une autre. . . . humeur, goûts, façons des gens" (natural talent . . . natural disposition towards one thing rather than another. . . . people's mood, tastes, manners); "Dans le système de la fable . . . , Dieu de la nature, qui a la force & vertu de produire toutes choses" (in the poetic economy of the fable . . . God of nature, who has the strength and power to produce all things); "en Poësie, sur tout d'une intelligence qu'on suppose attachée à une personne pour régler

sa conduite, l'aider dans ses entreprises, &c" (in poetry, is said of an intelligent being supposedly tied to a person in order to regulate their behaviour, help them in their endeavours, etc.).

68. Richelet 1680: "avec une bonne ou une méchante epitète veut dire *bon esprit*, ou *petit esprit*" (with a positive or negative qualifier means good wit, or petty wit); Trévoux 1740: "signifie plus ordinairement, l'esprit, ou la faculté de l'ame entant qu'elle pense, ou qu'elle juge" (means more commonly, the mind, or the faculty of the soul inasmuch at it thinks and judges). Saint Lambert 1757: "L'étendue de l'esprit, la force de l'imagination, & l'activité de l'ame, voilà le *génie*" (an expansive mind, a strong imagination, an active soul: that is genius).

69. On the relationship between *genius*, *genie* and *djinn*, see chapter 6. The regular translation of the Arabic *djinn* into the French *génie* or the English *genie* drew on pagan and Christian notions of the genius figure, but they are not connected etymologically.

70. Apuleius, *De deo Socratis*, 374–377, 15–4.

71. See chapter 1.

72. Trévoux 1740: "Le genie n'était pas l'âme de chaque homme, mais un dieu tutélaire qui lui était donné." (Genius was not each man's soul, but a tutelary god granted to each of them).

73. Richelet 1680: "Ce mot avec une bonne ou une méchante epitète veut dire *bon esprit*, ou *petit esprit*. (Ce n'est pas un grand genie que Mr un tel. C'est un petit genie. Pauvre genie.)" (This word with a good or bad epithet means a solid mind, or a small wit. [Mr so and so is not a great mind. He is a small wit, a poor wit.]).

74. Trévoux 1740: "Le Cardinal de Richelieu etoit le plus grand génie qui ait paru pour le gouvernement" (Cardinal Richelieu was the greatest genius ever in matters of governance).

75. Trévoux 1740, quoting from Bouhours and Saint-Évremond respectively.

76. Trévoux 1740 (quoted from Jean-Baptiste Morvan, abbé de Bellegarde 1761, 1:123): "De la discrétion, et de la retenue." This witticism became a commonplace.

77. The presence of Shaftesbury and Locke in this entry reflects the growing importance of English: see chapter 6.

78. For an example of the relevance of this context, see e.g. Yves Citton, "Spirits across the Channel. The Staging of Collective Mental Forces in Gabalistic Novels from Margaret Cavendish to Charles Tiphaigne de la Roche," *Comparatio. Zeitschrift für Vergleichende Literaturwissenschaft*, 1. 2 (2009): 291–319.

79. See Michael Moriarty, *Taste and Ideology in Seventeenth-Century France* (Cambridge: Cambridge University press, 1988).

5 German and Dutch

1. The relation between *Genie* and *ingenium* is found in Adelung 1787, 2:361–362.

2. Adelung 1793–1801, s.v.: "Kopf wäre vielleicht noch das einzige Deutsche Wort, welches das Französische mit der Zeit verdrängen könnte."

3. Adelung 1793–1801, s.v.: "Einige Neuere haben diesen Ausdruck für das Französ. Genie einführen wollen, welches aber dadurch nicht erschöpfet wird, weil zum Genie vornehmlich die obern Kräfte erfordert werden."

4. Cornelis Dekker, *The Origins of Old Germanic Studies in the Low Countries* (Leiden: Brill, 1999), 13–14. On the "otherness" fostered by Tacitus, consider the widely read *Germania* of Aeneas Silvius Piccolomini (later Pius II), the first seriously to employ Tacitus—Piccolomini praised Germanic tribes, but was taken to slight German culture because he defended Rome against northern concerns, out of concern for the Roman pontiffs; German humanists could point out that Tacitus held the ancient Germans as ethical superiors to the Romans, albeit primitive. See Else L. Etter, *Tacitus in der Geistesgeschichte des 16. und 17. Jahrhunderts* (Basel: Helbing and Lichtenhahn, 1966), 150.

5. Spellings vary during this period. These lists summarise a study of Dasypodius 1536, Frisius 1541, Calepino 1590, Calepino 1718, systematically comparing entries for the following terms: *astutus, astutia, facilitas, genius, genuinus, indoles, ingenium, ingeniosus, ingeniose, ingenuus, ingenuitas, inspiratio, intellectus, intelligens, natura, sagax, sagacitas, solers, sollertia, subtilitas.*

6. For example, see König 1668, where *ingenium* is "natürlicher Verstand." More generally see "Vernunft; Verstand," in *Historisches Wörterbuch der Philosophie*, ed. Joachim Ritter and Karlfried Gründer, vol. 3 (Darmstadt: Wissenschaftliche Buchgesellschaft, 1974), 748–863.

7. The great German dictionary of Stieler (1691) placed *Ver-nuft* under *ne[h]men*, reinforcing it as mental appropriation—to understand, see, or perceive. One of the most sophisticated Dutch lexicographers of this period, Lambert ten Kate (1723), also placed *vernunft* under *noemen*.

8. Stieler 1616 (German) and ten Kate 1723 (Dutch) respectively placed the word under *standen*.

9. Sommer, *Génie*; Schmidt, *Die Geschichte des Genie-Gedankens*; Bettina Brockmeyer, *Selbstverständnisse: Dialoge über Körper und Gemüt im frühen 19. Jahrhundert* (Göttingen: Wallstein, 2009).

10. This confirms and qualifies the recent argument of Stefanie Buchenau, *The Founding of Aesthetics in the German Enlightenment: The Art of Invention and the Invention of Art* (Cambridge: Cambridge University Press, 2013), that German aesthetics grew out of classical notions of *ingenium*. Cf. Frederick C. Beiser, *Diotima's Children: German Aesthetic Rationalism from Leibniz to Lessing* (Oxford: Oxford University Press, 2009).

11. E.g. Wolfhart Henckmann and Konrad Lotter, *Lexikon der Ästhetik* (Munich: Beck, 1992).

12. E.g. for *subtijl* see *inter alia* Madoets 1573 s.v. "Geest"; Arsy 1643 s.v. "Gheest," "Subtijl"; Hexham 1647 s.v. "Gheestigh"; van den Ende 1654 s.v. "Ingenieux." On words

about *Kunst*, but etymologically related to τέχνη during this period, see also Wilfried Seibicke, *Technik: Versuch einer Geschichte der Wortfamilie um* τέχνη *in Deutschland vom 16. Jahrhundert bis etwa 1830* (Düsseldorf: VDI-Verlag, 1968); see also Paul Taylor on the late emergence of *technique* in French, in "From Mechanism to Technique: Diderot, Chardin, and the Practice of Painting," in *Knowledge and Discernment in the Early Modern Arts*, ed. Sven Dupré and Christine Göttler (London: Routledge, 2017).

13. E.g. Ingeborg Hartmann-Werner, *"Gemüt" bei Goethe: Eine Wortmonographie* (Munich: W. Fink, 1976).

14. Jonathan West, *Lexical Innovation in Dasypodius' Dictionary, A Contribution to the Study of the Development of the Early Modern German Lexicon Based on Petrus Dasypodius' Dictionarium Latinogermanicum, Strassburg 1536* (Berlin, Boston: De Gruyter, 2013); more generally see Considine, *Dictionaries in Early Modern Europe*.

15. On the Germanic self-consciousness of Celtis and his circle, see Eckhard Bernstein, "From Outsiders to Insiders: Some Reflections on the Development of a Group Identity of the German Humanists between 1450 and 1530," in *In Laudem Caroli: Renaissance and Reformation Studies for Charles G. Nauert*, ed. James V. Mehl (Kirksville, MO: Truman State University Press, 1998).

16. For a way into the enormous literature on Gessner's lexicographically inflected natural history, see e.g. Florike Egmond and Sachiko Kusukawa, "Circulation of Images and Graphic Practices in Renaissance Natural History: The Example of Conrad Gessner," *Gesnerus* 73, no. 1 (2016): 29–72; Angela Fischel, "Collections, Images and Form in Sixteenth-Century Natural History: The Case of Conrad Gessner," in "Picturing Collections in Early Modern Europe," ed. Alexander Marr, special issue of *Intellectual History Review*, 20, no. 1 (2010): 147–64.

17. Gessner, in his preface to Maaler 1561, *6v. "decere videlicet homines eruditos, ut inter alia quibus patriam quisque suam ornat, linguae etiam maternae excolendae (ceu artium et scientiarum omnium, et totius civilis vitae, immo religionis etiam ac pietatis, instrumenti necessarii) rationem habeat."

18. Maaler 1561 also offers words that ring with similar meaning: *Klug* (clever, prudent: "Klüger mensch, artlich, zierlich, hurtig in allen dingen. Elegans homo"); and *Witzig* (knowing: "Ich bin Witzig und verstendig gnüg. Sat sapio").

19. See Introduction and chapter 1.

20. Plantin had Madoets collect this dictionary out of several other works including Estienne and especially Maaler. See Considine, *Dictionaries in Early Modern Europe*, 146, and Frans Claes, *De bronnen van drie woordenboeken uit de drukkerij van Plantin : Het Dictionarium tetraglotton (1562), de Thesaurus theutonicae linguae (1573) en Kiliaans Eerste Dictionarium teutonico-latinum (1574)* (Ghent: Koninklijke Vlaamse Academie, 1970), 143–266.

21. In the sixteenth century, there were three German grammars published in the 1570s, all in Latin: Albertus 1573; Ölinger 1574; Clajus 1578. The first known Dutch

grammar is Spieghel 1584. See Geert R.W. Dibbets, "Dutch Philology in the 16th and 17th Century," in *The History of Linguistics in the Low Countries*, ed. Jan Noordegraaf, C. H. M. Versteegh, and E. F. K. Koerner (Amsterdam: John Benjamins, 1992).

22. C. J. Wells, *German, a Linguistic History to 1945* (Oxford: Clarendon Press, 1985), 269. Out of the *Gesellschaft* emerged the first generation to make German literary in the same sense as the Du Bellays, Jacques Peletier du Mans and their circles had done for French a century earlier: e.g. Opitz 1617, which set out a standard of Germanic "purity" in poetry.

23. Goropius 1580 first suggested the notion of monosyllabic *primagenia*; this was theorised at length by Schottelius 1641. See William Jervis Jones, "Lingua teutonum victrix? Landmarks in German Lexicography (1500–1700)," *Histoire Épistémologie Langage* 13, no. 2 (1991): 131–52, at 139–40.

24. Considine, *Dictionaries in Early Modern Europe*, 144–5.

25. ten Kate, *Aenleiding*, 1.402.

26. Gerrit H. Jongeneelen, "Lambert ten Kate and the Origin of 19th-Century Historical Linguistics," in *The History of Linguistics in the Low Countries*, ed. Jan Noordegraaf, C. H. M. Versteegh, and E. F. K. Koerner (Amsterdam: John Benjamins, 1992).

27. Vivian Salmon, "Anglo-Dutch Linguistic Scholarship: A Survey of Seventeenth-Century Achievements," in *The History of Linguistics in the Low Countries*, ed. Jan Noordegraaf, C. H. M. Versteegh, and E. F. K. Koerner (Amsterdam: John Benjamins, 1992); Noel Edward Osselton, "Bilingual Lexicography with Dutch," in *Wörterbücher: Ein internationales Handbuch zur Lexikographie*, ed. Franz Josef Hausmann et al., vol. 3 (Berlin: Walter de Gruyter, 1991).

28. Osselton, "Bilingual Lexicography."

29. For more on Kramer, see Laurent Bray, *Matthias Kramer et la lexicographie du français en Allemagne au XVIIIe siècle: Avec une édition des textes métalexicographiques de Kramer* (Berlin: Walter de Gruyter, 2000).

30. E.g. Maaler 1561: "Wider sein alte Art widerumb an sich nemmen / thün wie von alter här. Ad ingenium reverti," which sets up an equivalence between a German and Latin saying.

31. Even in the late nineteenth-century, Friedrich Kluge (consulted in the English translation of Kluge 1891) observed possible connections with tilling and *ard*, the old Saxon for native dwelling, but continued that "it is more probable that *Art* is connected with Lat. *Ars*." Adelung notes as likely sources the verbs *ären*, *pflügen* (to plow, to till). This is the lineage attested in the current authoritative *Etymologisches Wörterbuch* (consulted online 9 September 2015), which makes no reference to the Latin.

32. Considine, *Dictionaries in Early Modern Europe*, 138.

33. See chapter 2.

34. The earlier entries of Kiliaan 1599 were, as we might expect from the foregoing narrative, modest and Latinate, revolving around *aerd* (*indoles, natura, ingenium,*

genius), the verb *aerden* (Indolem sive naturam sequi, *aerden nae den vader*; Patrissare; ingenium sive indolem patris referre), and related modifiers.

35. See also Gladov 1728, where *Naturel* is translated as "Geburts-Art, Zuneigung von Natur, Sinn, Humeur. . . ." (inborn kind, natural tendency, sense, humour).

36. On the relationship between style, art, and race, see Carlo Ginzburg, "Style: Inclusion and Exclusion," in *Wooden Eyes: Nine Reflections on Distance*, trans. Martin Ryle and Kate Soper (New York: Columbia University Press, 2001).

37. Wachter 1737: "Ab *erde* terra, quia res naturales ingenium eius soli, in quo natae sunt, referre solent, non fructus tantum, sed etiam incolae." Kramer 1676, under the Italian headword *genio*, also evoked the notion of the *genius linguae*: "il genio della lingua Italiana, *die Art und Eigenschafft der Italiänischen Sprache, etc.*"

38. Cf. Kiliaan 1599, who defines *aerd* as "indoles, natura, ingenium, genius," and then only follows with a second word, *aerde/eerde*, as *terra*.

39. ten Kate 1723, 634: "Daerenboven, gelijk de Heidenen in 't algemeen dat gene, waer van ze de grootste nuttigheid ontfingen, Goddelijke eere bewesen, zo hebben ook de Oud-Duitsche Voor-ouderen, volgens de getuigenis van Tacitus (van de zeden der Germannen cap. 40) HERTHA, dat is, DE AERDE gedient, als de Moder en Voedster van alle vleesch, die ons als op of in haren schoot koestern, verblijf, en rust verleent." See also chapter 6.

40. See chapter 1.

41. He first advertised the project in van Iperen 1755, and then followed up with examples seven years later (van Iperen 1762a and 1762b).

42. E.g. Estienne 1543. "Ingenium patris habet."

43. Wachter 1737, s.v. *Art*.

44. See Kant 2006, §40. This work collects lectures Kant used to teach; the first version of these lectures, dated 1772–73, is titled "Vom Witz und Scharfsinnigkeit." Later editions dropped the word "Scharfsinnigkeit." For the standard view, see Wolfgang Ritzel, "Kant über den Witz und Kants Witz," *Kant-Studien; Berlin* 82, no. 1 (1991): 102–109.

45. Kant 2006, §44. In this, he was influenced by developments in English philosophy concerning the distinction between *wit* and *judgment* (see chapter 6). Kant's response to English thinkers is enormously complex, both biographically and conceptually. E.g. Manfred Kuehn, *Scottish Common Sense in Germany, 1768–1800: A Contribution to the History of Critical Philosophy* (Kingston and Montreal: McGill-Queen's Press, 1987), and Daniel Garber and Béatrice Longuenesse, eds., *Kant and the Early Moderns* (Princeton NJ: Princeton University Press, 2008).

46. The word only arises in the *Dictionary of Untranslatables* as part of *Gefühl* (feeling), chiefly as part of eighteenth-century discourse of sentiment. See Dubost 2014.

47. See also Henisch 1616, who uses the term for the liberal arts.

48. See chapter 6.

49. Kiliaan 1599.

50. Maaler 1561: "Nach seinem Sinn, Fantasey, art und lust, läben. Ingenio suo vivere."

51. Maaler 1561: "Ein güter verstand. Facultas ingenii."

52. On this twinning, see chapter 2.

53. Madoets 1573: "Sinrijck, verstandich ende begrijpich in den sin. Riche d'entendement, ingenieux, capable de sens. Ingeniosus, capax, docilis, sagax, solers, industrius."

54. Thus in Kramer 1676, "sinnlich / id est, fleischlich, sensuale."

55. Kramer 1700: "Sinn-reich / f. Adj. Ricco di senno ò di sentimenti, cioè Sensato, Arguto, Sottile, Spiritoso, Ingenioso."

56. See chapter 2. Note that Kramer's German-French dictionary of 1712–15 followed the lead of its German-Italian predecessor.

57. E.g. Kiliaan 1599, Rodriguez 1624, Binnart 1649, Hexham 1647.

58. Hexham 1647.

59. Binnart 1649. (This is the second of at least 28 editions; the first edition was published 1635, and later editions seem to change very little.)

60. Kramer 1700-2. "ein sinnreicher Spruch / una sentenza ingeniosa. ein sinnreiches Gedicht / un poema spiritoso."

61. Stieler 1691: "aliquam corde machinari astutiam . . . fabricare fraudes."

62. Wachter 1737: "Proprie est *figura ingeniosa*, ex typo et lemmate composita, et dicitur ingeniosa a *sinn* ingenium . . . Pauca vero ingenia tam alte assurgunt, ut symbolum condere perfectum possint."

63. The first author to use *Sinnbild* as a synonym for emblem seems to be Hudemann, 162. For Hudemann's account of the genre, see Ingrid Höpel, *Emblem und Sinnbild: Vom Kunstbuch zum Erbauungsbuch.* (Frankfurt am Main: Athenäum, 1987), 116–7, 121. For a good example of how *Sinnbild* were expected to engage, and even shape, their reader's *Sinnen*, see e.g. the introduction to Camerarius 1671. The bibliography of *Sinnbild* as a genre is vast, with useful entries in Anthony John Harper and Ingrid Höpel, eds., *The German-Language Emblem in Its European Context: Exchange and Transmission* (Librairie Droz, 2000); Arthur Henkel and Albrecht Schöne, *Emblemata: Handbuch zur Sinnbildkunst des XVI. und XVII. Jahrhunderts*, 1st ed. 1976 (Berlin: Springer, 2013).

64. Weise 1678. The origins of such epigrammatic genres and their demise in the early eighteenth century are outlined by Thomas Neukirchen, *Inscriptio: Rhetorik und Poetik der Scharfsinnigen Inschrift im Zeitalter des Barock* (Tübingen: Niemeyer, 1999).

65. Sasbout 1576.

66. Madoets 1573. "Mijnen geest is geneycht om di te volgen. *Mon esprit est enclin à suivre ceulx là.* Hos ut sequare inclinat animus." This seems to come from Estienne, who gives "inclinare animum ad rem aliquam" s.v. *animus*. There are some exceptions, such as when "light of spirit" translates *esprit* and *ingenium* ("Licht van geest. *D'esprit legier et vif.* Ingenium vividum et acre, alacer"). But these cannot come from Estienne 1538 or 1571, which do not supply this example for *ingenium*.

67. See chapter 4.

68. The literature on this painting is substantial. See, most recently, Fiona Healy, "Vive l'Esprit. Sculpture as the Bearer of Meaning in Willem van Haecht's Art Cabinet

of Cornelis van Der Geest," in *Munuscula Amicorum: Contributions on Rubens and His Colleagues in Honour of Hans Vlieghe*, ed. K. van de Stighelen (Turnhout: Brepols, 2006); Charles M. Peterson, "The Five Senses in Willem II van Haecht's Cabinet of Cornelis van der Geest," in "Picturing Collections in Early Modern Europe," ed. Alexander Marr, special issue of *Intellectual History Review* 20, no. 1 (2010): 103–21; Alexander Marr, "Ingenuity and Discernment in *The Cabinet of Cornelis van der Geest*," *Nederlands kunsthistorisch Jaarboek* (forthcoming). On the "pictures of collections" genre, see Alexander Marr, "The Flemish 'Pictures of Collections' Genre: An Overview," in "Picturing Collections in Early Modern Europe," ed. Marr.

69. Panofsky favoured this English translation for the Latin phrase, which Dürer used in his engraved portrait of Wilibald Pirckheimer (1524). See Erwin Panofsky, *The Art and Life of Albrecht Dürer* (Princeton NJ: Princeton University Press 2005), 239.

70. See, for example, the depiction of *ingegno* in Ripa, 1613.

71. Bart Ramakers, "Embodied Wits: Personifications of Thinking, Sensing, Feeling and Willing in Rhetoricians' Drama," (forthcoming). We thank Ramakers for sharing an early version of this article. Notably, the painter Van Haecht's great uncle—also called Guillem—was a prominent member of this Chamber. On the *Violieren* and Antwerp's *liefhebbers*, see Zirka Zaremba Filipczak, *Picturing Art in Antwerp, 1550–1700* (Princeton NJ: Princeton University Press, 1986).

72. Walter S. Melion, *Shaping the Netherlandish Canon: Karel Van Mander's Schilder-Boeck* (Chicago: University of Chicago Press, 1991), 65. On this topos, we may note the entry *Na 't leven* in Marin 1696, where it is made the equivalent of *au naturel*: "Naturel, (m) *Aart, imborst* (m), *gemoeds-gesteldheid* (f); un bon naturel; être timide de son naturel; avoir un beau-naturel pour la poésie, *schoone bequaamheid hebben voor de dichtkunde;* au naturel (adv.) *na 't leven.*" For a discussion of *naer het leven* and *uyt den geest* see Claudia Swan, *Art, Science, and Witchcraft in Early Modern Holland: Jacques de Gheyn II (1565–1629)* (Cambridge: Cambridge University Press, 2005).

73. *Ibid.*, p. 66. For *geest* in Van Mander see also Hessel Miedema, *Kunst, kunstenaar en kunstwerk bij Karel van Mander: Een analyse van zijn levensbeschrijvingen* (Alphen aan den Rijn: Canaletto, 1981), 2–3, 165–168.

74. See chapter 2.

75. Binnart 1649: "Geest/ Engel. Quaet of goet/ Genius."

76. Binnart 1649: "Subtiliseren / scherpsinnigh maecken / de sinnen scherpen/ acuere, exacuere. ut *disputeren ende discoureren scherpt de sinnen ende den geest*, disputationes et discursus ingenium acuunt." Perhaps Binnard depended on Kiliaan 1599, who first offers *genius* as a direct translation, and then gives examples of "native potential" that typically belong to *ingenium*: "gheest. Genius. *Eenen ghest hebben*. Genium habere: nativa quadam indole et arcana ratione animos hominum movere, et ad sui amorem sponte sua trahere, vehementerque afficer."

77. Indeed, modern *geestig* has the sense of "funny," like *wit* in English or *witz* in German.

78. Ingeborg Hartmann-Werner, *"Gemüt" Bei Goethe: Eine Wortmonographie*, Münchner Germanistische Beiträge ; Bd. 23 (München: W. Fink, 1976); Hildegard Emmel, "Gemüt," in *Historisches Wörterbuch der Philosophie*, ed. Joachim Ritter and Karlfried Gründer, vol. 3 (Darmstadt: Wissenschaftliche Buchgesellschaft, 1974), 258–62; Klaus Junk, *Die Funktion von "Gemut" in der Literatur-theorie und—kritik (18. und 19. Jahrhundert)* (Erlangen, 1968). In particular, Goethe seems to make the word do the work of *ingegno* in Italian, saying that "Ohne Gemüt ist keine wahre Kunst denkbar." Cit. Otto Stelzer, *Goethe und die Bildende Kunst*, 1st ed. 1949 (Berlin: Springer, 2013), 138. See also Schlegel, in whose early work Jeffrey Librett finds *Gemüt* an "obsessional motif": Jeffrey S. Librett, *The Rhetoric of Cultural Dialogue: Jews and Germans from Moses Mendelssohn to Richard Wagner and Beyond* (Stanford: Stanford University Press, 2000), 109, 320n16. The topic of poetic *Gemüt-neigung* concludes the argument against artistic rules in Breitinger and Bodmer 1740, 466ff.

79. Maaler 1561 is slightly more helpful, offering only mental faculties: *Verstand* and *Vernunfft* as well as *animus* and *mens*.

80. Henisch 1616: "GEMÜTH . . . hertz, animus, mens, affectus, cor. GEMÜT: der sinn, die gedancken meijnung, sensus quo aliquid percipimus."

81. Tieck 1798, 1:188: "Meine innerlichen Bilder vermehren sich bei jedem Schritte, den ich tue, jeder Baum, jede Landschaft, jeder Wandersmann, Aufgang der Sonne und Untergang, die Kirchen, die ich besuche. . . . Meine Gemüt ist nunmehr so verwirrt, dass ich mich durchaus nicht unterstehen darf, selber an die Arbeit zu gehen." Cf. Brad Prager, *Aesthetic Vision and German Romanticism: Writing Images* (Rochester, NY: Camden House, 2007), 53.

82. Consider Dürer's reception in Goethe's circle.

83. The "aesthetic excursus" was based on a series of fourteen preparatory texts, composed ca. 1509-ca. 1513, which survive in Dürer 1956–9, 3:270–90. On the "aesthetic excursus" see, most recently, Jeffrey Ashcroft, "Art in German: Artistic Statements by Albrecht Dürer," *Forum for Modern Language Studies* 48, no. 4 (2012): 377–88; Peter Parshall, "Graphic Knowledge: Albrecht Dürer and the Imagination," *The Art Bulletin* 95, no. 3 (2013): 393–410. On Dürer and ingenuity see Alexander Marr, "Ingenuity in Nuremberg: Dürer and Stabius's Instrument Prints," *The Art Bulletin* (forthcoming 2018).

84. Eckhart 1993, 2:188 (Sermon 83): "das do mens heiset oder gemüote . . . die heisent die meistere [i.e. Augustine] ein sloz oder einen schrin in geistlicher formen oder formelicher bilde."

85. E.g. the anonymous late medieval tract *Theologia deutsch*, much praised by Martin Luther in an edition he introduced with a preface in 1518: Anon. 1908, 49, 52, 54.

86. See chapter 4.

87. See chapter 3.

88. Huarte 1659, 2–3.

89. Huarte 1659, 3.

90. E.g. Maaler 1561, s.v. *Geschwindigkeit*; Kramer 1676, s.v. *ingegno* and *genio*; Kramer 1700, s.v. *Sinn*.

91. Martin Franzbach, *Lessings Huarte-Übersetzung (1752): Die Rezeption und Wirkungsgeschichte des "Examen de ingenios para las ciencias" (1575) in Deutschland* (Hamburg: De Gruyter, 1965), 81–83.

92. Gottsched, *Neuesten aus der anmuthigen Gelehrsamkeit* (Feb., 1757), 152, cit. Franzbach, *Lessings Huarte-Übersetzung*, 81: "Unsere neuen Sprachverderber würden wohl gar schweren, daß er [der rezensierte Autor] das undeutsche Ding besitze, was sie Genie nennen; aber mit keiner deutschen Zunge ausgesprochen werden kann. Wir aber, die wir bey deutschen Sylben eben so viel als sie bey ausländischen denken, würden es Geist und Lebhaftigkeit des Witzes nennen" (The new corruptors of our language would gladly swear that [the reviewed author] possesses the unGerman thing, what they call Genie—but it cannot be pronounced by any German tongue. But we, who think in German syllables as much as they do in foreign ones, we would call it Geist and liveliness of wit).

93. See note 2 above.

6 English

1. Abercromby 1685, 6–7. Abercomby's text is discussed in detail in Thomas Colville, "Mental Capacity and the Pursuit of Knowledge in England 1650–1700," PhD diss. (King's College, London, 2017).

2. See Jonathan Bate, "Shakespeare and Original Genius," in *Genius*, ed. Murray; Lawrence Lipking, "Johnson and Genius," in *Samuel Johnson: The Arc of the Pendulum*, ed. Freya Johnston and Linda Mugglestone (Oxford: Oxford University Press, 2012); Simon Schaffer, "Fontenelle's Newton and the Uses of Genius," *L'Esprit Créateur* 55, no. 2 (2015): 48–61.

3. On the etymology and semantics of the Latin terms see above, chapter 1, and Vallini, "Genius/ingenium." In one of the earliest Anglo-Saxon dictionaries—Nowell 1567 (ca.)—we find entries for "Cunninȝ; Cunning, expert" and "Gleawnesse. Witte, discretion, cunning, circumspection." Minsheu 1617 offers *weten*. See also C. S. Lewis, *Studies in Words* (Cambridge: Cambridge University Press, 1967), 86.

4. See e.g. Geraldine Barnes, "Cunning and Ingenuity in the Middle English Floris and Blauncheflur," *Medium Aevum* 53, no. 1 (1984): 10–25. While Johnson 1755, identifies the Italian equivalent of *engine* as *ingegno*, the *ingenuity* sense is entirely absent from his definitions. See also the chapter "Mechanical" in Owen Barfield, *History in English Words* (London: Methuen, 1926).

5. Ben Jonson, *The Entertainment at Britain's Burse* (1609), ed. James Knowles, in *The Cambridge Edition of the Works of Ben Jonson Online*, ed. David Bevington, Martin Butler, and Ian Donaldson (Cambridge: Cambridge University Press, 2014), http://universitypublishingonline.org/cambridge/benjonson/.

6. The motto *Sal sapit omnia* was the motto of the Salterers' Company, but also a pun

on the title of Robert Cecil, Earl of Salisbury, who had supported the *Exchange*. On salt and ingenuity, see chapter 3.

7. See e.g. Bailey 1737: "*Engine*: [in a figurative sense] an artifice, contrivance or device." See also Kenneth K. Ruthven, *The Conceit* (London: Methuen, 1969).

8. The only instance of comparative analysis, grounded on statistical analysis of word pairings in digitized texts, is Colville, "Mental Capacity."

9. Lewis, *Studies in Words*.

10. Ruthven, *Conceit*, 2. See also Parker, "'Concept' and 'Conceit,'" which includes a discussion of the difficulties in attempting a comparative study of early modern *conceit*.

11. See e.g. the entries in Minsheu 1617, for *cunne, cunning* and *witte*.

12. On the early development of these trends in Latin, see above, chapter 1.

13. For *wit* in rhetoric see e.g. William G. Crane, *Wit and Rhetoric in the Renaissance: The Formal Basis of Elizabethan Prose Style* (New York: Columbia University Press, 1937); D. Judson Milburn, *The Age of Wit, 1650–1750* (New York: Macmillan, 1966); James Biester, *Lyric Wonder: Rhetoric and Wit in Renaissance English Poetry* (Ithaca, NY: Cornell University Press, 1997). For faculty psychology, including theories of the "five wits," see E. Ruth Harvey, *The Inward Wits: Psychological Theory in the Middle Ages and Renaissance* (London: The Warburg Institute, 1975).

14. Despite Johnson's evident importance, recent scholarship has revealed the significance of his eighteenth-century predecessors while noting that in certain respects his dictionary was not "especially innovative in many of its features" (Anne C. McDermott, "Introduction," in *Ashgate Critical Essays on Early English Lexicographers, Volume 5: The Eighteenth Century*, ed. Anne C. McDermott (Farnham: Ashgate, 2012), xv.

15. There was, however, widespread recognition of the need for improved dictionaries well before Johnson, notably in the Hartlib circle and the early Royal Society. See John Considine, "Introduction," in *Ashgate Critical Essays on Early English Lexicographers, Volume 4: The Seventeenth Century*, ed. John Considine (Farnham: Ashgate, 2012), xvii. On seventeenth-century English language planning see Rhodri Lewis, *Language, Mind and Nature: Artificial Languages in England from Bacon to Locke* (Cambridge: Cambridge University Press, 2007).

16. There is a substantial literature on English lexicography before Johnson. The classic account remains DeWitt T. Starnes and Gertrude E. Noyes, *The English Dictionary from Cawdrey to Johnson, 1604–1755* (Chapel Hill: North Carolina State University Print Shop, 1946), although recent work has offered significant revisions. For concise accounts with up-to-date bibliographies, see the essays by Bately, Cormier, and Assleton in *The Oxford History of English Lexicography*, ed. Anthony P. Cowie (Oxford: Oxford University Press, 2009). See also the essays collected in Ian Lancashire, ed., *Ashgate Critical Essays on Early English Lexicographers*, 5 vols. (Farnham: Ashgate, 2012), especially the editors' introductions to vols. 3–5. For bibliographies of period dictionaries see, in addition to the works cited above, Robin C. Alston, *A Bibliography of the English*

Language from the Invention of Printing to the Year 1800 (Leeds: Printed for the author, 1965–73); Jürgen Schäfer, *Early Modern English Lexicography* (Oxford: Clarendon Press, 1989); LEME.

17. While this chapter is concerned principally with "stand alone" printed dictionaries, it is important to note that in early modern England much lexicographical work appeared either in manuscript or in glossaries appended to publications that were not explicitly lexicographical in nature. See Considine, "Introduction," xvi.

18. With a few exceptions (for which see below), the ingenuity family of terms tends not to appear in specialized, "canting" dictionaries, on which see e.g. Maurizio Gotti, *The Language of Thieves and Vagabonds: 17th and 18th Century Canting Lexicography in England* (Berlin: De Gruyter, 1999). On the staggered development of dictionaries in the period see Considine, "Introduction," xvi and Roderick W. McConchie, "Introduction," in *Ashgate Critical Essays on Early English Lexicographers, Volume 3: The Sixteenth Century*, ed. Roderick W. McConchie (Farnham: Ashgate, 2012). McConchie (xli) makes the important point that commercial motives often lay behind the production of dictionaries in early modern England, which had a tendency to temper their evolution.

19. McDermott, "Introduction," xvi.

20. For the unfamiliarity of the English language abroad, see e.g. Peter Burke, *Languages and Communities in Early Modern Europe* (Cambridge: Cambridge University Press, 2004).

21. Edmund Weiner, "Early Modern English—An Overview," in OED Online (http://public.oed.com/aspects-of-english/english-in-time/early-modern-english-an-overview/; accessed 1.9.2016).

22. The OED cites Florio 1598, as the first instance, but a search in EEBO shows that it appears earlier, in Rainolds 1584, 196: "What sparkle of thankfulnes, but I let go thankfulnes, what sparkle of ingenuitie was there, and good nature, in *Marian Victorius*. . . ."

23. See McConchie, "Introduction," xiv; Considine, "Introduction," xxv-xxvi and, for an example of English ownership of continental dictionaries, xlix-xlx. For Estienne's influence see DeWitt T. Starnes, *Robert Estienne's Influence on Lexicography* (Austin: University of Texas Press, 1963). Likewise, there can be little doubt that major lexicographical achievements on the Continent, such as the *Vocabolario degli Accademici della Crusca* (1612 and subsequent editions) and the *Dictionnaire de l'Académie française* (1687–) spurred the developments in English lexicography leading to Johnson 1755 (see chapters 2 and 4). Equally, there is ample evidence that these large continental works informed English dictionary making, both structurally and in terms of content. For example, Torriano 1659, drew extensively on Crusca 1612. See Considine, "Introduction," xlii

24. McConchie, "Introduction," xiv-xv.

25. Considine, "Introduction," xvii.

26. Considine, "Introduction," xvi.

27. Lewis, *Studies in Words*, 89.

28. See McConchie, "Introduction," xxxvii.

29. To take just two examples, a free keyword search for *ingenuity* in the date range 1473–1750 produces 10,201 hits in 4,706 records, while a search for *ingenious* produces 21,397 hits in 7,449 records. For further statistical analysis of the ingenuity family of terms in English, see Colville, "Mental Capacity," 51.

30. For the word history of *genius* in English see Logan Pearsall Smith, *Words and Idioms: Studies in the English Language* (London: Constable and Co., 1925); Margaret L. Wiley, "Genius: A Problem in Definition," *Studies in English* 16 (1936): 77–83; David Bromwich, "Reflections on the Word *Genius*," *New Literary History* 17, no. 1 (1985): 141–64; Bate, "Shakespeare and Original Genius." Limited attention has been paid to its study as a concept in England before Romanticism, but see e.g. Joel E. Spingarn, *Critical Essays of the Seventeenth Century*, 3 vols. (Oxford: Clarendon Press, 1908); idem., *Creative Criticism: Essays on the Unity of Genius and Taste* (New York: Henry Holt, 1926), and the references at n. 41.

31. Wiley, "Genius." See especially the questions posed about words and meanings and the critique of the history of concepts (78–79).

32. Wiley, "Genius," 83.

33. This sense appears in Miège only as an example of usage in his definition of *genius/genie*: "*Do but observe the genius of the Age*, remarquez bien le genie du tems" and the French *porter*: "*Le tems le portoit ainsi,* such was the genius of the times."

34. This sense is, however, implicit in definitions that invoke the "genial" (birth-related) aspects of *genius*. For examples in other languages, see chapters 1 and 3.

35. For genius and the *daimon*, see McMahon, *Divine Fury*.

36. Notably, however, Elyot 1538 gives the headword *genii*, defined as "men, who do gyve all their studye to eatynge and drynkynge." This, presumably, is the "genial" aspect of the term, which is more pronounced in Latin and Italian (see chapters 1 and 2).

37. For Elyot's sources see Gabriele Stein, *Sir Thomas Elyot as Lexicographer* (Oxford: Oxford University Press, 2014). On this antiquarian impulse more generally, see chapter 1.

38. Chambers 1728: "Among the *Romans*, *Festus* observes, the Name *Genius* was given to the God who had the Power of doing all things, *Deum qui vim obtineret rerum omnium gerendarum*; which *Vossius, de Idololatriae* rather chuses to read *genendarum*, who has the Power of producing all Things; by reason *Censorinus* frequently uses *genere* for *gignere*."

39. Ibid.

40. Phillips 1658 notes in his entry for *Tages* that this deity was "the Grand-child of Jupiter, and son of Genius, he is said to have taught the Hetrurians the art of divining, when he was a boy of twelve year old." This may connect to the relationship of *genius* to prophesy.

41. E. C. Knowlton, "Genius as an Allegorical Figure," *Modern Language Notes* 39,

no. 2 (1924): 89–95; idem., "The Genii of Spenser," *Studies in Philology* 25, no. 4 (1928): 439–56; Wiley, "Genius"; DeWitt T. Starnes, "The Figure Genius in the Renaissance," *Studies in the Renaissance* 11 (1964): 234–44, all of whom wrote in response to the figure of Genius in Spenser. See also Morton, "Ingenious Genius," on playful conflations of the "natural" generation sense of *genius* with its traditional opposite—the artifice of "craft"—in the High Middle Ages.

42. E.g. Elyot 1538: "dilectation moved by nature"; Cooper 1584: "the grace and pleasantness of a thing." The latter derives from the entry for *angel* in Baret 1574, the Latin for which is identified as *genius*.

43. *Genial* appears somewhat earlier as a headword, but still in relation to the definitions of *genius*, e.g. Cotgrave 1611: "Geniall; belonging to lucke, or chance; or to a mans nature, disposition, inclination."

44. We should note that across the the vernacular bilingual dictionaries, *genius* plays a relatively minor role. *Genie, génie, genio*, and *génio* appear only very sporadically, largely prior to the eighteenth century, and tend to be defined simply either as a "good or bad angel," or "natural disposition."

45. "Genius*,* a good or evil spirit attending on particular men or places, also Nature, fancy or inclination." Coles 1676.

46. In this sense fancy was often used interchangeably with fantasy, as in e.g. Baret 1574: "*fantasie*: *Tuo frui genio. A vostre fantasie: a vostre plaisir.* To live as he will him selfe, or after his owne fantasie. *Ingenio suo vivere.*" We may note here Baret's inclusion of both *genio* and *ingenio*. On "fancy" in early modern·England, see e.g. Charles H. Hinnant, "Hobbes on Fancy and Judgment," *Criticism* 18 (1976): 15–26; Lisa T. Sarasohn, *The Natural Philosophy of Margaret Cavendish: Reason and Fancy During the Scientific Revolution* (Baltimore and London: The Johns Hopkins University Press, 2010).

47. Blount 1656, quoting Wotton 1624.

48. On this relationship in the context of *wit*, see Adam Zucker, *The Places of Wit in Early Modern English Comedy* (Cambridge: Cambridge University Press, 2011).

49. For which see e.g. Klibansky, Panofsky, and Saxl, *Saturn and Melancholy*; Gowland, *Worlds of Renaissance Melancholy*; Brann, *The Debate*. Notably, Dryden observed that the description of the "humours" was the particular "genius and talent" of Ben Jonson. Smith, *Words and Idioms*, 96.

50. E.g. Cockeram 1623: "A breathing into" and "An instruction from God." See, however, Sir William Alexander's opinion "That every author hath his own Genius, directing him by a secret Inspiration to that wherein he may most excel." Sir William Alexander, *Anacrisis* (ca. 1635), quoted in Smith, *Words and Idioms*, 98.

51. See John Mee, *Romanticism, Enthusiasm, and Regulation: Poetics and the Policing of Culture in the Romantic Period* (Oxford: Oxford University Press, 2003); Eron, *Inspiration*.

52. Florio 1611: "*Enthusiasmó*: A Poeticall or propheticall fury, a ravishment of

senses from above." We may note that Thomas 1587 defines *furor* as "A vehement concitation or stirring of the minde: fury, madnesse, rage, anger: exceeding desire of: a ravishing of the minde, a trance, a poeticall furie."

53. However, we may note Johnson's third sense of the word, quoting one of Sidney's references to the *furor poeticus* in the *New Arcadia*: "Fury . . . 3. Enthusiasm; exaltation of fancy. 'Taking up her lute, her wit began to be with a divine fury inspired; and her voice would, in so beloved an occasion second her wit'" (Johnson 1755).

54. See below and Isabel Rivers, *Reason, Grace, and Sentiment: A Study of the Language of Religion and Ethics in England, 1660–1780* (Cambridge: Cambridge University Press, 1991).

55. Thomas Hobbes, "Answer to the Preface to Gondibert" (1650), quoted in part in Smith, *Words and Idioms*. Kevin Pask, *The Fairy Way of Writing: Shakespeare to Tolkien* (Baltimore: The Johns Hopkins University Press, 2013), chapter 3, offers a useful overview of these issues.

56. This is presumably connected to definitions of *genius* in other languages as symmetry and harmony of parts. See chapters 1 and, but in a somewhat different sense, 3.

57. See e.g. Huloet 1552: "Grace or beauty in doing or speaking. *Decor. oris*"; Thomas, 1587: "Venustas . . . : Beautifull, faire, delectable, pleasant to the eye, sightly, having a good grace." See also Rivers, *Reason, Grace, and Sentiment*; Ita Mac Carthy, "Grace," in *Rennaisance Keywords*, ed. Ita Mac Carthy (Oxford: Legenda, 2013), 63–80.

58. Wotton 1624, 12. For the history of "beauty" as a concept see e.g. Władysław Tatarkiewicz, *A History of Six Ideas: An Essay in Aesthetics* (The Hague: Martinus Nijhoff, 1980). A similar set of relationships is at play in the definition of the *ingegnoso* as "very beautiful" (Crusca 1691; see chapter 2).

59. See also Smith, *Words and Idioms*, who notes Sidney's opinion in the *Apology* that "A Poet no industrie can make, if he own Genius be not carried unto it" (96).

60. See Gilberta Golinelli, "The Concept of Genius and Its Metamorphoses in Eighteenth-Century English Criticism," *Textus: English Studies in Italy* 18, no. 1 (2005): 189–204; Daniel Cook, "On Genius and Authorship: Addison to Hazlitt," *Review of English Studies* 64, no. 266 (2013): 610–29.

61. OED, *genius*, sense AII.9: "It is not recognised in Johnson's Dictionary." See also Paul W. Bruno, *Kant's Concept of Genius: Its Origin and Function in the Third Critique* (London: Continuum, 2010), 13.

62. See e.g. his statement in the "Life of Cowley": "The true Genius is a mind of large general powers, accidentally determined to some particular direction." Johnson 1779, 4. See Isobel Grundy, *Samuel Johnson and the Scale of Greatness* (Leicester: Leicester University Press, 1986); Lipking, "Johnson and Genius."

63. Bate, "Shakespeare and Original Genius," 77–78; Jonathan Bate, *The Genius of Shakespeare* (London: Picador, 2008).

64. Leman 1755, 314. On the early reviews of the *Dictionary* see Allen Reddick, *The*

Making of Johnson's Dictionary, 1746–1773 (Cambridge: Cambridge University Press, 1990), 83–7.

65. See Greene, "Whichcote, Wilkins, 'Ingenuity,' and the Reasonableness of Christianity." Ingenuity also has a prominent place in Bacon's thought. See Lewis, "Francis Bacon and Ingenuity."

66. On eighteenth-century lexicographers' attempts to clear up the confusion, see Susie I. Tucker, *Protean Shape: A Study in Eighteenth-Century Vocabulary and Usage* (London: Athlone Press, 1967), 119–20.

67. See Lewis, *Studies in Words*; Rhodri Lewis, "Impartiality and Disingenuousness in English Rational Religion," in *The Emergence of Impartiality*, ed. Anita Traninger and Kathryn Murphy (Leiden: Brill, 2013).

68. *Ingenious* appears as a synonym for *acute* in English and the foreign vernaculars in e.g. Florio 1598; Cotgrave 1611; Blount 1656.

69. It is presumably a false etymology to connect *Kenard* to *canard* ("An extravagant or absurd story circulated to impose on people's credulity; a hoax, a false report": OED, 1, 1864 >). However, in his entry for *canard*, Cotgrave 1611, gives "vendeur de canards à moitié. A cousener, guller, cogger, foister, lyer": precisely the kind of guile associated with sharp ingenuity in Baret 1574.

70. Boyle 1665, 177.

71. Kenny, *Curiosity*; idem., *The Uses of Curiosity*.

72. Lewis, *Studies in Words*, 112.

73. This explains why Lloyd-Wilkins 1668 could offer "fancy" as a definition of *ingenious*.

74. See chapter 2. There is, of course, a long tradition (beginning with Socrates) in which virtue (*arete*) is considered a branch of knowledge (*episteme*). See Lisa Raphals, *Knowing Words: Wisdom and Cunning in the Classical Traditions of China and Greece* (Ithaca, NY: Cornell University Press, 1992), 3.

75. With this in mind, we may note that for much of our period the word *expert* (a synonym for which is *cunning*) is defined as one skilled in a wide manner of pursuits.

76. Kersey 1702 defines *virtuoso* as "a Learned and Ingenious Person, more especially, well skill'd in natural Philosophy." On the early modern *virtuoso* see, most recently, Craig A. Hanson, *The English Virtuoso: Art, Medicine, and Antiquarianism in the Age of Empiricism* (Chicago: Chicago University Press, 2009).

77. See Schaffer and Shapin, *Leviathan and the Air Pump*; Jim Bennett, "Instruments and Ingenuity," in *Robert Hooke: Tercentennial Studies*, ed. Michael Cooper and Michael Hunter (Aldershot: Ashgate, 2006); Lewis, "Impartiality,"; Colville, "Mental Capacity."

78. Hilliard ca. 1600. See Alexander Marr, "Pregnant Wit: *Ingegno* in Renaissance England," *British Art Studies* 1 (2015): http://dx.doi.org/10.17658/issn.2058–5462/issue-01/amarr.

79. Ripa 1603. See chapter 2.

80. Notably, Kersey 1702 offers for *inventive* "ingenious, or apt to invent."

81. Similarly, the word *ingenio* appears also as a headword, from Blount 1656 on, as a word used in Barbados for a sugar-mill. See chapter 3.

82. See also Minsheu 1617, *Engine*: "Italian Ingegno, machina ... Latin ingenio."

83. Wolfe, *Humanism, Machinery and Renaissance Literature*; Sawday, *Engines of the Imagination*.

84. OED, II.4. The OED gives the date 1600, but the term appears in the first, 1599 version of the play. Shortly after, Richard Lynche used *ingenuity* to denote natural quick-wittedness in his free translation of selections from Annius of Viterbo's forged *Auctores vetustissimi*. Commenting on Annius's account of Noah's son, Shem, Lynche wrote: "many writers have affirmed that this *Cham* [sic] was a man of singular ingenuitie and sharpe capacitie, and that hee first found out the seven liberall Sciences, and had wrote many bookes of great worth, among which, his cheefest were of Negromancie." Nanni 1601, Diijr. This passage does not appear in the original, hence Lynche's use of *ingenuity* is not a translation from Latin.

85. Jonson 1599, III.3.64–74.

86. Jonson 1599, III.3.89–90; 102–05.

87. Harvey 1592, 17–18; Nashe 1592, I4v.

88. Here, the dictionaries are not fully aligned with period usage, since quite early on *ingenuity* could stand for some of the sense of *genius* we have already encountered. For example, in 1601 the lawyer and historian William Fullbeck used *ingenuity* for *genius*—in the sense of "spirit of a people"—when quoting Pompey's speech before the Senate: "if the Italian ingenuitie, and the heate of the Romane bloud be as yet warme within the Romanes, let them not marke upon what earth they stand." Fulbecke 1601, 134.

89. The literature on wit is especially large. Specifically word-historical studies include Alex Aronson, "Eighteenth Century Semantics of Wit," *ETC: A Review of General Semantics* 5, no. 3 (1948): 182–90; William Empson, "Wit in the *Essay on Criticism*," *The Hudson Review* 2, no. 4 (1950): 559–77; Lewis, *Studies in Words*; Klára Bikanovà, "The Dangers of Wit: Re-examining C.S. Lewis's Study of a Word," *Brno Studies in English* 35, no. 1 (2009): 87–102; Colville, "Mental Capacity." The most important studies of the wider meanings and applications of early modern *wit* are Spingarn, *Critical Essays*; W. Lee Ustick and Hoyt H. Hudson, "Wit, 'Mixt Wit,' and the Bee in Amber," *The Huntington Library Bulletin* 8 (1935): 103–30; George Williamson, *The Proper Wit of Poetry* (London: Faber and Faber, 1961); Milburn, *Age of Wit*; J. W. Van Hook, "'Concupiscence of Witt': The Metaphysical Conceit in Baroque Poetics," *Modern Philology* 84, no. 1 (1986): 24–38; John Sitter, *Arguments of Augustan Wit* (Cambridge: Cambridge University Press, 1991); James Biester, "Fancy's Images: Wit, the Sublime, and the Rise of Aestheticism," in *Wonders, Marvels, and Monsters in Early Modern Culture*, ed. Peter G. Platt (Newark: University of Delaware Press, 1999); Roger D. Lund, *Ridicule, Religion and the Politics of Wit in Augustan England* (Farnham: Ashgate, 2012); Klára Bikanovà, "From Rhetoric to Aesthetics: Wit and Esprit in the English and French Theoretical Writings of the Late

Seventeenth and Early Eighteenth Centuries," PhD diss. (Masaryk University, 2011). Bikanovà's dissertation is largely derivative—without acknowledgment—of Milburn, but it includes a useful account of the historiography (17–38).

90. On which see Scholar, *The Je-Ne-Sais-Quoi*.

91. Flecknoe 1664, unpaginated.

92. "[W]it, abstracted from its effects upon the hearer, may be more rigorously and philosophically considered as a kind of *discordia concors*; a combination of dissimilar images, or discovery of occult resemblances in things apparently unlike. Of wit, thus defined, they have more than enough. The most heterogeneous ideas are yoked by violence together; nature and art are ransacked for illustrations, comparisons, and allusions; their learning instructs, and their subtilty surprises; but the reader commonly thinks his improvement dearly bought, and, though he sometimes admires, is seldom pleased." Johnson 1779, 20–21. See David Perkins, "Johnson on Wit and Metaphysical Poetry," *ELH* 20, no. 3 (1953). On "false wit" see Roger D. Lund, "The Ghosts of Epigram, False Wit, and the Augustan Mode," *Eighteenth Century Life* 27, no. 2 (2003). On *discordia concors* and "metaphysical wit" see A. J. Smith, *Metaphysical Wit* (Cambridge: Cambridge University Press, 1991).

93. See Roger D. Lund, "Wit, Judgment, and the Misprisions of Similitude," *Journal of the History of Ideas* 65, no. 1 (2004). There is a substantial literature on *wit* in Hobbes. Important recent studies include Milburn, *Age of Wit*; Raman Selden, "Hobbes and Late Metaphysical Poetry," *Journal of the History of Ideas* 35, no. 2 (1974): 197–210; Quentin Skinner, *Reason and Rhetoric in the Philosophy of Hobbes* (Cambridge: Cambridge University Press, 1996). On the later eighteenth-century philosophical fortunes of the distinction between wit and judgment, see Stanley Corngold, "Wit and Judgment in the Eighteenth Century: Lessing and Kant," *MLN* 102, no. 3 (1987): 461–82.

94. Miège 1677 has for *sublime*: "haut, élevé, sublime, *exalted, high, or lofty*. Un esprit sublime, *a sublime (or exalted) wit*." For *wit* and the sublime, see Scott Elledge, "Cowley's Ode "Of Wit" and Longinus on the Sublime: A Study of One Definition of the Word Wit," *Modern Language Quarterly* 9 (1948); Biester, "Fancy's Images."

95. The physician and natural philosopher Walter Charleton noted that "some of our Glossaries derive [wit] from the Teutonic *Witz*, to understand; and others from the Latin *Videlicet*, contracted in to *viz.*, because instead thereof we say *to witt*." Walter Charleton, *A Brief Discourse Concerning the Different Wits of Men* (London: William Whitwood, 1669), 17. Charleton's discourse was composed in 1664, the same year in which he presented a paper on the anatomy of the brain to The Royal Society (The Royal Society Archive, MS 366/2/2). We are grateful to Sietske Fransen for bringing this reference to our attention.

96. The commonplace nature of this connection enabled George Herbert, in his forty-first epigram, to make the lexicographical pun "Wit's an unruly engine." See Sam Westgate, "George Herbert: 'Wit's an Unruly Engine,'" *Journal of the History of Ideas* 38, no. 2 (1977).

97. This range is evident already in Middle English, in which different kinds of *wit* were determined by qualifying adjectives. See Randolph Quirk, "Langland's Use of 'Kind Wit' and 'Inwit,'" *The Journal of English and Germanic Philology* 52, no. 2 (1953): 182–88.

98. See Crane, *Wit and Rhetoric*; Milburn, *Age of Wit* (esp. chapter 10); Biester, "Fancy's Images."

99. See Milburn, *Age of Wit*; Richard A. McCabe, "Wit, Eloquence, and Wisdom in 'Euphues: The Anatomy of Wit,'" *Studies in Philology* 81, no. 3 (1984): 299–324; Jocelyn Powell, *John Lyly and the Language of Play* (Farnham: Ashgate, 2011). The word *witticism* appears only once in the lexica, in Bailey 1737: "Turlupinade: a low, dry jest or witticism." The OED, however, notes its first use by Dryden in 1677.

100. The word *facetious* is regularly associated with *wit* in English, as in Phillips 1658: "Facetious, (lat[in]) wittily merry, or pleasant." On the impact of classical *facetiae* on early modern literature, see Barbara C. Bowen, "Ciceronian Wit and Renaissance Rhetoric," *Rhetorica* 16, no. 4 (1998).

101. See e.g. Hudson, *The Epigram*; Schäffer, *The Early Seventeenth-Century Epigram*. For alternative forms of *wit* in Elizabethan and Jacobean literature, see Arnold Stein, "On Elizabethan Wit," *Studies in English Literature, 1500–1900* 1, no. 1 (1961): 75–91; Judith Dundas, "Allegory as a Form of Wit," *Studies in the Renaissance* 11 (1964): 223–33.

102. Earlier Latin-English dictionaries attribute these qualities to *sententiae*. See e.g. the entries for *sententiose/us* in Thomas 1587.

103. See Ustick and Hudson, "Wit."

104. See, in general, Sitter, *Arguments of Augustan Wit*. For Dryden, see H. James Jensen, *A Glossary of John Dryden's Critical Terms* (Minneapolis: University of Minnesota Press, 1969); Sue W. Doederlein, "A Compendium of Wit: The Psychological Vocabulary of John Dryden's Literary Criticism," PhD diss. (University of Michigan, 1971). For Rochester, see Jeremy Treglown, ed., *Spirit of Wit: Reconsiderations of Rochester* (Oxford: Basil Blackwell, 1982). For Pope, chiefly in relation to the *Essay on Criticism*, see e.g. Empson, "Wit in the Essay"; Edward Niles Hooker, "Pope on Wit: the 'Essay on Criticism,'" in *The Seventeenth Century: Studies in the History of English Literature and Thought from Bacon to Pope*, ed. Richard Foster Jones et al. (Oxford: Oxford University Press, 1951). For Addison, see Elizabeth Kraft, "Wit and The Spectator's Ethics of Desire," *Studies in English Literature, 1500–1900* 45, no. 3 (2005): 625–46.

105. "The composition of all Poems is, or ought to be of wit; and wit in the poet, or wit writing (if you will give me leave to use a School distinction), is no other than the faculty of imagination in the writer. . . . Wit written is that which is well defin'd; the happy result of thought, or product of that imagination." Dryden 1667, 1 [A7r-v].

106. See Kenny, *Curiosity*.

107. The "drollery" element of *wit* is evident in a wide range of genres, including philosophical writings. Hobbes made pointed use of this mode, for which see Roger D. Lund, "The Bite of Leviathan: Hobbes and Philosophic Drollery," *ELH* 65, no. 4 (1998): 825–55.

108. On *wit* and incivility, see Phil Withington, "'Tumbled into the Dirt': Wit and Incivility in Early Modern England," *Journal of Historical Pragmatics* 12, nos. 1–2 (2011): 156–77.

109. For *wit* and gender, see Margaret D. Stein, "Margaret Cavendish: Engendering Restoration Wit," PhD diss. (University of Notre Dame, 1998); Raymond Stephanson, *The Yard of Wit: Male Creativity and Sexuality, 1650–1750* (Philadelphia: University of Pennsylvania Press, 2004). For "learned wit" see Martin Kallich, "The Association of Ideas and Critical Theory: Hobbes, Locke, and Addison," *ELH* 12, no. 4 (1945): 290–315; For "vulgar wit" see Colville, "Mental Capacity."

110. Detienne and Vernant, *Cunning Intelligence*, 2, for the polysemy of *métis*.

111. See also Hogarth 1689 "*Policy*: craft, especially in civil affairs; from the Greek Πολιτεία, idem; hence the English Politick, crafty, or cunning."

112. See Owen Davies, *Popular Magic: Cunning Folk in English History* (London: Hambledon, 2007).

113. On which see especially Stephanson, *The Yard of Wit*.

114. Walter Charleton, "Certain Differences observable betwixt the Brain of a Man and the Brain of all other Animals," (paper read 8 June 1664), The Royal Society Archive, MS 366/2/2, p. 5. For further discussion of received notions of female character, see Ian Maclean, *The Renaissance Notion of Woman: A Study in the Fortunes of Scholasticism and Medical Science in European Intellectual Life* (Cambridge: Cambridge University Press, 1980).

115. Willis 1683, 121. This case is discussed in Colville, "Mental Capacity," 137.

116. For the early modern period see e.g. Londa Schiebinger, *The Mind Has No Sex? Women in the Origins of Modern Science* (Cambridge, MA and London, 1989); Cynthia B. Bryson, "Mary Astell: Defender of the 'Disembodied Mind,'" *Hypatia* 13 (1998), 40–62; Paula Findlen, "Ideas in the Mind: Gender and Knowledge in the Seventeenth Century," *Hypatia* 17 (2002), 183–196; Kathleen P. Long (ed.), *Gender and Scientific Discourse in Early Modern Culture* (Farnham and Burlington: Ashgate, 2010); Colville, "Mental Capacity." There is a substantial literature on gender and genius in Romanticism. Key studies include Christine Battersby, *Gender and Genius: Towards a Feminist Aesthetics* (New York: The Women's Press, 1986); Marie-Hélène Huet, *Monstrous Imagination* (Cambridge, MA: Harvard University Press, 1993); Kari E. Lokke, *Tracing Women's Romanticism: Gender, History and Transcendence* (London and New York: Routledge, 2004); Jefferson, *Genius in France*.

117. See Juliet Fleming, "Dictionary English and the Female Tongue," in *Ashgate Critical Essays on Early English Lexicographers, Volume 4*, ed. Considine.

Conclusion

1. On ingenuity and the persona of the lexicographer see, besides the introduction above, Caspar Hirschi, "Compiler into Genius: The Transformation of Dictionary Writers in Eighteenth Century France and England," in *Scholars in Action: The Practice of Knowledge and the Figure of the Savant in the 18th Century*, ed. André Holenstein, Hubert Steinke, and Martin Stuber (Leiden: Brill, 2013). See also Benedict Anderson, *Imagined Communities: Reflections on the Origin and Spread of Nationalism* (London: Verso, 1992).

2. This language of inborn difference is the subject of Gambarota, *Irresistible Signs*.

3. Dürer famously described to the wonders of the New World in terms of *ingenia* that betrayed the skill—and thus humanity—of its inhabitants: "I saw among them amazing artistic objects, and I marvelled over the subtle *Ingenia* [ingenuity] of the men in these distant lands" (Dürer 1958, 101–2).

4. This work has been initiated by Russo in "An Artistic Humanity."

5. See chapter 4: fig.1.

6. On the latter, see Oosterhoff, Marcaida, and Marr, eds., *Ingenuity in the Making* (forthcoming).

7. Cardano 1550, 1r: "Est autem subtilitas ratio quaedam, qua sensilia a sensibus, intellegibilia a intellectu difficile comprahenduntur" (Subtlety is a certain cause by which sensible things and intellectual ones can only be grasped by the senses and the intellect with difficulty); later specified as material tenuity (Cardano 1550, 2r) and the knowledge of difficult things and the discovery of obscure ones (Cardano 1550, 2v).

8. Cardano 1550, 290r.

9. Cardano 1550, 292r (book 18, "On marvels, and on the way to represent diverse things in order to generate belief"): "ut filii ingeniosi procreantur—Arcent igitur cotonea vapores a cerebro: quo fit ut infantis substantia cerebri purior reddatur, unde ingenium clarius Galeno et medicis omnibus testibus proficiscur" (how to beget ingenious children—Wallwort fends off vapors from the brain so that the cerebral matter of the infant is made purer, from which a brighter *ingenium* can arise according to Galen and all the testimonies of physicians).

10. Cardano 1550, 6r (book 1, "On principles, matter, form, vacuum, the antipathies between bodies, and natural motion"): "Janellus Turrianus Cremonensis, vir magni ingenii in omnibus quae ad machinas pertinent" (Janello Torriani of Cremona, man of great wit when it came to anything relating to machines); 260r (book 16, "On the sciences"): in denoting the singer's voice, *ingenium* is a doublet for *natura* (nature). The *ingenium* as inborn nature also features in the description of animal behaviour (Cardano 1550, 201v).

11. Cardano 1550, 247v-254v (book 15, "On useless subtleties, or those of an uncertain nature"): "Sed haec, ut dixi ac similia ad ostentationem ingenii, utilitatem vero pene nullam inventa sunt."

12. Cardano 1550, 287r (book 18): "Audaciae haec potius portenta sunt quam ingenii."

13. Cardano 1550, 1v. Cardano makes this point explicit by bemoaning the inadequacy of existing words to denote his conceptual discovery.

14. Scaliger 1557, 1–2.

Bibliography

Primary Sources

MANUSCRIPTS

Anon. 1435–60. *Glossario.* In *Il glossario quattrocentesco latino-volgare della biblioteca universitaria di Padova (ms. 1329)*, edited by Massimo Arcangeli. Florence: Accademia della Crusca, 1997.

Charleton, Walter. 1664. "Certain Differences observable betwixt the Brain of a Man and the Brain of all other Animals." The Royal Society Archive, MS 366/2/2.

Hilliard, Nicholas. Ca. 1600. *The Arte of Limning*, edited by R. K. R. Thornton and T. G. S. Cain. Manchester: Carcanet, 1992.

Nowell, Laurence. Ca. 1567. "Vocabularium Saxonicum." Bodleian Library, Oxford. MS Selden Supra 63.

Pulci, Luigi. Ca. 1460–66. *Vocabolista.* In Guglielmo Volpi, "Il 'Vocabolista' di Luigi Pulci." *Rivista delle Biblioteche e degli Archivi* 19 (1908): 9–15, 21–128.

Rosal, Francisco del. Ca. 1601. *Origen y etimología de todos los vocablos originales de la lengua castellana*, edited by Enrique Gómez Aguado. Madrid: CSIC, 1992.

Tesauro, Emanuele. 2008. *Vocabulario Italiano*, edited by Marco Maggi. Florence: Leo S. Olschki.

Tranchedini, Nicodemo. 1455–70. *Vocabolario italiano-latino*, edited by Federico Pelle. Florence: Leo S. Olschki, 2001.

PRINTED BOOKS

Abercromby, David. 1685. *A Discourse of Wit.* London: John Weld, 1685.

Académie française. 1694. *Dictionnaire de l'Académie françoise.* Paris: Veuve Jean-Baptiste Coignard.

Adelung, Johann Christoph. 1787. *Über den Deutschen Styl.* Berlin: Christian Friedrich Voss and Son.

——. 1793–1801. *Grammatisch-kritisches Wörterbuch der Hochdeutschen Mundart.* 4 vols. 1st ed. 1774–1786. Leipzig: Breitkopf und Sohn.

Aisy, Jean d'. 1685. *Le génie de la langue françoise.* Paris: L. D'Houry.

Albertus, Laurentius. 1573. *Teutsch Grammatick oder Sprachkunst.* Würzburg: Michael Manger.

Aldrete, Bernardo de. 1606. *Del origen y principio de la lengua castellana o romance que hoy se usa en España.* Roma: Carlos Willeto.

Alsted, Johannes Heinrich. 1626. *Compendium philosophicum*. Herborn: G. Corvin and G. Muderspach.

Alunno, Francesco. 1539a. *Il Petrarca con le osservationi di messer Francesco Alunno*. Venice: Francesco Marcolini da Forlì.

Alunno, Francesco. 1539b. *Le osservationi sopra il Petrarca*. Venice: Francesco Marcolini da Forlì.

———. 1543. *Le ricchezze della lingua volgare*. Venice: Aldus.

———. 1557. *La fabrica del mondo*. Venice: Paolo Gherardo.

Andry de Boisregard, Nicolas. 1692. *Réflexions, ou remarques critiques sur l'usage présent de la langue françoise*. 1st ed. 1689. Paris: Laurent d'Houry.

Anonymous. 1500. *Ortus vocabulorum*. Westminster: Wynkyn de Worde.

Anonymous. 1533. *Quinque linguarum utilissimus vocabulista latine, tusche, galliche, hyspane et alemanice*. Augsburg: Philip Ulhart.

Anonymous. 1556. *Dictionarium quatuor linguarum, teutonicae, gallicae, latinae et hispanicae*. Louvain: Bartholomeus Gravius.

Anonymous. 1565. *Dictionaire, colloques, ou dialogues en quatre langues, flamen, françois, español et italien*. Antwerp: Iean Withaye.

Anonymous. "Genies (Architecture).'" In *Encyclopédie ou Dictionnaire raisonné des arts et des sciences*, ed. Denis Diderot, Jean-Baptiste Le Rond d'Alembert et al., 28 vols (Paris: Briasson, David, Le Beton, Durand), 7:584.

Anonymous. 1908. *Theologia deutsch*, edited by Hermann Mandel. Leipzig: Deickert'sche Verlagsbuchh andlung.

Antonini, Annibale. 1735. *Dictionnaire italien, latin et françois*. Paris: Jacques Vincent.

Arouet, Francois-Marie, dit Voltaire, "Esprit (Philosophie–Belles Lettres)," in *Encyclopédie ou Dictionnaire raisonné des arts et des sciences*, ed. Denis·Diderot, Jean-Baptiste Le Rond d'Alembert et al., 28 vols. Paris: Briasson, David, Le Breton, Durand. vol 5: pp. 973-4.

Bailey, Nathan. 1730. *Dictionarium brittanicum*. London: Thomas Cox.

———. 1737. *The Universal Etymological English Dictionary*. London: Thomas Cox.

Balbus de Janua, Joannes. 1506. *Catholicon*. Venice: Peter Liechtenstein.

Baldinucci, Filippo. 1681. *Vocabolario toscano dell'arte del disegno*. Florence: Santi Franchi.

Baret, John. 1574. *An Alveary or Triple Dictionary, in English, Latin, and French*. London: Henry Denham.

Barrientos, Bartolomé. 1570. *Synonymorum liber liberalium artium*. Salamanca: Simón Portonaris, Matías Mares.

Batman, Stephen. 1582. *Batman uppon Batholeme his dook De Proprietatibus Rerum*. London: Thomas East.

Bayle, Pierre. 1740. *Dictionnaire historique et critique*, edited by Des Maizeaux. 4 vols. 5th ed. 1st ed. 1697. Amsterdam: P. Brunel et al.; Leiden: Samuel Luchtmans; The Hague: P. Gosse et al.; Utrecht: Etienne Naulme.

B. E. 1699. *A New Dictionary of the Terms Ancient and Modern of the Canting Crew.* London: W. Hawes et al.

Bergantini, Gian Pietro. 1745. *Voci italiane d'autori approvati dalla Crusca.* Venice: Pietro Bassaglia.

Bevilacqua, Luca Antonio. 1567. *Vocabulario volgare et latino non solamente di tutte le voci italiane, ma ancora de i nomi moderni, et antichi delle provincie, città.* Venice: Niccolò Bevilacqua.

Binnart, Martin. 1649. *Biglotton sive dictionarium teuto-latinum novum,* 1st ed. 1535. Antwerp: Cnobbarts.

Blount, Thomas. 1656. *Glossographia.* London: Humphrey Moseley.

Bluteau, Raphael. *Vocabulario portuguez et latino.* 8 vols. Coimbra: Collegio das Artes da Companhia de Jesus, 1712.

Bouhours, Dominique. 1671. *Les Entretiens d'Ariste et d'Eugène.* Paris: Sebastien Mabre-Cramoisy.

———. 1675. *Remarques nouvelles sur la langue françoise.* Paris: Sebastien Mabre-Cramoisy.

———. 1692. *Suite des remarques nouvelles sur la langue françoise.* Paris. G. L. Josse.

Bovelles, Charles de. 1553. *De differentia vulgarium linguarum.* Paris: Robert Estienne.

Boyle, Robert. 1665. *Occasional Reflections upon Several Subjects.* London: Henry Herringman.

Bravo, Bartolomé. 1599. *Thesaurus verborum ac phrasium ad orationem ex hispana latinam efficiendam et ornandam plurimis locis.* Salamanca: Andreas Renaut.

Breitinger, Johann Jakob, and Johann Jacob Bodmer. 1740. *Critische Dichtkunst: Worinnen die poetische Mahlerey in Absicht auf die Erfindung im Grunde untersuchet und mit Beyspielen aus den berühmtesten Alten und Neuern erläutert wird.* Zurich: Conrad Orell & Company.

Brocardo, Antonio. 1558. *Nuovo modo de intendere la lingua zerga, cioe parlare forbesco* Venice: [no publisher].

Bullokar, John. 1616. *An English Expositor.* London: John Legate.

Buti, Francesco da. 1934. *Commento . . . sopra la Divina Commedia di Dante.* Pisa: Nistri, 1858.

Butrón, Juan de. 1626. *Discursos apologéticos, en que se defiende la ingenuidad del arte de la pintura.* Madrid: Luis Sánchez.

Calderino Mirani, Cesare. 1586. *Dictionarium.* Venice: Felice Valgrisi.

Calepino, Ambrogio. 1502. *Dictionarium.* Bergamo: Regii Longobardiae.

———. 1553. *Il Dittionario di Ambrogio Calepino dalla lingua latina nella volgare brevemente ridotto.* Translated by Lucio Minerbi. Venice: Marco Trivisano.

———. 1555. *Dictionarium.* Venice: Giovanni I Griffio.

———. 1590. *Dictionarium undecim linguarum.* Basel: Sebastian HeinrichPeter.

———. 1718. *Dictionarium septem linguarum.* Pavia: Jacopo Facciolati.

Camerarius, Joachim. 1671. *Vierhundert Wahl-Sprüche und Sinnen-Bilder, durch welche beygebracht und außgelegt werden die angeborne Eigenschafften, wie auch lustige Historien und Hochgelährter Männer weiße Sitten-Sprüch*. Meintz: Borgeat.

Canal, Pierre. 1603. *Dittionario francese e italiano*. 2 vols. 1st ed. 1598. Geneva: Giacopo Choveto.

Capmany y de Montpalau, Antonio de. 1812. *Filosofía de la eloquencia*. London: Longman Hurst, Rees Orme and Brown.

Cardano, Girolamo. 1550. *De subtilitate libri XXI*. Paris: Michel Fezandat and Robert Granjon.

Casas, Cristóbal de las. 1570. *Vocabulario de las dos lenguas toscana y castellana*. Seville: Alonso Escrivano.

Castiglione, Baldassare. 1528. *Il libro del cortegiano*. Venice: Aldus Manutius.

———. 1556. *Il libro del cortegiano*. Venice: Girolamo Scoto.

Caussin, Nicolas. 1623. *De Symbolica aegyptorum sapientia*. 1st ed. 1618. Cologne: Johann Kinck.

Cawdrey, Robert, 1604. *A Table Alphabeticall*. I. Roberts for Edmund Weaver.

———. 1617. *A Table Alphabetical*. London: T. S. for Edmund Weaver.

Censorinus. 2007. *The Birthday Book*. Translated by Holt N. Parker. Chicago: University of Chicago Press.

Cervantes, Miguel de. 1612. *The History of the Valorous and Wittie Knight-Errant, Don-Quixote of the Mancha. Translated out of the Spanish*. Translated by Thomas Shelton. London: Printed by William Stansby, for Edward Blount and William Barret.

———. 1687. *The History of the most Renowned Don Quixote of Mancha And his Trusty Squire Sancho Pancha. Now made English according to the Humour of our Modern Language. And Adorned with several Copper Plates*. Translated by John Phillips. London: Printed by Thomas Hodgkin, and sold by William Whitwood.

———. 1998. *Don Quijote de La Mancha*, edited by Francisco Rico. Barcelona: Instituto Cervantes; Crítica.

Chambers, Ephraim. 1728. *Cyclopaedia, or, An Universal Dictionary of Arts and Sciences*. 2 vols. London: James and John Knapton et al.

———. 1747–9. *Dizionario universale delle arti e delle scienze*. 9 vols. Venice: Giambattista Pascquali.

Charleton, Walter. 1669. *A Brief Discourse Concerning the Different Wits of Men*. London: William Whitwood.

Chauvin, Stéphane. 1692. *Lexicon rationale sive thesaurus philosophicus*. Rotterdam: Petrus van der Slaart.

Clajus, Johannes. 1578. *Grammatica germanicae linguae*. Leipzig: Hans Rambau.

Cockeram, Henry. 1623. *The English Dictionarie*. London: Edmund Weaver.

Coles, Elisha. 1676. *An English Dictionary*. London: Samuel Crouch.

Collective. 1740. *Dictionnaire universel françois et latin*, 5th edn. Nancy: Pierre Antoine.
Conti, Natale. 1567. *Mythologiae*. Venice: [al segno della Fontana].
Cooper, Thomas. 1578. *Thesaurus linguae Romanae et Brittanicae*. London: Henry Wykes.
———. 1584. *Thesaurus linguae Romanae et Brittanicae*. London: Henry Bynneman.
Coote, Edmund. 1596. *The English Schoole-maister*. London: Ralph Jackson and Robert Dexter.
Corneille, Thomas. 1694. *Le dictionnaire des arts et des sciences*. 2 vols. Paris: Veuve Jean-Baptiste Coignard.
Cotgrave, Randle. 1611. *A Dictionarie of the French and English Tongues*. London: Adam Islip.
Covarrubias, Sebastián de. 1610. *Emblemas Morales*. Madrid: Luis Sánchez.
———. 1611. *Tesoro de la lengua castellana, o española*. Madrid: Luis Sánchez.
———. 1674. *Primera parte y segunda parte del Tesoro de la lengua castellana o española*, edited by Benito Remigio Noydens. Madrid: Melchor Sánchez.
———. 2006. *Tesoro de la lengua castellana o española. Edición integral e ilustrada*, edited by Ignacio Arellano and Rafael Zafra. Madrid: Iberoamericana.
Cowell, John. 1607. *The Interpreter*. Cambridge: John Legate.
Crusca [Accademici della]. 1612. *Vocabolario degli Accademici della Crusca*. Venice: Giovanni Alberti.
———. 1623. *Vocabolario degli Accademici della Crusca*. Venice: Bastiano de' Rossi.
———. 1691. *Vocabolario degli Accademici della Crusca*. Florence: Accademia della Crusca.
———. 1729–38. *Vocabolario degli Accademici della Crusca*. 7 vols. Florence: Domenico Maria Manni.
Dasypodius, Petrus. 1536. *Dictionarium latinogermanicum*. 1st ed. 1535. Strassburg: Wendelin Rihelium.
Defoe, Benjamin. 1735. *A New English Dictionary*. Westminster: John Brindley et al.
Descartes, René. 1897–1913. *Oeuvres de Descartes*, edited by Charles Adam and Paul Tannery. 12 vols. Paris: Léopold Cerf.
———. 1984–91. *The Philosophical Writings of Descartes*, edited and translated by John Cottingham et al. 3 vols. Cambridge: Cambridge University Press.
Diderot, Denis, Jean-Baptiste Le Rond d'Alembert et al. 1751–72. *Encyclopédie ou Dictionnaire raisonné des arts et des sciences*. 28 vols. Paris: Briasson, David, Le Breton, Durand.
Dryden, John. 1667. *Annus mirabilis*. London: Henry Herringman.
Du Fresnes, Charles, seigneur du Cange. 1678. *Glossarium mediae et infimae latinitatis*. Paris: Gabriel Martin.
Du Laurens, André. 1600. *Historia anatomica humani corporis, e singularum eius partium multis controversiis et observationibus novi*. Paris: M. Orry.
Dürer, Albrecht. 1956–9. *Schriftlicher Nachlass*, edited by Hans Rupprich. 3 vols. Berlin: Deutscher Verein für Kunstwissenschaft.

---. 1958. *The Writings of Albrecht Dürer*, translated by William Martin Conway. London: Peter Owen.

Elyot, Thomas. 1538. *The Dictionary of Syr Thomas Elyot*. London: Thomas Berthelet.

Erasmus. 2011. *Les Adages*, edited by Jean-Christophe Saladin et al. Paris: Les Belles Lettres.

Estienne, Robert. 1530. *Sententiae et proverbia ex omnibus Plauti & Terentii comoediis*. Paris: Robert Estienne.

---. 1531. *Dictionarium, seu latinae linguae thesaurus*. Paris: Robert Estienne.

---. 1536. *Dictionarium, seu latinae linguae thesaurus: non singulas modo dictiones continens, sed integras quoque latine et loquendi, & scribendi formulas—ex Catone, Cicerone, Plinio avunculo, Terentio, Varrone, Livio, Plinio secundo, Virgilio, Caesare, Columella, Plauto, Martiale—cum latine grammaticorum, tum varii generis scriptorum interpretatione*. Paris: Robert Estienne.

---. 1538. *Dictionarium latinogallicum, thesauro nostro ita ex adverso respondens, ut extra pauca quaedam aut obsoleta: aut minus in usu necessaria vocabularia, & quas consulto praetermisimus, authorum appellationes, in hoc eadem sint omnia, eodem ordine, sermone patrio explicata*. Paris: Robert Estienne.

---. 1543. *Dictionarium latinogallicum*. Paris: Robert Estienne.

---. 1571. *Dictionarium latinogallicum*. Paris: Sebastianus Honoratus.

Feijoo, Benito Jerónimo. 1726–1740. *Teatro crítico universal*. Madrid: Lorenzo Francisco Mojados; Francisco del Hierro.

---. 1773. *Cartas eruditas y curiosas. Nueva impresión*. Madrid: Pedro Marín.

Fenice, Giovanni Antonio. 1584. *Dictionnaire françois et italien, profitable et necessaire à ceux qui prenent plaisir en ces deux langues.* Paris: Nicolas Nivelle.

Fernández de Palencia, Alfonso. 1490. *Universal vocabulario en latín y en romance*. Seville: Paulus de Colonia Alemanus et al.

Fernández de Santaella, Rodrigo. 1499. *Vocabularium ecclesiasticum per ordinem alphabeti*. Seville: Johannes Pegnitzer, Magnus Herbst, Thomas Glockner.

Flecknoe, Richard. 1664. "A Short Discourse of the English Stage." In *Love's Kingdom*. London: s.n.

Florio, John. 1598. *A Worlde of Wordes*. London: Edward Blount.

---. 1611. *Queen Anna's New World of Words*. London: Edward Blount and William Baret.

Franciosini, Lorenzo. 1620. *Vocabolario español e italiano*. Rome: Iuan Angel Rufineli, Angel Manni.

Freylas, Alonso de. 1606. *Si los Melanchólicos pueden saber lo que está por venir con la fuerça de su ingenio o soñando*. Jaén: Fernando Díaz de Montoya.

Frisius, Johannes. 1541. *Dictionarium latinogermanicum*. Zurich: Christoph Froschauer.

Fulbecke, William. 1601. *A Historicall Collection of the Continuall Factions, Tumults, and Massacres of the Romans and Italians*. London: William Ponsonby.

Furetière, Antoine. 1690. *Dictionaire universel*. 3 vols. The Hague & Rotterdam: Arnoud and Reinier Leers.

Geoffrey the Grammarian. 1499. *Promptorium parvulorum*. London: Frederic Egmondt and Petrus Post.
Gessner, Conrad. 1555. *Mithridates, de differentii linguarum*. Zurich: Froschauer.
Girard, Gabriel. 1718. *La justesse de la langue françoise ou les differentes significations des mots qui passent pour synonimes*. Paris: Laurent d'Houry.
Gladov, Friedrich. 1728. *A la Mode-Sprach der Teutschen Oder Compendieuses Hand-LEXICON: In welchem die meisten aus fremden Sprachen entlehnten Wörter und gewöhnliche Redens-Arten, So in denen Zeitungen, Briefen und täglichen Conversation vorkommen*. Nuremberg: Buggel and Seitz.
Glocenius, Rodolph. 1613. *Lexicon philosophicum quo tanquam clave philosophiae fores aperiuntur*. Frankfurt: Widow M. Becker.
Gómez Miedes, Bernardino. 2003. *Comentarios sobre la sal*, edited and translated by Sandra Inés Ramos Maldonado. 3 vols. Madrid: Alcáñiz.
Goropius Becanus, Johannes. 1569. *Origines Antwerpianae, sive Cimmeriorum becceselana novem libros complexa*. Antwerp: Christophe Plantin.
Gournay Marie de. 1641. *Les Advis, ou les presens de la demoiselle de Gournay*. 1st ed. 1634. Paris: Jean du Bray.
Gracián, Baltasar. 1637. *El Héroe*. Huesca: Francisco de Larumbe.
———. 1642. *Arte de ingenio. Tratado de la Agudeza*. Madrid: Juan Sánchez.
———. 1648. *Agudeza y arte de ingenio*. Huesca: Juan Nogués.
———. 1969. *Agudeza y arte de ingenio*, edited by Evaristo Correa Calderón. 2 vols. Madrid: Castalia.
———. 1997. *El Discreto*, edited by Aurora Egido. Madrid: Alianza.
———. 2003. *El Héroe. Oráculo manual y arte de prudencia*, edited by Antonio Bernat Vistarini and Abraham Madroñal. Madrid: Castalia.
———. 2011. *The Pocket Oracle and the Art of Prudence*, edited and translated by Jeremy Robbins. London: Penguin Books.
Guez de Balzac, Jean-Louis. 1933–34. *Les premières lettres de Guez de Balzac: 1618–1627*, edited by Henri Bibas and Kathleen T. Butler. 2 vols. Paris: Droz.
Gutiérrez de los Ríos, Gaspar. 1600. *Noticia general para la estimacion de las artes, y de la manera en que se conocen las liberales de las que son mecanicas*. Madrid: Pedro Madrigal.
Harvey, Gabriel. 1592. *Foure Letters and Certaine Sonnets*. London: John Wolfe.
Henisch, Georg. 1616. *Teütsche Sprach und Weißheit: Thesaurus linguae et sapientiae Germanicae*. Augsburg: David Frank.
Herrera, Fernando de. 1580. *Obras de Garcilaso de la Vega con anotaciones de Fernando de Herrera*. Seville: Alonso de la Barrera.
Hexham, Henry. 1647. *Het Groot woorden-boek gestelt in 't Engelsch ende Nederduytsch*. Rotterdam: Arnout Leers.
Hille, Karl Gustav von. 1647. "Alamodischer Brief," in *Der teutsche Palmbaum*. Nuremberg: Endter.
Hofmann, Johann Jacob. 1677. *Lexicon universale*. Basel: J. H. Widerhold.

Hogarth, Richard. 1689. *Gazophylacium anglicanum*. London: Randall Taylor.
Hollyband, Claude. 1593. *A Dictionarie French and Englishe*. London: Thomas Woodcock.
Hornkens, Henricus. 1599. *Recueil de dictionaires francoys, espaignolz et latins*. Brussels: Rutgeer Velpius.
Howell, James. 1660. *Lexicon Tetraglotton, an English-French-Italian-Spanish Dictionary*. London: J.G. for Samuel Thomson.
Huarte de San Juan, Juan. 1575. *Examen de ingenios, para las sciencias*. Baeza: Juan Baptista de Montoya.
——. 1594. *Examen de ingenios, para las sciencias . . . Agora nuevamente enmedado*. Baeza: Juan Baptista de Montoya.
——. 1622. *Scrutinium ingeniorum, pro iis qui excellere cupiunt*. Translated by Joachim Cesar. Leipzig: Kote.
——. 1659. *Onderzoek der byzondere vernuftens*. Translated by Henryk Takama. Amsterdam: Jean van Ravesteyn.
——. 1698. *Examen de Ingenios, Or, The Tryal of Wits Discovering the Great Difference of Wits among Men, and What Sort of Learning Suits Best with Each Genius, published Originally in Spanish by Doctor Juan Huartes; and Made English from the Most Correct Edition by Mr. Bellamy*. Translated by Edward Bellamy. London: Richard Sare.
——. 1989. *Examen de ingenios para las ciencias*, edited by Guillermo Serés. Madrid: Cátedra.
Hudemann, Heinrich. 1626. *Hirrnschleiffer. Das ist: Außerlesene teutsche Emblemata oder Sinnbilder: welche zu schärffung der Verstands/ besserung des sündlichen Lebens/ unnd Erlustigung des gantzen Menschen mit Verssen gezieret/ unnd in dieser Sprach hiebevor nicht außkommen seynd*. s.l., s.n.
Huloet, Richard. 1552. *Abcedarium anglico latinum*. London: William Riddel.
Iperen, Josua van. 1755. *Proeve van taalkunde, als eene wetenschap behandeld*. Amsterdam: G. Rykman.
——. 1762(a). "Uitnoodiginge der Liefhebbers en Kenners van Onze Modertale, tot het helpen toestellen van een oordeelkundig Nederduitsch Woordenboek." *Maendelijksche By-Dragen ten opbouw van Neer-land's Tael- en Dicht-kunde* 47, no. 9: 509–14.
——. 1762(b). "Schetze van woorden-scharinge en Zin-bepalinge, die men hoopt te volgen, in het toestellen van een oordeelkundig nederduitsch woordenboek." *Maendelijksche By-Dragen ten opbouw van Neer-land's Tael- en Dicht-kunde* 48, no. 10: 541–51.
Isidore of Seville. 1911. *Etymologiarum sive Originum libri XX*, edited by W. M. Lindsay. Oxford: Oxford University Press.
Jaucourt, Louis, Chevalier de. "Génie (Mythologie-Littérature-Antiquité),"in *Ency-

clopédie ou Dictionnaire raisonné des arts et des sciences, ed. Denis Diderot, Jean-Baptiste Le Rond d'Alembert et al., 28 vols. Paris: Briasson, David, Le Breton.

Johnson, Samuel. 1755. *A Dictionary of the English Language*. 2 vols. London: James and Paul Knapton et al.

———. 1779. *The Lives of the English Poets*, vol. 1. London: [Mr] Whitestone et al.

Jonson, Ben. 1599. *Every Man Out of His Humour*, edited by Randall Martin. In *The Cambridge Edition of the Works of Ben Jonson Online*, edited by David Bevington, Martin Butler, and Ian Donaldson. Cambridge: Cambridge University Press, 2014. http://universitypublishingonline.org/cambridge/benjonson/.

———. 1609. *The Entertainment at Britain's Burse*, edited by James Knowles. In *The Cambridge Edition of the Works of Ben Jonson Online*, edited by David Bevington, Martin Butler, and Ian Donaldson. Cambridge: Cambridge University Press, 2014. http://universitypublishingonline.org/cambridge/benjonson/.

Jouvancy, Joseph de. 1690. *De arte docendi et dicendi*. Paris: Barbou Frères.

Junius, Hadrianus. 1606. *Nomenclator octilinguis: omnium rerum propria nomina continens. Ab Adriano Junio antehac collectus, nunc vero renovatus*. Paris: David Douceur.

Kahl, Johann. 1600. *Lexicon juridicum juris Caesarei simul, et canonici: feudalis item, civilis, criminalis, theoretici, ac practici*. Frankfurt: Wechel, Marnius & Aubrius.

Kant, Immanuel. 2000. *Critique of the Power of Judgment*. Translated by Paul Guyer. Cambridge: Cambridge University Press.

———. 2006. *Anthropology from a Pragmatic Point of View*. Translated by Robert B. Louden. Cambridge: Cambridge University Press.

Kate, Lambert ten. 1723. *Aenleiding tot de Kennisse van het verhevene deel der nederduitsche sprake*. Amsterdam: Rudolph and Gerard Wetstein.

Kersey, John. 1702. *A New English Dictionary*. London: Henry Bonwick and Robert Knaplock.

———. 1713. *A New English Dictionary*. London: Robert Knaplock and R. and J. Bonwick.

Kiliaan, Cornelis. 1562. *Dictionarium tetraglotton, seu voces latinae omnes, et graecae eis respondentes, cum gallica & teutonica (quam passim flandricam vocant) earum interpretatione*. Antwerp: Christoph Plantin.

———. 1588. *Dictionarium teutonico-latinum, praecipuas teutonicae linguae dictiones latine interpretatas complectens*, 1st ed. 1574. Antwerp: Christophe Plantin.

———. 1599. *Etymologicum teutonicae linguae, sive dictionarium teutonico-latinum, praecipuas teutonicae linguae dictiones et phrases latine interpretatas, et cum aliis nonnullis linguis obiter collatas complectens*. Antwerp: Officina Plantiniana.

König, Georg Matthias. 1668. *Gazophylacium latinitatis sive lexicon novum latino-germanicum*. Nuremberg: Endter.

Kramer, Matthias. 1676–78. *Il nuovo dizzionario delle due lingue italiana-tedesca et tedesca-italiana / Das neue Dictionarium oder Wort-Buch, in Italiänischer-Teutscher und Teutsch-Italiänischer Sprach*. 2 vols. Nuremberg: Endter.

———. 1700–02. *Das herrlich grosse Teutsch–Italiänische Dictionarium/Il gran dittionario reale, Tedesco–Italiano*. 2 vols. Nuremberg: Endter.

———. 1712–15. *Le vraiment parfait dictionnaire roial, radical, etimologique, sinonimique, phraseologique, & syntactique, françois-allemand / Das recht vollkommen-Königliche Dictionarium Radicale, Etymologicum, Synonymicum, Phraseologicum et Syntacticum, Frantzösisch-Teutsch*. 4 vols. Nuremberg: Endter.

La Bruyère, Jean de. 1688. *Les Caractères*, edited by Emmanuel Bury. Paris: Librairie générale française, 1995.

Landino, Cristoforo. 1487. *Comento . . . sopra la Comedia di Dante Alighieri*. Brescia: Bonimum de Boninis.

La Perrière, Guillaume de. 1544. *Le theatre des bons enginsen*. 1st ed. 1540. Paris: Denis Janot.

Le Blond, Guillaume. 1757. "Genie (art militaire)." In *Encyclopédie ou Dictionnaire raisonné des arts et des sciences*, ed. Denis Diderot, Jean-Baptiste Le Rond d'Alembert et al., 28 vols. (Paris: Briasson, David, Le Breton, Durand), 7:584.

Leman, Tanfield. 1755. Review of Johnson's *Dictionary*. *The Monthly Review*, XII (April 1755): 292–324.

Lever, Ralph. 1573. *The Art of Reason*. London: H. Bynneman.

Liburnio, Niccolò. 1526. *Le tre fontane di Messer Nicolò Liburnio in tre libri divise, sopra la grammatica, et eloquenza di Dante, Petrarcha, et Boccaccio*. Venice: Gregorio de Gregorii.

Livy. 1600. *The Romane Historie*. Translated by Philemon Holland. London: Adam Islip.

Lloyd, William. 1668. "An Alphabetical Dictionary." In John Wilkins, *An Essay Towards a Real Character, And a Philosophical Language*. London: Samuel Gellibrand and John Martyn. [Lloyd-Wilkins 1668]

Luna, Fabricio. 1536. *Vocabulario di cinquemila vocabuli toschi non men oscuri che utili e necessarij del Furioso, Bocaccio, Petrarcha e Dante*. Naples: Giovanni Sultzbach.

Maaler, Josua. 1561. *Die teutsch Spraach. Alle Wörter, Namen und Arten zu reden in Hochteutscher Spraach*. Zurich: Froschauer.

Madoets, André. 1573. *Thesaurus theutonicae linguae. Schat der Neder-Duytscher Spraken*. Antwerp: Christophe Plantin.

Marin, Pierre. 1696. *Dictionnaire portatif hollandais et français of Nederdeutsch en Fransch woordenboekje*. Amsterdam: Jan van Eyl.

Marinelli, Giovanni. 1565. *Dittionario di tutte le voci italiane usate da migliori scrittori antichi, et moderni*. Venice: Niccolò Bevilacqua.

Martin, Benjamin. 1749. *Lingua britannica reformata*. London: J. Hodges et al.

Megiser, Hieronymus. 1603. *Thesaurus polyglottus*. Frankfurt am Main: By the Author.

Meister Eckhart. 1993. *Werke*, edited by Niklaus Largier. 2 vols. Frankfurt am Main: Deutscher Klassiker Verlag.

Ménage, Gilles. 1650. *Les Origines de la langue françoise*. Paris: Augustin Courbé.

———. 1669. *Le origini della lingua italiana*. Paris: Sebastien Mabre-Cramoisy.

———, et al. 1750. *Dictionnaire étymologique ou origines de la langue française*, edited by Augustin-François Jault. 2 vols. Paris: Briassou.

Micraelius, Johannes. 1653. *Lexicon philosophicum*. Jena: Caspar Freyschmidt.

Miège, Guy. 1677. *A New Dictionary French and English*. London: Thomas Basset.

Minerbi, Lucilio. 1535. *Il decamerone di M. Giovanni Boccaccio col vocabulario di Lucilio Minerbi*. Venice: Bernardino di Vidali.

Minsheu, John. 1599. *A Dictionarie in Spanish and English, first published into the English tongue by Ric. Percivale*. London: Edmund Bollifant.

———. 1617. *Ductor in linguas. The Guide into Tongues*. London: John Browne.

Morhof, Daniel Georg. 1688. *Polyhistor sive de notitia auctorum et rerum commentarii*. Lübeck: Böckmannus.

Morvan, Jean-Baptiste, abbé de Bellegarde. 1761. *Réflexions sur la politesse des moeurs avec des maximes pour la société civile, suite des Réflexions sur le ridicule*. Vol. 1 in *Oeuvres de Monsieur l'Abbé de Bellegarde*. 15 vols. 1st ed. 1698. The Hague: Pierre Gosse Junior.

Moxon, James. 1679. *Mathematicks made Easie: Or, a Mathematical Dictionary*. London: Joseph Moxon.

Nanni, Giovanni. 1601. *An Historical Treatise of the Travels of Noah into Europe*. Translated by Richard Lynche. London: Adam Islip.

Nashe, Thomas. 1592. *Strange Newes*. London: J. Danter.

Navarro, Miguel. 1599. *Libro muy útil y provechoso para aprender la latinidad*. Madrid: Imprenta Real.

Nebrija, Elio Antonio de. 1492. *Lexicon hoc est Dictionarium ex sermone latino in hispaniensem*. Salamanca.

———. Ca. 1495. *Dictionarium hispanarum in latinum sermonem*. Salamanca.

———. 1545. *Dictionarium Aelii Antonii Nebrissensis iam denuo innumeris dictionibus locupletatum*. Antwerp: Jean Steelsius.

Nicot, Jean. 1599. *Le grand dictionaire françois-latin*. Geneva: Jacob Stoer.

———. 1606. *Thresor de la langue françoyse, tant ancienne que moderne*. Paris: David Douceur.

Ogier de Gombauld, Jean. 1658. *Les épigrammes*. Paris: Augustin Courbé.

Ölinger, Albertus. 1574. *Underricht der Hoch Teutschen Spraach: Grammatica seu Institutio verae germanicae linguae*. Strassburg: Nicolaus Vuyriot.

Opitz, Martin. 1617. *Aristarchus, sive De contemptu linguae Teutonicae: Und Buch von der deutschen Poeterey*. Beuthen: Dörfer.

Oudin, César. 1607. *Tesoro de las dos lenguas francesa y española*. Paris: Marc Orry.

———. 1616. *Tesoro de las dos lenguas francesa y española*. Paris: Marc Orry.

Palomino, Antonio. 1715–24. *El museo pictórico y escala óptica*. 3 vols. Madrid: Lucas Antonio de Bedmar.

Pascal, Blaise. 1656–57. *Lettres provinciales*. In *Provinciales, Pensées et opuscules divers*, edited by Philippe Sellier and Gérard Ferreyrolles. Paris: La Pochothèque, 2004.

Percivale, Richard. 1591. *Bibliotheca hispanica. Containing a Grammar with a Dictionarie in Spanish, English and Latine*. London: John Jackson, Richard Watkins.

Pereira, Bento. 1653. *Prosodia in vocabularium trilingual*. Lisbon: Paul Craesbeek, 1653 [first ed. 1634].

Pergamino, Giacomo. 1602. *Il Memoriale della lingua*. Venice: Giacomo Pergamino.

Perotti, Niccolò. 1496. *Cornu copia . . . emendatissimum*. Venice: Giovanni Tacuino.

———. 1513. *Cornucopiae*. Venice: Aldus Manutius.

———. 1525. *Cornu copiae*. 1st ed. 1489. Venice: Aldus Manutius.

Persio, Antonio. 1576. *Trattato dell'ingegno dell'huomo*. Venice: Aldus.

———. 1999. *Trattato dell'ingegno dell'huomo*, edited by Luciano Artese. Pisa: Istituti Editoriali e Poligrafici Internazionali.

Petrarca, Francesco. 1859. *Epistolae de rebus familiaribus et variae*, edited by Giuseppe Fracassetti. Florence: Le Monnier.

———. 2003. *Invectives*, edited by David Marsh. Cambridge, MA: Harvard University Press.

Phillips, Edward. 1658. *The New World of English Words*. London: Nathaniel Brooke.

———. 1706. *The New World of English Words*, edited by John Kersey. London: J. Phillips et al. [Phillips-Kersey 1706].

———. 1720. *The New World of English Words*, edited by John Kersey. London: J. Phillips et al. [Phillips-Kersey 1720].

Pineda, Pedro. 1740. *Nuevo Diccionario, español e ingles e ingles y español: que contiene la etimologia, de la propria, y metaphorica significacion de las palabras, terminos de artes y sciencias*. London: F. Gyles, T. Woodward, T. Cox.

Politi, Adriano. 1614. *Dittionario toscano*. Rome: Giovanni Angelo Ruffinelli.

Porcacchi, Tommaso. 1584. "Vocabulario nuovo del Porcacchi." In Francesco Alunno, *Della fabrica del mondo*. Venice: Giovanni Battista Porta.

Porta, Arnaldo de la. 1659. *Nueuo dictionario, o thesoro de la lengua española y flamenca*. Amsterdam: Verdussen.

Possevino, Antonio. 1603. "De cultura ingeniorum." In *Bibliotheca selecta Antonii Possevinii, Examen Ingeniorum Io. Huartis expenditur*, vol. 1. Venice: Altobellum Salicatium.

Preston, Henry. Ca. 1674. *Brief Directions for True Spelling*. London: William Bishop.

Priscianese, Francesco. 1579. *Dictionarium Ciceronianum*. Venice: Giovanni Antonio Bertano.

Rabelais, François. 1532. *Pantagruel*. Lyon: Claude Nourry.

Rainolds, John. 1584. *The Summe of the Conference betwene John Rainoldes and John Hart Touching the Head and the Faith of the Church*. London: George Bishop.

Ray, John. 1674. *A Collection of English Words*. London: T. Burrell.

Real Academia Española. 1726–39. *Diccionario de la lengua castellana, en que se explica el verdadero sentido de las voces, su naturaleza y calidad, con las phrases o modos de hablar, los proverbios o refranes, y otras cosas convenientes al uso de la lengua*. 6 vols. Madrid: Imprenta de Francisco del Hierro.

———. 1743. *Reglas, que formó la Academia en el año de 1743, y mandó observassen los señores Académicos, para trabajar con uniformidad en la correccion, y suplemento del Diccionario*: s.l., s.n.

Reuchlin, Johannes. 1478. *Vocabularius breviloquus*. Basel: Amerbach.

Richelet, César-Pierre. 1680. *Dictionnaire françois contenant les mots et les choses, plusieurs nouvelles remarques sur la langue françoise, ses expressions propres, figurées & burlesques . . . avec les termes les plus connus des arts et des sciences . . . tiré de l'usage des bons auteurs de la langue françoise*. 2 vols. Geneva: Jean Herman Widerhold.

Rider, John. 1589. *Bibliotheca scholastica*. Oxford: John Barnes.

Ripa, Cesare. 1603. *Iconologia*. Rome: Lepido Facii.

Roboredo, Amaro de. 1621. *Raizes da lingva latina mostradas em hum trattado, e diccionario*. Lisbon: Pedro Craesbeeck.

Ruscelli, Girolamo. 1588. *Vocabolario delle voci latine dichiarate con l'italiane scelte da' migliori scrittori*. Venice: Haeredi di Valerio Bonello.

Saint-Lambert, Jean-François de. 1757. "Génie (Philosophie-Littérature)." In *Encyclopédie ou Dictionnaire raisonné des arts et des sciences*, ed. Denis Diderot, Jean-Baptiste Le Rond d'Alembert et al., 28 vols. (Paris: Briasson, David, Le Breton, Durand). 7:582–84.

Saint-Réal, César Vichard de. 1726. *Cesarion, ou entretiens sur divers sujets*. 1st ed. 1684. The Hague: Alexandre de Rogissart.

Sánchez de la Ballesta, Alonso. 1587. *Dictionario de vocablos castellanos*. Salamanca: Juan Renaut, Andrés Renaut.

Sansovino, Francesco. 1568. *Orthographia delle voci della lingua nostra o vero Dittionario volgare et latino*. Venice: Francesco Sansovino.

Sasbout, Mathias. 1576. *Dictionaire flameng-francois tres-ample et copieux*. Antwerp: Jan I van Waesberghe.

Savary des Bruslons, Jacques, and Philemon-Louis Savary. 1723, 1726, 1730. *Dictionnaire universel du commerce*. 3 vols. Amsterdam: Jansons.

Scaliger, Julius Caesar. 1557. *Exotericarum exercitationum liber quintus decimus*. Paris: Michel Vascosan.

Scobar, Cristoforo de. 1519. *Vocabularium nebrissense ex siciliensi sermone in latinum, L. Christophoro Scobare Bethico interprete traductum*. Venice: Domini Dominici de Nesi et al.

Scoppa, Lucio Giovanni. 1511/15. *Spicilegium*. In Fatima Stefania Sorrentino, "Lo Spicilegium di Lucio Giovanni Scoppa," 28–319. PhD diss., Università degli Studi di Napoli Federico II, 2011.

Servius, Maurus Honoratus. 1471. *In tria Virgilii expositio*. Florence: Bernardo Cennini.

Sewel, Willem. 1691. *A New Dictionary English and Dutch . . . Nieuw Woordenboek der Engelsche en Nederduytsche Taale*. Amsterdam: Weduwe van Steven Swart.

Soarez, Cypriano. 1562. *De arte rhetorica libri tres ex Aristotele, Cicerone & Quintiliano praecipue deprompti*. Coimbra: Ioannis Barrerium.

Spiegel, Jakob. 1538. *Lexicon iuris civilis*. Strassburg: Schott.

———. 1554. *Lexicon iuris civilis*. Basel: Johann Hervagius.

Spieghel, Hendrik Laurenszoon. 1584. *Twe-spraack vande Nederdeutsche letterkunst*. Leiden: Christophe Plantin.

Stevens, John. 1706. *A New Spanish and English Dictionary*. London: George Sawbridge.

Stieler, Kaspar. 1691. *Der Teutschen Sprache Stammbaum und Fortwachs oder Teutscher Sprachschatz / Teutonicae linguae semina et germina, sive lexicon Germanicum*. Nuremberg: Johann Hofmann.

Tesauro, Emanuele. 1654. *Il cannocchiale aristotelico o sia idea delle argutezze heroiche vulgarmente chiamate imprese et di tutta l'arte simbolica et lapidaria*. Turin: Giovanni Sinibaldo.

Thesaurus linguae Latinae. Leipzig and Berlin: Teubner & De Gruyter, 1894–.

Thomas, Thomas. 1587. *Dictionarium linguae Latinae et Anglicanae*. Cambridge: Thomas Thomas.

Tieck, Johann Ludwig. 1798. *Franz Sternbalds Wanderungen: Eine altdeutsche Geschichte*. 2 vols. Berlin: Johann Friedrich Unger.

Torriano, Giovanni. 1659. *Dizionario Italiano & Inglese*. London: John Martin et al.

Ulloa, Alfonso de. 1553. *Introdutione del Signor Alphonso di Uglioa nella quale s'insegna pronuntiare la lingua Spagnuola. Con una espositione da lui fatta nella Italiana di parecchi vocaboli Hispagnuoli dificili, contenuti quasi tutti nella Tragicomedia di Calisto e Melibea o Celestina*. Venice: Gabriel Giolito.

Valdés, Juan de. 1969. *Diálogo de la lengua*, edited by Juan M. Lope Blanch. Madrid: Castalia.

Vaugelas, Claude Favre de. 1647. *Remarques sur la langue françoise: utiles à ceux qui veulent bien parler et bien escrire*. Paris: Veuve J. Camusat et P. Le Petit.

Venuti, Filippo. 1561. *Dittionario volgare, et latino*. Venice: Giovanni Andrea Valvassori.

———. 1576. *Dittionario volgare & latino*. Rome: Nelle Case del Popolo Romano.

Vittori, Girolamo. 1609. *Tesoro de las tres lenguas francesa, italiana y española*. Geneva: Philippe Albert, Alexandre Pernet.

Vives, Juan Luis. 1538. *De anima et vita libri tres*. Basel: Robert Winter.

Volckmar, Henning. 1675. *Dictionarium Philosophicum, hoc est enodatio terminorum ac distinctionum celebriorum in philosophia occurentium*. Frankfurt: Jakob Gottfred Seyler.

Wachter, Johann Georg. 1737. *Glossarium germanicum: continens origines et antiquitates totius linguae germanicae, et omnium pene vocabulorum, vigentium et desitorum*. 2 vols. Leipzig: Johann Friedrich Gleditsch.

Weise, Christian. 1678. *De poesi hodiernorum politicorum sive de argutis inscriptionibus*. Jena: Matthias Berckner.
Willis, Thomas. 1683. *Two Discourses Concerning the Soul of Brutes*. London: Thomas Dring, Charles Harper and John Leigh.
Wilson, Thomas. 1612. *A Christian Dictionarie, Opening the signification of the chiefe wordes dispersed generally through Holie Scriptures of the Old and New Testament, tending to increase Christian knowledge. Whereunto is annexed: A perticular Dictionary For the Reuelation of S. Iohn. For the Canticles, or Song of Salomon. For the Epistle to the Hebrues*. London: William Jaggard.
Wotton, Henry. 1624. *The Elements of Architecture*. London: John Bill.

Secondary Sources

Acero Durántez, Isabel. "La lexicografía plurilingüe del español." In *Lexicografía española*, edited by Antonia María Medina Guerra, 175–204. Barcelona: Ariel, 2003.
Aguzzi-Barbagli, Danilo. "*Ingegno, acutezza*, and *meraviglia* in the Sixteenth Century Great Commentaries to Aristotle's Poetics." In *Petrarch to Pirandello: Studies in Italian Literature in Honour of Beatrice Corrigan*, edited by Julius A. Molinaro. Toronto: University of Toronto Press, 1973.
Ahumada, Ignacio, ed. *Cinco siglos de lexicografía del español*. Jaén: Universidad de Jaén, 2000.
Alston, Robin C. *A Bibliography of the English Language from the Invention of Printing to the Year 1800*. Leeds: Printed for the author, 1965–73.
Alvar, Manuel. "Nebrija, Lexicógrafo." In *Cinco siglos de lexicografía del español*, edited by Ignacio Ahumada, 179–201. Jaén: Universidad de Jaén, 2000.
Alvar Ezquerra, Manuel. "El largo camino hasta el diccionario monolingüe." *Voz y Letras* 5 (1994): 47–66.
———. "Dictionaries of Spanish in their Historical Context." In *Lexicography. Reference Works across Time, Space and Languages*, edited by Reinhard R. K. Hartmann, 343–74. London: Routledge, 2003.
Álvarez de Miranda, Pedro. "La lexicografía académica de los siglos XVIII y XIX." In *Cinco siglos de lexicografía del español*, edited by Ignacio Ahumada, 35–61. Jaén: Universidad de Jaén, 2000.
Amelang, James S. *The Flight of Icarus: Artisan Autobiography in Early Modern Europe*. Stanford: Stanford University Press, 1998.
Anderson, Benedict. *Imagined Communities: Reflections on the Origin and Spread of Nationalism*. London: Verso, 1992.
Andreu Celma, José María. *Gracián y el arte de vivir*. Zaragoza: Institución Fernando el Católico, 1998.

Aracil, Alfredo. *Juego y artificio. Autómatas y otras ficciones en la cultura del Renacimiento a la Ilustración.* Madrid: Cátedra, 1998.

Ardila, J. A. G., ed. *The Cervantean Heritage. Reception and Influence of Cervantes in Britain.* Oxford: Legenda, 2009.

Arellano, Ignacio. *El ingenio de Lope de Vega. Escolios a las "Rimas humanas y divinas del licenciado Tomé de Burguillos."* New York: IDEA/IGAS, 2012.

Aricò, Denise. "Prudenza e ingegno nella Filosofia morale di Emanuele Tesauro." *Studi Secenteschi* 42 (2001): 187–208.

Aronson, Alex. "Eighteenth Century Semantics of Wit." *ETC: A Review of General Semantics* 5 (1948): 182–90.

Arrizabalaga, Jon. "Huarte en la medicina de su tiempo." In *Huarte au XXIe siècle*, edited by Véronique Duché-Gavet, 65–98. Anglet: Atlantica, 2003.

Arvisu Dumol, Paul. *The Metaphysics of Reading Underlying Dante's Commedia: The Ingegno.* New York: Peter Lang, 1998.

Ashcroft, Jeffrey. "Art in German: Artistic Statements by Albrecht Dürer." *Forum for Modern Language Studies* 48, no. 4 (2012): 377–88.

Ayala, Jorge Manuel. "El 'ingenio' en Huarte de San Juan y otros escritores españoles." In *Actas del VI Seminario de Historia de la Filosofía Española e Iberoamericana*, edited by Antonio Heredia Soriano, 211–24. Salamanca: Universidad de Salamanca, 1990.

———. "Introducción." In *Agudeza y arte de ingenio*, by Baltasar Gracián, 13–113. Edited by Ceferino Peralta, Jorge M. Ayala, and José María Andreu. Zaragoza: Prensas Universitarias de Zaragoza, 2004.

Ayres-Bennett, Wendy, and Magali Seijido. *Remarques et observations sur la langue française: histoire et évolution d'un genre.* Paris: Classiques Garnier, 2011.

Azara, Pedro. "Furor divino. Contribución a la historia de la teoría del arte." PhD diss., Universitat Politècnica de Catalunya, 1986.

Azorín Fernández, Dolores. "Sebastián de Covarrubias y el nacimiento de la lexicografía española monolingüe." In *Cinco siglos de lexicografía del español*, edited by Ignacio Ahumada, 3–34. Jaén: Universidad de Jaén, 2000.

Barfield, Owen. *History in English Words.* London: Methuen, 1926.

Barnes, Geraldine. "Cunning and Ingenuity in the Middle English Floris and Blauncheflur." *Medium Aevum* 53, no. 1 (1984): 10–25.

Barolsky, Paul. *Infinite Jest: Wit and Humor in Italian Renaissance Art.* Columbia: University of Missouri Press, 1978.

———, and Andrew Ladis. "The Pleasurable Deceits of Bronzino's So-called London 'Allegory.'" *Source: Notes in the History of Art* 10, no. 3 (1991): 32–36.

Bartra, Roger. *Melancholy and Culture: Essays on the Diseases of the Soul in Golden Age Spain.* Cardiff: University of Wales Press, 2008.

Bate, Jonathan. "Shakespeare and Original Genius." In *Genius: The History of An Idea*, edited by Penelope Murray, 76–97. Oxford: Basil Blackwell, 1989.

———. *The Genius of Shakespeare*. London: Picador, 2008.
Battersby, Christine. *Gender and Genius: Towards a Feminist Aesthetics*. New York: The Women's Press, 1986.
Baxandall, Michael. "A Dialogue on Art from the Court of Leonello d'Este: Angelo Decembrio's De Politia Litteraria Pars LXVIII." *Journal of the Warburg and Courtauld Institutes* 26, no. 3/4 (1963): 304–26.
———. *Giotto and the Orators: Humanist Observers of Painting in Italy and the Discovery of Pictorial Composition*. Oxford: Oxford University Press, 1986.
Beer, Susanna de, K. A. E. Enenkel, and David Rijser, eds. *The Neo-Latin Epigram: A Learned and Witty Genre*. Leuven: Leuven University Press, 2009.
Beiser, Frederick C. *Diotima's Children: German Aesthetic Rationalism from Leibniz to Lessing*. Oxford: Oxford University Press, 2009.
Bennett, Jim. "Instruments and Ingenuity." In *Robert Hooke: Tercentennial Studies*, edited by Michael Cooper and Michael Hunter, 65–76. Aldershot: Ashgate, 2006.
Bernard, Cerquiglini. *Une langue orpheline*. Paris: Editions de Minuit, 2007.
Bernstein, Eckhard. "From Outsiders to Insiders: Some Reflections on the Development of a Group Identity of the German Humanists between 1450 and 1530." In *In Laudem Caroli: Renaissance and Reformation Studies for Charles G. Nauert*, edited by James V. Mehl, 45–64. Kirksville, MO: Truman State University Press, 1998.
Biester, James. *Lyric Wonder: Rhetoric and Wit in Renaissance English Poetry*. Ithaca, NY: Cornell University Press, 1997.
———. "Fancy's Images: Wit, the Sublime, and the Rise of Aestheticism." In *Wonders, Marvels, and Monsters in Early Modern Culture*, edited by Peter G. Platt, 294–327. Newark: University of Delaware Press, 1999.
Bikanová, Klára. "The Dangers of Wit: Re-examining C.S. Lewis's Study of a Word." *Brno Studies in English* 35, no. 1 (2009): 87–102.
———. "From Rhetoric to Aesthetics: Wit and Esprit in the English and French Theoretical Writings of the Late Seventeenth and Early Eighteenth Centuries." PhD diss., Masaryk University, 2011.
Blair, Ann. *Too Much to Know: Managing Scholarly Information before the Modern Age*. New Haven: Yale University Press, 2010.
Blanco, Emilio. "Introducción." In *Arte de ingenio, tratado de la agudeza*, by Baltasar Gracián, edited by Emilio Blanco, 9–88. Madrid: Cátedra, 1998.
Blanco, Mercedes. *Les rhétoriques de la pointe. Baltasar Gracián et le conceptisme en Europe*. Geneva: Slatkine, 1992.
———. *Góngora o la invención de una lengua*. León: Universidad de León, 2012.
Bocchi, Andrea. "I Florio contro la Crusca." In *La nascita del vocabolario. Convegno di studio per i quattrocento anni del Vocabolario della Crusca*, edited by Antonio Daniele and Laura Nascimben, 51–80. Padua: Esedra, 2014.
Bombart, Mathilde. *Guez de Balzac et la querelle des lettres: écriture, polémique et critique dans la France du premier XVIIe siècle*. Paris: Champion, 2007.

Bonfait, Olivier. "'Ingegno divino' o 'beauté du génie': Bellori, Félibien e il 'super-artist' nel Seicento." In *Begrifflichkeit, Konzepte, Definitionen. Schreiben über Kunst und ihre Medien in Giovanni Pietro Belloris Viten und der Kunstliteratur der Frühen Neuzeit,* edited by Elisabeth Oy-Marra, Marieke von Bernstorff, and Henry Keazor, 105–25. Wiesbaden: Harrassowitz, 2014.

Bonnefoy, Yves, ed. *Dictionnaire des mythologies et des religions des sociétés traditionnelles et des mondes antiques.* Paris: Flammarion, 1981.

Bouza, Fernando, and Francisco Rico. "Digo que yo he compuesto un libro intitulado *El ingenioso hidalgo de la mancha.*" *Cervantes: Bulletin of the Cervantes Society of America* 29, no. 1 (2009): 13–30.

Bowen, Barbara C. "Ciceronian Wit and Renaissance Rhetoric." *Rhetorica* 16, no. 4 (1998): 409–29.

Brann, Noel L. *The Debate Over the Origin of Genius During the Italian Renaissance: The Theories of Supernatural Frenzy and Natural Melancholy in Accord and in Conflict on the Threshold of the Scientific Revolution.* Leiden: Brill, 2002.

Bredekamp, Horst. *Der Künstler als Verbrecher: Ein Element der frühmodernen Rechts- und Staatstheorie.* Munich: Carl Friedrich von Siemens Stiftung, 2008.

Brinker, Menachem, and Rita Sabah. "Le 'naturel' et le 'conventionnel' dans la critique et la théorie." *Littérature* 57 (1985): 17–30.

Brockmeyer, Bettina. *Selbstverständnisse: Dialoge über Körper und Gemüt im frühen 19. Jahrhundert.* Göttingen: Wallstein, 2009.

Bromwich, David. "Reflections on the Word *Genius.*" *New Literary History* 17, no. 1 (1985): 141–64.

Bruno, Paul W. *Kant's Concept of Genius: Its Origin and Function in the Third Critique.* London: Continuum, 2010.

Bruster, Douglas. *Shakespeare and the Question of Culture: Early Modern Literature and the Cultural Turn.* New York: Palgrave Macmillan, 2003.

Bryson, Cynthia B. "Mary Astell: Defender of the 'Disembodied Mind.'" *Hypatia* 13 (1998), 40–62.

Buchenau, Stefanie. *The Founding of Aesthetics in the German Enlightenment: The Art of Invention and the Invention of Art.* Cambridge: Cambridge University Press, 2013.

Burioni, Matteo. *Die Renaissance der Architekten. Profession und Souveränität des Baukünstlers in Giorgio Vasaris Viten.* Berlin: Gebr. Mann Verlag, 2008.

Burke, Peter. *Languages and Communities in Early Modern Europe.* Cambridge: Cambridge University Press, 2004.

Cacho Casal, Rodrigo. *La esfera del ingenio. Las silvas de Quevedo y la tradición europea.* Madrid: Biblioteca Nueva, 2012.

Cámara, Alicia. *Los ingenieros militares de la Monarquía Hispánica en los siglos XVII y XVIII.* Madrid: Centro de Estudios Europa Hispánica, 2005.

Cámara, Alicia, and Bernardo Revuelta, eds. *Ingenieros del Renacimiento.* Madrid: Fundación Juanelo Turriano, 2014.

Cameron, Euan. *Enchanted Europe: Superstition, Reason, and Religion, 1250–1750*. Oxford: Oxford University Press, 2010.
Campi, Riccardo. "*Ingenio* ed *esprit* tra Gracián e Bouhours: una questione di metodo." *Studi di estetica* 25, no. 16 (1997): 185–209.
Campione, Francesco Paolo. *La regola del Capriccio. Alle origini di una idea estetica*. Palermo: Centro Internazionale Studi di Estetica, 2011.
Cantarino, Elena, and Emilio Blanco, eds. *Diccionario de conceptos de Baltasar Gracián*. Madrid: Cátedra, 2005.
Carrera, Elena. "Madness and Melancholy in Sixteenth- and Seventeenth-Century Spain: New Evidence, New Approaches." *Bulletin of Spanish Studies* 87, no. 8 (2010): 1–15.
Carriazo, José Ramón, and María Jesús Mancho. "Los comienzos de la lexicografía monolingüe." In *Lexicografía española*, edited by Antonia María Medina Guerra, 205–34. Barcelona: Ariel, 2003.
Casteix, Jean-Gerald. "Réduire la gravure en art et en principes: lecture et réception du *Traité des manières de graver a l'eau-forte* d'Abraham Bosse." In *Réduire en art: la technologie de la Renaissance aux Lumières*, edited by Pascal Dubourg-Glatigny and Hélène Vérin, 235–48. Paris: Éditions de la maison des sciences de l'homme, 2008.
Cavaillé, Jean-Pierre. *Dis/simulations: Jules-César Vanini, François La Mothe Le Vayer, Gabriel Naudé, Louis Machon et Torquato Accetto: religion, morale et politique Au XVIIe siècle*. Paris: Honoré Champion, 2002.
———. "Le paladin de la République des lettres contre l'épouvantail des sciences sociales," *Les Dossiers du Grihl* (2007): URL: http://dossiersgrihl.revues.org/278; DOI: 10.4000/dossiersgrihl.278.
———. *Postures libertines. La culture des esprits forts*. Toulouse: Anacharsis, 2011.
Cave, Terence. *Pré-histoires*. 2 vols. Geneva: Droz, 1999–2001.
———. *Retrospectives: Essays in Literature, Poetics and Cultural History*, edited by Neil Kenny and Wes Williams. Oxford: Legenda, 2009.
Chambers, David S., ed. *Patrons and Artists of the Italian Renaissance*. London: Macmillan, 1970.
Chevalier, Maxime. *Quevedo y su tiempo. La agudeza verbal*. Barcelona: Crítica, 1992.
Citton, Yves. "Spirits across the Channel. The Staging of Collective Mental Forces in Gabalistic Novels from Margaret Cavendish to Charles Tiphaigne de la Roche." *Comparatio. Zeitschrift für Vergleichende Literaturwissenschaft* 1, no. 2 (2009): 291–319.
Claes, Frans. *De bronnen van drie woordenboeken uit de drukkerij van Plantin: Het Dictionarium tetraglotton (1562), de Thesaurus theutonicae linguae (1573) en Kiliaans Eerste dictionarium teutonico-latinum (1574)*. Ghent: Koninklijke Vlaamse Academie, 1970.
Clarke, Georgia. "'La più bella e meglio lavorata opera': Beauty and Good Design in Italian Renaissance Architecture." In *Concepts of Beauty in Renaissance Art*, edited by Francis Ames-Lewis and Mary Rogers, 107–23. Aldershot: Ashgate, 1998.

Codoñer, Carmen. "Ingenio y agudeza. Reflexiones léxicas." In *Baltasar Gracián IV Centenario (1601–2001). Actas I Congreso Internacional "Baltasar Gracián: pensamiento y erudición,"* edited by Aurora Egido, Fermín Gil Encabo, and José Enrique Laplana Gil, 203–33. Huesca-Zaragoza: Instituto de Estudios Altoaragoneses; Institución Fernando el Católico, 2003.

Cole, Michael W. *Cellini and the Principles of Sculpture.* New York: Cambridge University Press, 2002.

———. "The Demonic Arts and the Origin of the Medium." *The Art Bulletin* 84, no. 4 (2002): 621–40.

Colón, Germán. *Las primeras traducciones europeas del Quijote.* Barcelona: Universitat Autònoma de Barcelona, 2005.

Connor, Desmond. *A History of Italian and English Bilingual Dictionaries.* Florence: Leo S. Olschki, 1990.

Considine, John. *Dictionaries in Early Modern Europe: Lexicography and the Making of Heritage.* Cambridge: Cambridge University Press, 2008.

———, ed. *Ashgate Critical Essays on Early English Lexicographers, Volume 4: The Seventeenth Century.* Farnham: Ashgate, 2012.

———. *Academy Dictionaries 1600–1800.* Cambridge: Cambridge University Press, 2014.

Conticelli, Valentina. "Prometeo, Natura e il Genio sulla volta dello Stanzino di Francesco I: fonti letterarie, iconografiche e alchemiche." *Mitteilungen Des Kunsthistorischen Institutes in Florenz* 46, no. 2/3 (2002): 321–56.

Cook, Daniel. "On Genius and Authorship: Addison to Hazlitt." *Review of English Studies* 64, no. 266 (2013): 610–29.

Corngold, Stanley. "Wit and Judgment in the Eighteenth Century: Lessing and Kant." *MLN* 102, no. 3 (1987): 461–82.

Corominas, Joan, and José A. Pascual. *Diccionario crítico etimológico castellano e hispánico.* 6 vols. Madrid: Gredos, 1991.

Cowie, Anthony P., ed. *The Oxford History of English Lexicography.* Oxford: Oxford University Press, 2009.

Crane, William G. *Wit and Rhetoric in the Renaissance: The Formal Basis of Elizabethan Prose Style.* New York: Columbia University Press, 1937.

Cronk, Nicholas. *The Classical Sublime: French Neoclassicism and the Language of Literature.* Charlottesville, VA: Rookwood, 2002.

Cruz Suárez, Juan Carlos. *Ojos con mucha noche. Ingenio, poesía y pensamiento en el Barroco español.* Bern: Peter Lang, 2014.

Cunchillos Jaime, Carmelo. "Traducciones inglesas del 'Quijote' (1612–1800)." In *De clásicos y traducciones. Versiones inglesas de clásicos españoles (ss. XVI-XVII)*, edited by Julio-César Santoyo and Isabel Verdaguer, 89–114. Barcelona: Promociones y Publicaciones Universitarias, 1987.

Dauzat, Albert. *Le génie de la langue française.* Paris: Payot, 1943.

Davies, Owen. *Popular Magic: Cunning Folk in English History*. London: Hambleden, 2007.
Dekker, Cornelis. *The Origins of Old Germanic Studies in the Low Countries*. Leiden: Brill, 1999.
Detienne, Marcel, and Jean-Pierre Vernant. *Cunning Intelligence in Greek Culture and Society*. Translated by Janet Lloyd. Chicago: University of Chicago Press, 1991.
Dibbets, Geert R. W. "Dutch Philology in the 16th and 17th Century." In *The History of Linguistics in the Low Countries*, edited by Jan Noordegraaf, C. H. M. Versteegh, and E. F. K. Koerner, 39–61. Amsterdam: John Benjamins, 1992.
Doederlein, Sue W. "A Compendium of Wit: The Psychological Vocabulary of John Dryden's Literary Criticism." PhD diss., University of Michigan, 1971.
Domenici, Davide. "Missionary Gift Records of Mexican Objects in Early Modern Italy." In *The New World in Early Modern Italy, 1492–1750*, edited by Elizabeth Horodowich and Lia Markey, 86–102. Cambridge: Cambridge University Press, 2017.
Drusi, Riccardo. *La lingua "cortigiana romana." Note su un aspetto della questione cinquecentesca della lingua*. Venice: Il Cardo, 1997.
Dubourg-Glatigny, Pascal, and Hélène Vérin, eds. *Réduire en art. la technologie de la Renaissance aux Lumières*. Paris: Éditions de la maison des sciences de l'homme, 2008.
Duché-Gavet, Véronique, ed. *Juan Huarte au XXIe siècle*. Anglet: Atlantica, 2003.
Dundas, Judith. "Allegory as a Form of Wit." *Studies in the Renaissance* 11 (1964): 223–33.
Dunkelgrün, Theodor. "The Multiplicity of Scripture: The Confluence of Textual Traditions in the Making of the Antwerp Polyglot (1568–1573)." PhD diss., University of Chicago, 2012.
Duro, Paul. "'The Surest Measure of Perfection': Approaches to Imitation in Seventeenth-century French Art and Theory." *Word & Image* 25, no. 4 (2009): 363–83.
Dursteler, Eric R. "Speaking in Tongues: Language and Communication in the Early Modern Mediterranean." *Past & Present* 217, no. 1 (2012): 47–77.
Egido, Aurora. "De las academias a la Academia." In *The Fairest Flower. The Emergence of Linguistic National Consciousness in Renaissance Europe*, 85–94. Florence: Presso L'Accademia, 1985.
———. "Arte y literatura: lugares e imágenes de la memoria en el Siglo de Oro." In *El Siglo de Oro de la pintura española*, edited by Alfoso E. Pérez Sánchez et al., 273–95. Madrid: Mondadori, 1991.
———. Introduction to *El discreto*, by Baltasar Gracián, edited by Aurora Egido, 7–128. Madrid: Alianza, 1997.
———. *Las caras de la prudencia y Baltasar Gracián*. Madrid: Editorial Castalia, 2000.
———. *Humanidades y dignidad del hombre en Baltasar Gracián*. Salamanca: Universidad de Salamanca, 2001.

———. "Estudio preliminar." In *Arte de ingenio, tratado de la agudeza. Edición facsímil (Madrid, Juan Sánchez, 1642)*, by Baltasar Gracián, VII–CXLVIII. Zaragoza: Institución Fernando el Católico, 2005.

———. "Estudio preliminar." In *Agudeza y arte de ingenio (Huesca, Juan Nogués, 1648) Edición facsímil*, by Baltasar Gracián. Zaragoza: Institución Fernando el Católico, 2007.

———. *Bodas de arte e ingenio. Estudios sobre Baltasar Gracián*. Barcelona: Acantilado, 2014.

———. "La fuerza del ingenio y las lecciones cervantinas." *Boletín de la Real Academia Española*, t. 96, cuaderno 314 (2016): 771–94.

———, and María Carmen Martín, eds. *Baltasar Gracián. Estado de la cuestión y nuevas perspectivas*. Zaragoza: Institución Fernando el Católico, 2001.

Egmond, Florike, and Sachiko Kusukawa. "Circulation of Images and Graphic Practices in Renaissance Natural History: The Example of Conrad Gessner." *Gesnerus* 73, no. 1 (2016): 29–72.

Eiche, Sabine, ed. *I Gheribizzi di Muzio Oddi*. Urbino: Accademia Raffaello, 2005.

Elledge, Scott. "Cowley's Ode 'Of Wit' and Longinus on the Sublime: A Study of One Definition of the Word Wit." *Modern Language Quarterly* 9 (1948): 185–98.

Else, Felicia M. "Ammanati's Shield of Achilles: Making a Virtue out of Necessity." *Source: Notes in the History of Art* 28, no. 1 (2008): 30–38.

Emison, Patricia A. "Grazia." *Renaissance Studies* 5, no. 4 (1997): 427–60.

———. *Creating the "Divine" Artist: From Dante to Michelangelo*. Leiden: Brill, 2004.

Emmel, Hildegard. "Gemüt." In *Historisches Wörterbuch der Philosophie*, edited by Joachim Ritter and Karlfried Gründer, vol. 3, 258–62. Darmstadt: Wissenschaftliche Buchgesellschaft, 1974.

Empson, William. "Wit in the *Essay on Criticism*." *The Hudson Review* 2, no. 4 (1950): 559–77.

Engell, James. *The Creative Imagination: Enlightenment to Romanticism*. Cambridge, MA: Harvard University Press, 1981.

Eron, Sarah. *Inspiration in the Age of Enlightenment*. Newark: University of Delaware Press, 2014.

Esparza Torres, Miguel Ángel, and Hans-Josef Niederehe. *Bibliografía nebrisense. Las obras completas del humanista Antonio de Nebrija desde 1481 hasta nuestros días*. Amsterdam and Philadelphia: John Benjamins Publishing Company, 1999.

Etter, Else L. *Tacitus in der Geistesgeschichte des 16. und 17. Jahrhunderts*. Basel: Helbing and Lichtenhahn, 1966.

Evans, R. J. W., and Alexander Marr, eds. *Curiosity and Wonder from the Renaissance to the Enlightenment*. Aldershot: Ashgate, 2006.

Farina, Caterina Mongiat. *Questione della lingua: L'ideologia del dibattita sul italiano nel Cinquecento*. Ravenna: Longo, 2014.

Fattori, Marta, and Massimo Bianchi, eds. *Spiritus. IV° Colloquio Internazionale del Lessico Intellettuale Europeo*. Rome: Edizioni dell'Ateneo, 1984.

Febvre, Lucien. *A New Kind of History, from the Writings of Lucien Febvre*, edited by Peter Burke, translated by K. Folca. London: Routledge & Kegan Paul, 1973.

Filipczak, Zirka Zaremba. *Picturing Art in Antwerp, 1550–1700*. Princeton: Princeton University Press, 1986.

Findlen, Paula. "Ideas in the Mind: Gender and Knowledge in the Seventeenth Century." *Hypatia* 17 (2002): 183–196.

Fischel, Angela. "Collections, Images and Form in Sixteenth-Century Natural History: The Case of Conrad Gessner." In "Picturing Collections in Early Modern Europe," edited by Alexander Marr. Special issue of *Intellectual History Review* 20, no. 1 (2010): 147–64.

Fleming, Juliet. "Dictionary English and the Female Tongue." In *Ashgate Critical Essays on Early English Lexicographers, Volume 4*, edited by John Considine. Farnham: Ashgate, 2012.

Fransen, Sietske, and Niall Hodson, eds. *Translating Early Modern Science*. Leiden: Brill, 2017.

Franzbach, Martin. *Lessings Huarte-Übersetzung (1752): Die Rezeption und Wirkungsgeschichte des "Examen de ingenios para las ciencias" (1575) in Deutschland*. Hamburg: De Gruyter, 1965.

Freifrau von Gemmingen, Barbara. "Los inicios de la lexicografía española." In *Lexicografía española*, edited by Antonia María Medina Guerra, 151–74. Barcelona: Ariel, 2003.

Freixas, Margarita. "Las autoridades en el primer diccionario de la Real Academia española." PhD diss., Universidad Autónoma de Barcelona, 2003.

Fumaroli, Marc. *L'âge de l'éloquence: rhétorique et res litteraria en France de la Renaissance au seuil de l'âge classique*. 1st ed. 1980. Paris: Droz, 2002.

———. *Le genre des genres littéraires français: la conversation—The Zaharoff Lecture, 1990–1*. Oxford: Clarendon Press, 1992.

———. "Le génie de la langue francaise." In *Trois institutions littéraires*. Paris: Gallimard, 1994.

———. *Quand L'Europe parlait français*. Paris: Fallois, 2001.

———. *La diplomatie de l'esprit: de Montaigne à La Fontaine*. Paris: Gallimard, 2002.

Furno, Martine. *Le* Cornu copiae *de Niccolò Perotti: culture et méthode d'un 'humaniste qui aimait les mots*. Geneva: Droz, 1995.

Gállego, Julián. *El pintor de artesano a artista*. Granada: Universidad de Granada, 1976.

Gambarota, Paola. *Irresistible Signs: The Genius of Language and Italian National Identity*. Toronto: University of Toronto Press, 2011.

Gambin, Felice. *Azabache. El debate sobre la melancolía en la España de los Siglos de Oro*. Madrid: Biblioteca Nueva, 2008.

García Berrio, Antonio. *Formación de la teoría literaria. Volumen 1.* Madrid: Cupsa, 1977.
——. *Formación de la teoría literaria. Volumen 2.* Murcia: Universidad de Murcia, 1980.
García Diego, José Antonio. *Juanelo Turriano, Charles V's Clockmaker. The Man and His Legend.* Wadhurst: Antiquarian Horological Society, 1986.
García Tapia, Nicolás. *Ingeniería y arquitectura en el renacimiento español.* Valladolid: Universidad de Valladolid, 1990.
García Tapia, Nicolás, and Jesús Carrillo. *Tecnología e imperio: Turriano, Lastanosa, Herrera, Ayanz.* Madrid: Nivola, 2012.
Garrod, Raphaële with Alexander Marr, eds. *Descartes and the* Ingenium: *The Embodied Soul in Cartesianism.* Oxford: Oxford University Press, forthcoming.
Geiger, Ludwig. *Johann Reuchlin, sein Leben und seine Werke.* Leipzig: Duncker & Humblot, 1871.
Gensini, Stefano. "L'ingegno e le metafore: alle radici della creatività linguistica fra Cinque e Seicento." *Studi di estetica* 25, no. 16 (1997): 135–62.
——. "Ingenium/ingegno fra Huarte, Persio e Vico: le basi naturali dell'inventività umana." *Ingenium propria hominis natura*, edited by Stefano Gensini and Arturo Martone, 29–70. Naples: Liguori, 2002.
——, and Arturo Martone, eds. *Ingenium propria hominis natura.* Naples: Liguori, 2002.
Gili Gaya, Samuel. *Tesoro lexicográfico, 1492–1726.* Madrid: CSIC, 1947.
Gilson, Simon. *Dante and Renaissance Florence.* Cambridge: Cambridge University Press, 2005.
——. *Reading Dante in Renaissance Italy: Florence, Venice and the Divine Poet.* Cambridge: Cambridge University Press, 2018.
Ginzburg, Carlo. "Style: Inclusion and Exclusion," in *Wooden Eyes: Nine Reflections on Distance*, translated by Martin Ryle and Kate Soper, 109–38. New York: Columbia University Press, 2001.
Giovanardi, Claudio. *La teoria cortigiana e il dibattito linguistico nel primo Cinquecento.* Rome: Bulzoni, 1998.
Golinelli, Gilberta. "The Concept of Genius and Its Metamorphoses in Eighteenth-Century English Criticism." *Textus: English Studies in Italy* 18, no. 1 (2005): 189–204.
Goodey, C. F. *A History of Intelligence and Intellectual "Disability": The Shaping of Psychology in Early Modern Europe.* Farnham: Ashgate, 2011.
Gotti, Maurizio. *The Language of Thieves and Vagabonds: 17th and 18th Century Canting Lexicography in England.* Berlin: De Gruyter, 1999.
Göttler, Christine, and Wolfgang Neuber, eds. *Spirits Unseen: The Representation of Subtle Bodies in Early Modern European Culture.* Leiden: Brill, 2008.
Gowland, Angus. *The Worlds of Renaissance Melancholy: Robert Burton in Context.* Cambridge: Cambridge University Press, 2006.

Grafton, Anthony, and Lisa Jardine. *From Humanism to the Humanities: Education and the Liberal Arts in Fifteenth- and Sixteenth-Century Europe*. Cambridge, MA: Harvard University Press, 1986.

Graviana, Teresa. "Breve storia della parola 'genio.'" *Lingua nostra* 28, no. 2 (1967): 37–43.

Graziosi, Elisabetta. *Questioni di lessico: l'ingegno, le passioni, il linguaggio*. Modena: Mucci Editore, 2004.

Green, Otis H. "El Ingenioso Hidalgo." *Hispanic Review* 25, no. 3 (1957): 175–93.

———. *The Literary Mind of Medieval and Renaissance Spain*. Lexington: University Press of Kentucky, 1970.

Greene, Robert A. "Whichcote, Wilkins, 'Ingenuity,' and the Reasonableness of Christianity." *Journal of the History of Ideas* 42, no. 2 (1981): 227–52.

Gregory, Sharon, and Sally Anne Hickson, eds. *Inganno—The Art of Deception*. Aldershot: Ashgate, 2012.

Gründler, Hana. "'Gloriarsi della mano e dell'ingegno': Hand, Geist und pädagogischer Eros bei Vasari und Bellori." In *Begrifflichkeit, Konzepte, Definitionen. Schreiben über Kunst und ihre Medien in Giovanni Pietro Belloris Viten und der Kunstliteratur der Frühen Neuzeit*, edited by Elisabeth Oy-Marra, Marieke von Bernstorff, and Henry Keazor, 77–103. Wiesbaden: Harrassowitz, 2014.

Grundy, Isobel. *Samuel Johnson and the Scale of Greatness*. Leicester: Leicester University Press, 1986.

Gundersheimer, Werner L. "Bartolommeo Goggio: A Feminist in Renaissance Ferrara." *Renaissance Quarterly* 33, no. 2 (1980): 175–200.

Hale, John R. *Renaissance Fortification: Art or Engineering?* London: Thames & Hudson, 1977.

Hall, Robert A. *The Italian 'Questione della Lingua': An Interpetative Essay*. Chapel Hill: University of North Carolina Press, 1942.

Hallyn, Fernand. *Descartes: Dissimulation et ironie*. Geneva: Droz, 2006.

Hanson, Craig Ashley. *The English Virtuoso: Art, Medicine, and Antiquarianism in the Age of Empiricism*. Chicago: University of Chicago Press, 2009.

Harper, Anthony John, and Ingrid Höpel, eds. *The German-Language Emblem in Its European Context: Exchange and Transmission*. Geneva: Droz, 2000.

Harris, Jim. "Lorenzo Ghiberti and the Language of Praise." *Sculpture Journal* 26, no. 1 (2017): 107–18.

Hartmann-Werner, Ingeborg. *"Gemüt" bei Goethe: Eine Wortmonographie*. Munich: W. Fink, 1976.

Harvey, E. Ruth. *The Inward Wits: Psychological Theory in the Middle Ages and Renaissance*. London: The Warburg Institute, 1975.

Hastings, Robert. "Questione della lingua." In *The Oxford Companion to Italian Literature*, edited by Peter Hainsworth and David Robey, 495–96. Oxford: Oxford University Press, 2002.

Healy, Fiona. "*Vive l'Esprit.* Sculpture as the Bearer of Meaning in Willem van Haecht's Art Cabinet of Cornelis van Der Geest." In *Munuscula Amicorum: Contributions on Rubens and His Colleagues in Honour of Hans Vlieghe*, edited by K. van de Stighelen, 423–41. Turnhout: Brepols, 2006.

Hempel, Wido. "Zur Geschichte von *spiritus, mens* und *ingenium* in den romanischen Sprachen." *Romanistisches Jahrbuch* 16 (1965): 21–33.

Henckmann, Wolfhart, and Konrad Lotter, *Lexikon der Ästhetik*. Munich: Beck, 1992.

Hidalgo-Serna, Emilio. "'Ingenium' and Rhetoric in the Work of Vives." *Philosophy & Rhetoric* 16, no. 4 (1983): 228–41.

———. *El pensamiento ingenioso en Baltasar Gracián*. Barcelona: Anthropos, 1993.

Hinnant, Charles H. "Hobbes on Fancy and Judgment." *Criticism* 18 (1976): 15–26.

Hooker, Edward Niles. "Pope on Wit: the 'Essay on Criticism.'" In Richard Foster Jones et al., *The Seventeenth Century: Studies in the History of English Literature and Thought from Bacon to Pope*, 225–46. Oxford: Oxford University Press, 1951.

Höpel, Ingrid. *Emblem und Sinnbild: Vom Kunstbuch zum Erbauungsbuch*. Frankfurt am Main: Athenäum, 1987.

Hudson, Hoyt H. *The Epigram in the English Renaissance*. Princeton: Princeton University Press, 1947.

Huet, Marie-Hélène. *Monstrous Imagination*. Cambridge, MA: Harvard University Press, 1993.

Hutchings, C. M. "The *Examen de Ingenios* and the Doctrine of Original Genius." *Hispania* 19, no. 2 (1936): 273–82.

Iriarte, Mauricio de. *El doctor Huarte de San Juan y su Examen de ingenios. Contribución a la historia de la psicología diferencial*. Madrid: CSIC, 1948.

Jakobs, Frederika H. "Woman's Capacity to Create: The Unusual Case of Sofonisba Anguissola." *Renaissance Quarterly* 47, no. 1 (1994): 74–101.

Jefferson, Ann. *Genius in France: An Idea and Its Uses*. Princeton: Princeton University Press, 2014.

———, and Jean-Alexandre Perras, eds. *Thinking Genius, Using Genius / Penser le génie à travers ses usages*. Special issue of *L'Esprit Créateur* 55, no. 2 (2015).

Jensen, H. James. *A Glossary of John Dryden's Critical Terms*. Minneapolis: University of Minnesota Press, 1969.

Jones, William Jervis. "*Lingua teutonum victrix*? Landmarks in German Lexicography (1500–1700)." *Histoire Épistémologie Langage* 13, no. 2 (1991): 131–52.

Jongeneelen, Gerrit H. "Lambert Ten Kate and the Origin of 19th-Century Historical Linguistics." In *The History of Linguistics in the Low Countries*, edited by Jan Noordegraaf, C. H. M. Versteegh, and E. F. K. Koerner, 201–20. Amsterdam: John Benjamins, 1992.

Jouhaud, Christian. *Les pouvoirs de La littérature: histoire d'un paradoxe*. Paris: Gallimard, 2000.

Junk, Klaus. *Die Funktion von "Gemüt" in der Literatur-theorie und—kritik (18. und 19. Jahrhundert)*. Erlangen: Schmitt and Meyer, 1968.

Kallich, Martin. "The Association of Ideas and Critical Theory: Hobbes, Locke, and Addison." *English Literary History* 12, no. 4 (1945): 290–315.

Kanz, Roland. *Die Kunst des Capriccio. Kreativer Eigensinn in Renaissance und Barock*. Munich: Deutscher Kunstverlag, 2002.

Keazor, Henry. "'Spirito abile' ed 'elevatissimo ingegno.' Giovan Pietro Bellori e Carlo Cesare Malvasia." *Mitteilungen des Kunsthistorischen Instituts in Florenz* 52, no. 1 (2008): 73–82.

Kemp, Martin. "From 'Mimesis' to 'Fantasia': The Quattrocento Vocabulary of Creation, Inspiration and Genius in the Visual Arts." *Viator* 8 (1977): 347–98.

———. "'Equal excellences': Lomazzo and the Explanation of Individual Style in the Visual Arts." *Renaissance Studies* 1, no. 1 (1987): 175–200.

———. "The 'Super-Artist' as Genius: The Sixteenth-Century View." In *Genius: History of an Idea*, edited by Penelope Murray, 32–53. Oxford: Basil Blackwell, 1989.

Kenny, Neil. *Curiosity in Early Modern Europe: Word Histories*. Wiesbaden: Harrassowitz, 1998.

———. *The Uses of Curiosity in Early Modern France and Germany*. Oxford: Oxford University Press, 2004.

———, and Wes Williams. "Introduction." In Terence Cave, *Retrospectives: Essays in Literature, Poetics and Cultural History*, edited by Neil Kenny and Wes Williams, 1–8. Oxford: Legenda, 2009.

Kessler, Eckhard. "The Intellective Soul." In *The Cambridge History of Renaissance Philosophy*, edited by Charles B. Schmitt et al., 485–539. Cambridge: Cambridge University Press, 1988.

Kim, David Young. *The Traveling Artist in the Italian Renaissance*. New Haven and London: Yale University Press, 2014.

Klein, Jürgen. "Genius, Ingenium, Imagination: Aesthetic Theories of Production from the Renaissance to Romanticism." In *The Romantic Imagination: Literature and Art in England and Germany*, edited by Frederick Burwick and Jürgen Klein, 16–62. Amsterdam: Rodopi, 1996.

Klein, Robert. *Form and Meaning: Writings on the Renaissance and Modern Art*, translated by Madeline Jay and Leon Wiseltier. Princeton: Princeton University Press, 1979.

———. "Spirito Pelegrino." In *Form and Meaning: Writings on the Renaissance and Modern Art*. Translated by Madeline Jay and Leon Wiseltier, 62–85. Princeton: Princeton University Press, 1979.

Klibansky, Raymond, Erwin Panofsky, and Fritz Saxl. *Saturn and Melancholy: Studies in the History of Natural Philosophy, Religion and Art*. London: Thomas Nelson and Sons, 1964.

Knowlton, E. C. "The Allegorical Figure Genius." *Classical Philology* 15, no.4 (1920): 380–84.
———. "Genius as an Allegorical Figure." *Modern Language Notes* 39, no. 2 (1924): 89–95.
———. "The Genii of Spenser." *Studies in Philology* 25, no. 4 (1928): 439–56.
Kraft, Elizabeth. "Wit and The Spectator's Ethics of Desire." *Studies in English Literature, 1500–1900* 45, no. 3 (2005): 625–46.
Kristeller, Paul O. "The Modern System of the Arts." In *Renaissance Thought and the Arts: Collected Essays*, by Paul O. Kristeller. Princeton: Princeton University Press, 1990.
Kuehn, Manfred. *Scottish Common Sense in Germany, 1768–1800: A Contribution to the History of Critical Philosophy*. Kingston and Montreal: McGill-Queen's Press, 1987.
Kunkel, Hille. *Der römische Genius*. Heidelberg: Kerl, 1974.
Labarre, Albert. *Bibliographie du Dictionarium d'Ambrogio Calepino (1502–1779)*. Baden-Baden: Koerner, 1975.
Laird, W. Roy. "Archimedes amongst the Humanists." *Isis* 82, no. 4 (1991): 629–38.
Laurens, Pierre. *L'abeille dans l'ambre: célébration de l'épigramme de l'époque alexandrine à la fin de la Renaissance*. Paris: Belles Lettres, 1989.
Lavin, Irving. "David's Sling and Michelangelo's Bow: A Sign of Freedom." In *Past-Present: Essays on Historicism in Art from Donatello to Picasso*, edited by Irving Lavin, 29–62. Berkeley: University of California Press, 1993.
Lázaro Carreter, Fernando. *Crónica del Diccionario de Autoridades (1713–1740)*. Madrid: Real Academia Española, 1972.
Lecointe, Jean. *L'idéal et la différence: la perception de la personnalité littéraire à la Renaissance*. Geneva: Librairie Droz, 1993.
Ledo, Jorge, and Harm den Boer, eds., *Moria de Erasmo Roterodamo. A Critical Edition of the Early Modern Spanish Translation of Erasmus's Encomium Moriae*. Leiden: Brill, 2014.
Lee, Rensselaer W. *Ut Pictura Poesis: The Humanistic Theory of Painting*. New York and London: Norton, 1967.
LEME. *Lexicons of Early Modern English*, edited by Ian Lancashire. https://leme.library.utoronto.ca
Levy, Alison, ed. *Playthings in Early Modernity: Party Games, Word Games, Mind Games*. Kalamazoo, MI: Medieval Institute Publications, 2017.
Lewis, C. S. *Studies in Words*. Cambridge: Cambridge University Press, 1967.
Lewis, Rhodri. *Language, Mind and Nature: Artificial Languages in England from Bacon to Locke*. Cambridge: Cambridge University Press, 2007.
———. "Impartiality and Disingenuousness in English Rational Religion." In *The Emergence of Impartiality*, edited by Anita Traninger and Kathryn Murphy, 224–45. Leiden: Brill, 2013.
———. "Francis Bacon and Ingenuity." *Renaissance Quarterly* 67, no. 1 (2014): 113–63.

Librett, Jeffrey S. *The Rhetoric of Cultural Dialogue: Jews and Germans from Moses Mendelssohn to Richard Wagner and Beyond.* Stanford, CA: Stanford University Press, 2000.

Lipking, Lawrence. "Johnson and Genius." In *Samuel Johnson: The Arc of the Pendulum,* edited by Freya Johnston and Linda Mugglestone, 83–94. Oxford: Oxford University Press, 2012.

Lokke, Kari E. *Tracing Women's Romanticism: Gender, History and Transcendence.* London and New York: Routledge, 2004.

Long, Kathleen P., ed. *Gender and Scientific Discourse in Early Modern Culture.* Farnham and Burlington: Ashgate, 2010.

Long, Pamela O. *Openness, Secrecy, Authorship. Technical Arts and the Culture of Knowledge from Antiquity to the Renaissance.* Baltimore and London: Johns Hopkins University Press, 2001.

Lund, Roger D. "The Bite of Leviathan: Hobbes and Philosophic Drollery." *English Literary History* 65, no. 4 (1998): 825–55.

———. "The Ghosts of Epigram, False Wit, and the Augustan Mode." *Eighteenth Century Life* 27, no. 2 (2003): 67–95.

———. "Wit, Judgment, and the Misprisions of Similitude." *Journal of the History of Ideas* 65, no. 1 (2004): 53–74.

———. *Ridicule, Religion and the Politics of Wit in Augustan England.* Farnham: Ashgate, 2012.

Mac Carthy, Ita. "Grace and the 'Reach of Art' in Castiglione and Raphael." *Word and Image* 25, no. 1 (2009): 33–45.

———. "Grace." In *Renaissance Keywords*, edited by Ita Mac Carthy, 63–80. Oxford: Legenda, 2013.

Maclean, Ian. *The Renaissance Notion of Woman: A Study in the Fortunes of Scholasticism and Medical Science in European Intellectual Life.* Cambridge: Cambridge University Press, 1980.

Maldonado de Guevara, Francisco. "Del 'Ingenium' de Cervantes al de Gracián." *Revista de estudios políticos* 100 (1958): 147–66.

Mannini, Maria Pia. "Chimere, capricci, ghiribizzi e altre cose. Esempi periferici di grottesca del tardo Cinquecento." *Annali / Fondazione di Studi di Storia dell'Arte Roberto Longhi* 1 (1984): 71–86.

Manrique Ara, María Elena. "De memoriales artísticos zaragozanos (I), Una defensa de la ingenuidad de la pintura presentada a Cortes de Aragón en 1677." *Artigrama: Revista del Departamento de Historia del Arte de la Universidad de Zaragoza* 13 (1998): 277–94.

Marcaida, José Ramón. "Examen de ingenios en la pintura de género de Murillo." In *Murillo ante su IV Centenario: Perspectivas historiográficas y culturales*, edited by Benito Navarrete Prieto. Seville: Instituto de la Cultura y las Artes de Sevilla, forthcoming, 2018.

Marías, Fernando. "El género de *Las Meninas*: los servicios de la familia." In *Otras Meninas*, edited by Fernando Marías, 247–78. Madrid: Siruela, 1995.

———. *El Greco. Biografía de un pintor extravagante*. Madrid: Editorial Nerea, 1997.

Marr, Alexander. "Understanding Automata in the Late Renaissance." *Journal de la Renaissance* 2 (2004): 205–22.

———. "The Flemish 'Pictures of Collections' Genre: An Overview." In "Picturing Collections in Early Modern Europe," edited by Alexander Marr, 5–25. *Intellectual History Review* 20, no. 1 (2010).

———. *Between Raphael and Galileo: Mutio Oddi and the Mathematical Culture of Late Renaissance Italy*. Chicago: University of Chicago Press, 2011.

———. "Pregnant Wit: *Ingegno* in Renaissance England." *British Art Studies* 1 (2015). Accessed May 1, 2017, http://dx.doi.org/10.17658/issn.2058-5462/issue-01/amarr.

———. "Ingenuity and Discernment in *The Cabinet of Cornelis van der Geest*." *Nederlands Kunsthistorisch Jaarboek* (forthcoming).

Marrazini, Claudio. "Questione della lingua." In *Enciclopedia dell'Italiano* (2011). Accessed May 1, 2017. http://www.treccani.it/enciclopedia/questione-della-lingua_(Enciclopedia-dell%27Italiano)/

Marshall, David L. *Vico and the Transformation of Rhetoric in Early Modern Europe*. Cambridge: Cambridge University Press, 2010.

Martín Herrero, Cristina. "El léxico de los ingenios y máquinas en el Renacimiento." PhD diss., Universidad de Salamanca, 2013.

Martínez Bogo, Enrique. *Retórica y agudeza en la prosa satírico-burlesca de Quevedo*. Santiago de Compostela: Universidad de Santiago de Compostela, 2010.

May, Terence. *Wit of the Golden Age. Articles on Spanish Literature*. Kassel: Reichenberger, 1986.

McCabe, Richard A. "Wit, Eloquence, and Wisdom in 'Euphues: The Anatomy of Wit.'" *Studies in Philology* 81, no. 3 (1984): 299–324.

McConchie, Roderick W., ed. *Ashgate Critical Essays on Early English Lexicographers, Volume 3: The Sixteenth Century*. Farnham: Ashgate, 2012.

———. "Introduction." In *Ashgate Critical Essays on Early English Lexicographers, Volume 3: The Sixteenth Century*, edited by Roderick W. McConchie. Farnham: Ashgate, 2012.

McDermott, Anne C., ed. *Ashgate Critical Essays on Early English Lexicographers, Volume 5: The Eighteenth Century*. Farnham: Ashgate, 2012.

———. "Introduction." In *Ashgate Critical Essays on Early English Lexicographers, Volume 5: The Eighteenth Century*, edited by Anne C. McDermott. Farnham: Ashgate, 2012.

McMahon, Darrin M. *Divine Fury: A History of Genius*. New York: Basic Books, 2013.

Medina Guerra, Antonia María. "Rodrigo Fernández de Santaella, *Vocabularium Ecclesiasticum*." *Analecta Malacitana* 13, no. 2 (1990): 329–42.

———. "Modernidad del *Universal vocabulario* de Alfonso Fernández de Palencia." *Estudios de Lingüística* 7 (1991): 45–60.

———, ed. *Lexicografía española*. Barcelona: Ariel, 2003.

Mee, John. *Romanticism, Enthusiasm, and Regulation: Poetics and the Policing of Culture in the Romantic Period*. Oxford: Oxford University Press, 2003.

Meerhoff, Kees. "La Ramée et Peletier Du Mans: Une deffence du 'naturel usage.'" *Nouvelle revue du XVIe Siècle* 18, no. 1 (2000): 77–93.

Melion, Walter S. *Shaping the Netherlandish Canon: Karel Van Mander's Schilder-Boeck*. Chicago: University of Chicago Press, 1991.

Meschonnic, Henri. *De la langue française: essai sur une clarté obscure*. Paris: Hachette, 1997.

Meschonnic, Henri, and Jean-Louis Chiss, eds. *Et le génie des langues?* Saint-Denis: Presses Universitaires de Vincennes, 2000.

Mestre-Zaragozá, Marina. "La '*Philosophía antigua poética*' de Alonso López Pinciano, un nuevo estatus para la prosa de ficción." *Criticón*, no. 120–21 (2014): 57–71.

Michel, Christian. "La peinture peut-elle être réduite en art?" In *Réduire en art: la technologie de la Renaissance aux Lumières*, edited by Pascal Dubourg-Glatigny and Hélène Vérin, 133–48. Paris: Éditions de la maison des sciences de l'homme, 2008.

Milburn, D. Judson. *The Age of Wit, 1650–1750*. New York: Macmillan, 1966.

Mirollo, James V. *The Poet of the Marvelous, Giambattista Marino*. New York: Columbia University Press, 1963.

Moretti, Franco. *Distant Reading*. London: Verso, 2013.

Moriarty, Michael. *Taste and Ideology in Seventeenth-Century France*. Cambridge: Cambridge University Press, 1988.

Morreale, Margherita. *Castiglione y Boscán: el ideal cortesano en el Renacimiento español*. 2 vols. Anejos del Boletín de la Real Academia Española. Madrid: S. Aguirre Torre, 1959.

Morton, Jonathan. "Ingenious Genius: Invention, Creation, Reproduction in the High Middle Ages." *L'Esprit Créateur* 55, no. 2 (2015): 4–19.

Moss, Ann. *Renaissance Truth and the Latin Language Turn*. Oxford: Oxford University Press, 2003.

Motta, Uberto. "La 'questione della lingua' nel primo libro del Cortegiano: dalla seconda alla terza redazione." *Aevum* 72, no. 3 (1998): 693–732.

Müller, Cristina. *Ingenio y melancolía. Una lectura de Huarte de San Juan*. Madrid: Biblioteca Nueva, 2002.

Mulsow, Martin. "Antiquarianism and Idolatry: The *Historia* of Religions in the Seventeenth Century." In *Historia: Empiricism and Erudition in Early Modern Europe*, edited by Gianna Pomata and Nancy G. Siraisi, 181–210. Cambridge, MA: MIT Press, 2005.

Murray, Penelope. "Introduction." In *Genius: The History of An Idea*, edited by Penelope Murray, 1–8. Oxford: Basil Blackwell, 1989.

Nancy, Jean-Luc. *The Discourse of the Syncope: Logodaedalus*, trans. Saul Anton. Stanford: Stanford University Press, 2008.

Nebes, Liane. *Der furor poeticus im italianischen Renaissanceplatonismus. Studien zu Kommentar und Literaturtheorie bei Ficino, Landino und Patrizi*. Marburg: Tectum-Verlag, 2001.

Neukirchen, Thomas. *Inscriptio: Rhetorik und Poetik der Scharfsinnigen Inschrift im Zeitalter des Barock*. Tübingen: Niemeyer, 1999.

Newman, Karen, and Jane Tylus, eds. *Early Modern Cultures of Translation*. Philadelphia: University of Pennsylvania Press, 2015.

Nieto Jiménez, Lidio. "Repertorios lexicográficos españoles menores en el siglo XVI." In *Cinco siglos de lexicografía del español*, edited by Ignacio Ahumada, 203–24. Jaén: Universidad de Jaén, 2000.

Nieto Jiménez, Lidio, and Manuel Alvar Ezquerra, eds. *Nuevo tesoro lexicográfico del español (s. XIV-1726)*, 11 vols. Madrid: Arco-Libros, 2007.

Nitzsche, Jane C. *The Genius Figure in Antiquity and the Middle Ages*. New York: Columbia University Press, 1975.

Norman, Larry F. *The Public Mirror: Molière and the Social Commerce of Depiction*. Chicago: University of Chicago Press, 1999.

Orobitg, Christine. *L'humeur noire. Mélancolie, écriture et pensée en Espagne au XVIe et au XVIIe siècle*. Bethesda: International Scholars Press, 1997.

———. "Del 'Examen de ingenios' de Huarte a la ficción cervantina, o cómo se forja una revolución literaria." *Criticón* 120 (2014): 23–39.

Ossa-Richardson, Anthony. *The Devil's Tabernacle: The Pagan Oracles in Early Modern Thought*. Princeton: Princeton University Press, 2013.

Osselton, Noel Edward. "Bilingual Lexicography with Dutch." In *Wörterbücher: Ein internationales Handbuch zur Lexikographie*, edited by Franz Josef Hausmann et al., vol. 3, 3034–39. Berlin: Walter de Gruyter, 1991.

Ott, Christine. "Terribile meraviglia: animismo artistico, empatia ed ekplexis nella poesia di Giovan Battista Marino." *Modern Language Notes* 130, no. 1 (2015): 63–85.

Paasche Grudin, Michaela, and Robert Grudin, *Boccaccio's Decameron and the Ciceronian Renaissance*. New York: Palgrave, 2012.

Pade, Marianne. "Niccolò Perotti's *Cornu Copiae*: Commentary on Martial and Encyclopedia." In *On Renaissance Commentaries*, edited by Marianne Pade, 49–63. Hildesheim: Olms, 2005.

———. "Niccolò Perotti's *Cornu Copiae*: The Commentary as a Repository of Knowledge." In *Neo-Latin Commentaries and the Management of Knowledge in the Late Middle Ages and the Early Modern Period (1400 -1700)*, edited by Karl A. E. Enenkel and Henk Nellen, 241–62. Leuven: Leuven University Press, 2013.

Panofsky, Erwin. "Artist, Scientist, Genius: Notes on the Renaissance *Dämmerung*." In *The Renaissance: Six Essays*, edited by Walace K. Ferguson et al., 121–82. New York: Harper & Row, 1962.

———. *The Art and Life of Albrecht Dürer*. Princeton: Princeton University Press 2005.

Pardo Tomás, José. "Ancora su Michel de Montaigne e Huarte de San Juan. Ricezione dei lettori e comunità interpretative tagli *Essais* e l'*Examen de ingenios para las sciencias*." In *Michel de Montaigne e il termalismo*, edited by Anna Bettoni, Massimo Rinaldi, and Maurizio Rippa Bonati, 95–119. Florence: Leo S. Olschki, 2010.

Parker, Alexander A. "'Concept' and 'Conceit': An Aspect of Comparative Literary History." *The Modern Language Review* 77, no. 4 (1982): xxi-xxxv.

Parker, Deborah. *Commentary and Ideology: Dante in the Renaissance*. Durham, NC: Duke University Press, 1993.

Parshall, Peter. "Graphic Knowledge: Albrecht Dürer and the Imagination." *The Art Bulletin* 95, no. 3 (2013): 393–410.

Pascual Rodríguez, José Antonio. "El comentario lexicográfico: tres largos paseos por el laberinto del diccionario." In *Lexicografía española*, edited by Antonia María Medina Guerra, 353–85. Barcelona: Ariel, 2003.

Pask, Kevin. *The Fairy Way of Writing: Shakespeare to Tolkien*. Baltimore and London: Johns Hopkins University Press, 2013.

Pelle, Federico. *Vocabolario italiano-latino. Edizione del primo lessico del volgare. Secolo XV*. Florence: Leo S. Olschki, 2001.

Pepe, Inoria, and José María Reyes. "Introducción." In *Anotaciones a la poesía de Garcilaso*, by Fernando de Herrera. Edited by Inoria Pepe and José María Reyes. Madrid: Cátedra, 2001.

Pepper, Simon, and Nicholas Adams. *Firearms and Fortifications: Military Architecture and Siege Warfare in Sixteenth-Century Siena*. Chicago: University of Chicago Press, 1986.

Pérez Fernández, José María. "Translation, *Sermo Communis*, and the Book Trade." In *Translation and the Book Trade in Early Modern Europe*, edited by José María Pérez Fernández and Edward Wilson-Lee, 40–60. Cambridge: Cambridge University Press, 2014.

Pérez Lasheras, Antonio. "Arte de ingenio y agudeza y arte de ingenio." In *Baltasar Gracián. Estado de la cuestión y nuevas perspectivas*, edited by Aurora Egido and María Carmen Martín, 71–88. Zaragoza: Institución Fernando el Católico, 2001.

Perkins, David. "Johnson on Wit and Metaphysical Poetry." *English Literary History* 20, no. 3 (1953): 200–17.

Perkinson, Stephen. "Engin and Artifice: Describing Agency at the Court of France, ca. 1400." *Gesta* 41, no. 1 (2002): 51–67.

Perras, Jean-Alexandre. *L'Exception exemplaire: Inventions et usages du génie (XVIe-XVIIIe siècle)*. Paris: Classiques Garnier, 2016.

Peset, José Luis. *Genio y desorden*. Valladolid: Cuatro, 1999.

———. *Las melancolías de Sancho. Humores y pasiones entre Huarte y Pinel*. Madrid: Asociación Española de Neuropsiquiatría, 2010.

Peterson, Charles M. "The Five Senses in Willem II van Haecht's Cabinet of Cornelis van Der Geest." In "Picturing Collections in Early Modern Europe," edited by Alexander Marr. Special issue of *Intellectual History Review* 20, no. 1 (2010): 103–21.

Philippe, Gilles. *Le français, dernière des langues. Histoire d'un procès littéraire*. Paris: Presses Universitaires de France, 2010.

Pigeaud, Jackie. "Fatalisme des tempéraments et liberté spirituelle dans '*l'Examen des esprits*' de Huarte de San Juan." *Littérature, Médicine et Société* 1 (1979): 115–59.

———. *Melancholia. Le malaise de l'individu*. Paris: Payot, 2008.

Pons, Alain. "Ingenium." In *Dictionary of Untranslatables: A Philosophical Lexicon*, edited by Barbara Cassin, Steven Rendall, and Emily S. Apter, 485–89. Princeton: Princeton University Press, 2014.

Prager, Brad. *Aesthetic Vision and German Romanticism: Writing Images*. Rochester, NY: Camden House, 2007.

Quirk, Randolph. "Langland's Use of 'Kind Wit' and 'Inwit.'" *The Journal of English and Germanic Philology* 52, no. 2 (1953): 182–88.

Raimondi, Ezio. "Ingegno e metafora nella poetica del Tesauro." *Verri* 2, no. 2 (1958): 53–75.

Ramakers, Bart. "Embodied wits. The representation of deliberative thought in rhetoricians' drama." *Renaissance Studies* 32, no. 1 (2018): 85–105.

Raphals, Lisa. *Knowing Words: Wisdom and Cunning in the Classical Traditions of China and Greece*. Ithaca, NY: Cornell University Press, 1992.

Reddick, Allen. *The Making of Johnson's Dictionary, 1746–1773*. Cambridge: Cambridge University Press, 1990.

Renouard, Antoine Auguste. *Annales de l'imprimerie des Estienne; ou, Histoire de la famille des Estienne et de ses editions*. Paris: J. Renouard et cie, 1843.

Riahi, Pari. *Ars et ingenium. The Embodiment of Imagination in Francesco di Giorgio Martini's Drawings*. Abingdon: Routledge, 2015.

Richardson, Brian. "The Concept of a *lingua comune* in Renaissance Italy." In *Languages of Italy: Histories and Dictionaries*, edited by Arturo Tosi and Anna Laura Lepschy. Ravenna: Longo, 2007.

Ritzel, Wolfgang. "Kant über den Witz und Kants Witz." *Kant-Studien; Berlin* 82, no. 1 (1991): 102–9.

Rivers, Isabel. *Reason, Grace, and Sentiment. A Study of the Language of Religion and Ethics in England, 1660–1780*. Cambridge: Cambridge University Press, 1991.

Robbins, Jeremy. *Arts of Perception. The Epistemological Mentality of the Spanish Baroque, 1580–1720*. Abingdon: Routledge, 2007.

Roberts, Lissa L., Simon Schaffer, and Peter Dear, eds. *The Mindful Hand: Inquiry and*

Invention from the Late Renaissance to Early Industrialisation. Chicago: University of Chicago Press, 2008.

Rodríguez de la Flor, Fernando. *Era melancólica. Figuras del imaginario barroco.* Palma de Mallorca: José J. de Olañeta, 2007.

Rodríguez Pardo, José Manuel. *El alma de los brutos en el entorno del Padre Feijoo.* Oviedo: Pentalfa Ediciones, 2008.

Roses Lozano, Joaquín. "Sobre el ingenio y la inspiración en la edad de Góngora." *Criticón* 49 (1990): 31–49.

Ruhstaller, Stefan. "Las autoridades del Diccionario de autoridades." In *Tendencias en la investigación lexicográfica del español: El diccionario como objeto de estudio lingüístico y didáctico,* edited by Stefan Ruhstaller and Josefina Prado Aragonés, 193–225. Huelva: Universidad de Huelva, 2000.

———. "Las obras lexicográficas de la Academia." In *Lexicografía española,* edited by Antonia María Medina Guerra, 235–61. Barcelona: Ariel, 2003.

Russo, Alessandra. "An Artistic Humanity. New Positions on Art and Freedom in the Context of Iberian Expansion, 1500–1600." *RES: Anthropology and Aesthetics* 65/66 (2014/15): 352–63.

Ruthven, Kenneth K. *The Conceit.* London: Methuen, 1969.

Salillas, Rafael. *Un gran inspirador de Cervantes: El Dr. Juan Huarte y su "Examen de ingenios."* Madrid: Eduardo Arias, 1905.

Salmon, Vivian. "Anglo-Dutch Linguistic Scholarship: A Survey of Seventeenth-Century Achievements." In *The History of Linguistics in the Low Countries,* edited by Jan Noordegraaf, C. H. M. Versteegh, and E. F. K. Koerner, 129–53. Amsterdam: John Benjamins, 1992.

Sarasohn, Lisa T. *The Natural Philosophy of Margaret Cavendish: Reason and Fancy During the Scientific Revolution.* Baltimore and London: Johns Hopkins University Press, 2010.

Sauer, Hans. "Glosses, Glossaries, and Dictionaries in the Medieval Period." In *The Oxford History of English Lexicography,* edited by Anthony Paul Cowie, vol. 1, 17–40. Oxford: Clarendon Press, 2009.

Sawday, Jonathan. *Engines of the Imagination: Renaissance Culture and the Rise of the Machine.* London: Routledge, 2007.

Schaffer, Simon. "Fontenelle's Newton and the Uses of Genius." *L'Esprit Créateur* 55, no. 2 (2015): 48–61.

Schäffer, Tatjana. *The Early Seventeenth-Century Epigram in England, Germany and Spain: A Comparative Study.* Oxford: Peter Lang, 2004.

Schiebinger, Londa. *The Mind Has No Sex? Women in the Origins of Modern Science.* Cambridge, MA and London: Harvard University Press, 1989.

Schleiner, Winfried. *Melancholy, Genius, and Utopia in the Renaissance.* Wiesbaden: Otto Harrassowitz, 1991.

Schmidt, Jochen. *Die Geschichte des Genie-Gedankens in der deutschen Literatur, Philosophie und Politik, 1750–1945*. Darmstadt: Wissenschaftliche Buchgesellschaft, 1985.

Scholar, Richard. *The Je-Ne-Sais-Quoi in Early Modern Europe: Encounters with a Certain Something*. Oxford: Oxford University Press, 2005.

———. "The New Philologists." In *Renaissance Keywords*, edited by Ita Mac Carthy, 1–10. Oxford: Legenda, 2013.

Scott Soufas, Teresa. *Melancholy and the Secular Mind in Spanish Golden Age Literature*. Columbia: University of Missouri Press, 1990.

Schreiber, Fred. *The Estiennes*. New York: E.K. Schreiber, 1982.

Schwenger, Peter. "Crawshaw's Perspectivist Metaphor." *Comparative Literature* 28, no. 1 (1976): 65–74.

Seco, Manuel. *Estudios de lexicografía española*. Madrid: Paraninfo, 1987.

Seibicke, Wilfried. *Technik: Versuch einer Geschichte der Wortfamilie Vom 16. Jahrhundert Bis Etwa 1830*. Heidelberg: Verein Deutscher Ingenieure, 1968.

Selden, Raman. "Hobbes and Late Metaphysical Poetry." *Journal of the History of Ideas* 35, no. 2 (1974): 197–210.

Serés, Guillermo. "Introducción." In *Examen de ingenios para las ciencias*, by Juan Huarte de San Juan, edited by Guillermo Serés, 13–131. Madrid: Cátedra, 1989.

———. "El ingenio de Huarte y el de Gracián. Fundamentos teóricos." *Ínsula* 655–56 (2001): 51–53.

———. "El ingenio en Gracián: de la invención a la elocución." In *Baltasar Gracián IV Centenario (1601–2001). Actas I Congreso Internacional "Baltasar Gracián: pensamiento y erudición,"* edited by Aurora Egido, Fermín Gil Encabo, and José Enrique Laplana Gil, 235–56. Huesca-Zaragoza: Instituto de Estudios Altoaragoneses; Institución Fernando el Católico, 2003.

———. "Don Quijote, ingenioso." In *Los rostros de Don Quijote. IV Centenario de la publicación de su primera parte*, edited by Aurora Egido, 9–36. Zaragoza: IberCaja, 2004.

Shapin, Steven, and Simon Schaffer. *Leviathan and the Air Pump: Hobbes, Boyle, and the Experimental Life*. Princeton: Princeton University Press, 1985.

Silva Suárez, Manuel. "Sobre Técnica e Ingeniería: en torno a un excursus lexicográfico." In *Técnica e ingeniería en España I. El Renacimiento. De la técnica imperial y la popular*, edited by Manuel Silva Suárez, 27–66. Zaragoza: Real Academia de Ingeniería, Institución Fernando el Católico, Prensas Universitarias de Zaragoza, 2008.

Siouffi, Gilles. *Le génie de la langue française. Étude sur les structures imaginaires de la description linguistique à l'âge classique*. Paris: Champion, 2010.

Sitter, John. *Arguments of Augustan Wit*. Cambridge: Cambridge University Press, 1991.

Skinner, Quentin. *Reason and Rhetoric in the Philosophy of Hobbes*. Cambridge: Cambridge University Press, 1996.

Smith, A. J. *Metaphysical Wit*. Cambridge: Cambridge University Press, 1991.

Smith, Logan Pearsall. *Words and Idioms: Studies in the English Language*. London: Constable and Co., 1925.
Snyder, Jon R. *La estética del Barroco*. Madrid: Antonio Machado Libros, 2014.
Sohm, Philip. "Gendered Style in Italian Art Criticism from Michelangelo to Malvasia." *Renaissance Quarterly* 48, no. 4 (1995): 759–808.
———. *Style in the Art Theory of Early Modern Italy*. Cambridge: Cambridge University Press, 2001.
Sommer, Hubert. *Génie: zur Bedeutungsgeschichte des Wortes von der Renaissance zur Aufklärung*. Frankfurt am Main: Peter Lang, 1999.
Spear, Richard. *The "Divine" Guido: Religion, Sex, Money, and Art in the World of Guido Reni*. New Haven and London: Yale University Press, 1997.
Spingarn, Joel E. *Critical Essays of the Seventeenth Century*, 3 vols. Oxford: Clarendon Press, 1908.
———. *Creative Criticism: Essays on the Unity of Genius and Taste*. New York: Henry Holt, 1917.
Starnes, DeWitt T. *Robert Estienne's Influence on Lexicography*. Austin: University of Texas Press, 1963.
———. "The Figure Genius in the Renaissance." *Studies in the Renaissance* 11 (1964): 234–44.
———, and Gertrude E. Noyes. *The English Dictionary from Cawdrey to Johnson, 1604–1755*. Chapel Hill: North Carolina State University Print Shop, 1946.
Stefano, Fiorella di. "Ingegno et Sprezzatura du Cortegiano dans la traduction de l'Abbé Duhamel (1690)." In *Traduire en français à l'âge classique. Génie national. Génie des langues*, edited by Y. M. Tran-Gervat, 139–51. Paris: Presses Sorbonne Nouvelle, 2013.
Stein, Arnold. "On Elizabethan Wit." *Studies in English Literature, 1500–1900* 1, no. 1 (1961): 75–91.
Stein, Gabriele. *Sir Thomas Elyot as Lexicographer*. Oxford: Oxford University Press, 2014.
Stein, Margaret D. "Margaret Cavendish: Engendering Restoration Wit." PhD diss., University of Notre Dame, 1998.
Stelzer, Otto. *Goethe und die Bildende Kunst*. 1st ed. 1949. Berlin: Springer, 2013.
Stephanson, Raymond. *The Yard of Wit: Male Creativity and Sexuality, 1650–1750*. Philadelphia: University of Pennsylvania Press, 2004.
Steptoe, Andrew. "Artistic Temperament in the Italian Renaissance: A Study of Giorgio Vasari's Lives." In *Genius and the Mind: Studies of Creativity and Temperament*, edited by Andrew Steptoe, 253–70. Oxford: Oxford University Press, 1998.
Sumillera, Rocío G. "Introduction." In *The examination of men's wits*, by Juan Huarte de San Juan, translated by Richard Carew, 1–67. London: Modern Humanities Research Association, 2014.

———. "From Inspiration to Imagination: The Physiology of Poetry in Early Modernity." *Parergon* 33, no. 3 (2017): 17–42.

Summers, David. *Michelangelo and the Language of Art.* Princeton: Princeton University Press, 1981.

———. *The Judgment of Sense: Renaissance Naturalism and the Rise of Aesthetics.* Cambridge: Cambridge University Press, 1987.

———. "Pandora's Crown: On Wonder and Imitation in Western Art." In *Wonders, Marvels, and Monsters in Early Modern Culture*, edited by Peter G. Platt, 45–75. Newark: University of Delaware Press, 1999.

Tatarkiewicz, Władysław. *A History of Six Ideas: An Essay in Aesthetics.* The Hague: Martinus Nijhoff, 1980.

Taylor, Paul. "From Mechanism to Technique: Diderot, Chardin, and the Practice of Painting." In *Knowledge and Discernment in the Early Modern Arts*, edited by Sven Dupré and Christine Göttler, 296–316. London: Routledge, 2017.

Tocanne, Bernard. *L'idée de nature en France dans la seconde moitié du XVIIe siècle: contribution à l'histoire de la pensée classique.* Paris: Klincksieck, 1978.

Torello-Hill, Giulia, and Andrew J. Turner, eds. *Terence Between Late Antiquity and the Age of Printing.* Leiden: Brill, 2015.

Torre, Esteban. *Ideas lingüísticas y literarias del Doctor Huarte de San Juan.* Seville: Universidad de Sevilla, 1977.

Tosi, Arturo, and Anna Laura Lepschy. *Languages of Italy: Histories and Dictionaries.* Ravenna: Longo, 2007.

Trabant, Jürgen. "Du génie aux gènes des langues." In *Et le génie des langues?*, edited by Henri Meschonnic, 80–102. Saint Denis: Presses Universitaires de Vincennes, 2000.

Treglown, Jeremy, ed. *Spirit of Wit: Reconsiderations of Rochester.* Oxford: Basil Blackwell, 1982.

Trovato, Mario. "The Semantic Value of Ingegno and Dante's Ulysses in the Light of the *Metalogicon.*" *Modern Philology* 84, no. 3 (1987): 258–66.

Tucker, Susie I. *Protean Shape: A Study in Eighteenth-Century Vocabulary and Usage.* London: Athlone Press, 1967.

Urquízar, Antonio. "La ingenuidad de la pintura y la teoría jurídica y social de los clásicos." In *Siete memoriales españoles en defensa del arte de la pintura*, edited by Antonio Sánchez Jiménez and Adrián J. Sáez, 15–28. Madrid and Frankfurt: Iberoamericana Vervuert, 2018.

———. "La profesión de pintor (principiante, aprovechado o perfecto) en la teoría artística de Francisco Pacheco." *Studi Ispanici* 43 (2018): 183–99.

Ushijima, Nobuaki. "Sobre los títulos del Quijote. La función del ingenio." In *Actas del Tercer Coloquio Internacional de la Asociación de Cervantistas*, 325–29. Barcelona: Anthropos, 1993.

Ustick, W. Lee, and Hoyt H. Hudson. "Wit, 'Mixt Wit,' and the Bee in Amber." *The Huntington Library Bulletin* 8 (1935): 103–30.

Valdovinos, José Manuel. "El fuero y el huevo. La liberalidad de la pintura: textos y pleitos." In *"Sacar de la sombra lumbre." La teoría de la pintura en el Siglo de Oro (1560–1724)*, edited by José Riello, 173–202. Madrid: Abada, 2012.

Vallini, Cristina. "Genius/ingenium: derive semantiche." In *Ingenium propria hominis natura*, edited by Stefano Gensini and Arturo Martone, 3–26. Naples: Liguori, 2002.

Van Hook, J. W. "'Concupiscence of Witt': The Metaphysical Conceit in Baroque Poetics." *Modern Philology* 84, no. 1 (1986): 24–38.

Vélez Posada, Andrés. "Ingenia: puissances d'engendrement. Philosophie naturelle et pensée géographique à la Renaissance." PhD diss., École des Hautes Études en Sciences Sociales, Paris, 2013.

Verdonk, Robert. "La importancia del '*Recueil*' de Hornkens para la lexicografía bilingüe del Siglo de Oro." *Boletín de la Real Academia Española* 70 (1990): 69–109.

Vérin, Hélène. *La gloire des ingénieurs. L'intelligence technique du XVIe au XVIIIe siècle*. Paris: Albin Michel, 1993.

Vialla Alain. *Naissance de l'écrivain. Sociologie de la littérature à l'âge classique*. Paris: Editions De Minuit, 1985.

Vitale, Maurizio. *La questione della lingua*. 2nd ed. Palermo: Palumbo, 1970.

Vivanco, Verónica. "*El Quijote* y la industria." *Destiempos. Revista de Curiosidad Cultural* 4, no. 20 (2009): 26–40.

"Vernunft; Verstand." In *Historisches Wörterbuch der Philosophie*, edited by Joachim Ritter and Karlfried Gründer, vol. 3, 748–863. Darmstadt: Wissenschaftliche Buchgesellschaft, 1974.

Waquet, Françoise. *Latin, or the Empire of the Sign: From the Sixteenth to the Twentieth Century*, translated by John Howe. New York and London: Verso, 2001.

Warnke, Martin. *The Court Artist: On the Ancestry of the Modern Artist*. Cambridge: Cambridge University Press, 1993.

Weijers, Olga. "Lexicography in the Middle Age." *Viator* 20 (1989): 139–53.

———. *Dictionnaires et répertoires au moyen âge: une étude du vocabulaire*. Turnhout: Brepols, 1991.

Weiner, Edmund. "Early Modern English—An Overview." In OED Online. Accessed May 1, 2017. http://public.oed.com/aspects-of-english/english-in-time/early-modern-english-an-overview/

Weinrich, Harald. *Das Ingenium Don Quijotes. Ein Beitrag zur literarischen Charakterkunde*. Munich: Aschendorf, 1956.

Wells, C. J. *German, a Linguistic History to 1945*. Oxford: Clarendon Press, 1985.

West, Jonathan. *Lexical Innovation in Dasypodius' Dictionary, A Contribution to the Study of the Development of the Early Modern German Lexicon Based on Petrus Dasypodius' Dictionarium Latinogermanicum, Strassburg 1536*. Berlin: De Gruyter, 2013.

Westgate, Sam. "George Herbert: 'Wit's an Unruly Engine.'" *Journal of the History of Ideas* 38, no. 2 (1977): 281–96.

White, Veronica Maria. "*Serio Ludere*: Baroque *Invenzione* and the Development of the *Capriccio*." PhD diss., Columbia University, 2009.
Wiley, Margaret L. "Genius: A Problem in Definition." *Studies in English* 16 (1936): 77–83.
Williams, Raymond. *Keywords: A Vocabulary of Culture and Society*. Oxford: Oxford University Press, 1983.
Williams, Robert. *Art, Theory, and Culture in Sixteenth-Century Italy. From Techne to Metatechne*. Cambridge: Cambridge University Press, 1997.
Williamson, George. *The Proper Wit of Poetry*. London: Faber and Faber, 1961.
Withington, Phil. "'Tumbled into the Dirt': Wit and Incivility in Early Modern England." *Journal of Historical Pragmatics* 12, nos. 1–2 (2011): 156–77.
Wittkower, Rudolf, and Margot Wittkower. *Born under Saturn: The Character and Conduct of Artists: A Documented History from Antiquity to the French Revolution*. New York: Norton, 1969.
Wolfe, Jessica. *Humanism, Machinery and Renaissance Literature*. Cambridge: Cambridge University Press, 2004.
Woods, Michael J. *Gracián Meets Góngora: The Theory and Practice of Wit*. Warminster: Aris & Phillips, 1995.
Zanetti, Cristiano. *Janello Torriani and the Spanish Empire*. Leiden: Brill, 2017.
Zárate Ruiz, Arturo. *Gracián, Wit, and the Baroque Age*. New York: Peter Lang, 1996.
Zilsel, Edgar. *Die Entstehung des Geniebegriffes: ein Beitrag zur Ideengeschichte der Antike und des Frühkapitalismus*. Tübingen: J. B. C. Mohr, 1926.
Zucker, Adam. *The Places of Wit in Early Modern English Comedy*. Cambridge: Cambridge University Press, 2011.

Index

Note: Page numbers in *italics* refer to illustrations.

aard/aerdt (D.), 155, 156, 165–67, 169–71, 184, 189
Abercromby, David: *A Discourse of Wit*, 193–94
ability, 100, 103, 105, 112, 122, 135, 150, 185, 189, 229; artistic, 30, 57, 68, 112, 143–44, 201; cognitive, 19, 35, 51, 57, 103, 113, 148, 186, 222; generative, 32; to learn, 20, 49; natural, 9, 103, 207, 229; poetic, 224; practical, 88, 91, 218; superior, 1, 2, 197, 201, 207–8, 237
abstract reasoning, 149, 189
Académie française, 97, 126, 136; *Dictionnaire de l'Académie françoise*, 15, 24, 97, *123*, 126, 127, 129, 132–36, 139, 140, 146, 147–48
academies, 62, 84, 162, 246n46
Accademia della Crusca, 14, 60, 62, 64, 84, 96; *Vocabolario della lingua italiana*, 13, 24, 55, 57–59, 60–64, *61*, 67, 71–76, 78–80, 84, 97
acquired ability, 6, 68, 211, 215, 229. *See also* study; training
acutezza (I.), 55, 58, 72, 75, 78, 81, 180
Addison, Joseph, 208–9, 222, 224
Adelung, Johann Christoph: *Grammatisch-kritisches Wörterbuch der hochdeutschen Mundart*, 153–54, 155–56, 185, 190
aesthetics, 17, 155–57, 185, 187–88, 198, 222, 223; of the marvelous, 76, 80; Renaissance, 35, 58
affectation, 134, 137, 140, 142–43

agudeza (S.), 17, 72, 90, 92–93, 114–18 (114–15), 119; — *de ingenio* (S.), 92, 115–16, 277n93
agudo (S.), 101–2, 114–17, 185
Alberti, Leon Battista, 58, 70
Alexander of Aphrodisias, 50, 255n96
allegory, 34, 146, 147
Alsted, Johann: *Compendium philosophicum*, 25, 50, 255n95
Alunno, Francesco: *La fabrica del mondo*, 60, 66, 71; *Il Petrarca con le osservationi di messer Francesco Alunno*, 79
angels, 251n45; English, 197, 204; French, 145; Italian, 56, 66; Latin, 20, 28, 33–34; Spanish, 91, 103, 111–12
anima (L.), 131, 132, 252n48, 280n22
animals, 46, 76, 141, 182, 233, 304n10
animus (L.), 181–82, 186, 197, 210, 290n66, 292n79
anthropology, 34, 92, 145–47, 156, 238
antiquarianism, 30, 34, 51, 91, 110, 124–29, 144–47, 204; classical, 20, 28; dictionaries/lexica, 14, 28, 30, 35, 39, 47, 125–26; medieval, 22, 25
aphorisms, 10–11, 30, 223–24, 245n37
archaism, 59–60, 63–64, 129
architecture, 58, 70–71, 76–77, 81, 126–27, 130, 146
arete (Gk.), 167, 299n74
Ariosto, 60, 66
Aristotle, 32, 48, 49, 76, 111, 255n95, 76
Art (G.), 153, 155, 156, 159–60, 161, 164–71 (165–67)

347

art/*ars*, 9, 21, 91, 143, 148, 164–65, 168, 173, 206, 236; *ars et ingenium*, 11, 68–69, 72, 260n48; artistic expression, 142; the arts, 150, 215; reception, 106; theory, 16, 48, 55, 57–58, 64, 70, 92, 157, 182–83, 187–88, 238
artifice, 79, 112, 142, 173, 195, 229; as tool/invention, 81, 99, 109
artifice (F.), 142–43
artificers, 195, 215, 218, 233
artificio (S.), 91, 99, 104, 106, 112–13
artificium (L.), 109, 230
artisans, 9, 58, 70, 80–81, 88, 214–15, 229
artists, 194, 218; definition of, 68, 72, 75, 165, 218; mechanical, 84–85, 218; patronage of, 8, 182; qualities of, 56, 70, 75–76, 79, 88, 114, 143, 187, 205, 216
astutia (I.), 21, 69, 72, 79, 81
Augustine, 35, 74, 188, 204
author, 124, 137–40, 151

Bacon, Francis, 13, 150
Bailey, Nathan: *Dictionarium brittanicum*, 198, 199, 207; *The Universal Etymological English Dictionary*, 218, 204, 226
Balbus de Janua, Joannes: *Catholicon*, 22, 38–40, 248n12
Baldinucci, Filippo: *Vocabolario toscano dell'arte del disegno*, 64, 80
Baret, John: *An Alveary or Triple Dictionary, in English, Latin, and French*, 211, 222
Bayle, Pierre: *Dictionnaire historique et critique*, 121–22, 124, 126, 190
B. E.: *A New Dictionary of the Terms Ancient and Modern of the Canting Crew*, 207, 231–32, 225
beauty, 149, 187, 206
behavior, 24, 37, 39, 44–45, 50, 92, 134–35, 143, 214
bel esprit, 121–22, 124, 132–33, 135–39, 142–43, 148; false/*faux*, 137, 138
Bembo, Pietro: *Prose della volgar lingua*, 60

Binnart, Martin: *Biglotton sive dictionarium teuto-latinum novum*, 170, 176, 184
birth, 40–41, 44–45, 104, 141, 250n32
Blount, Thomas: *Glossographia*, 199, 205–7, 214, 215, 225, 231
Boccaccio, 13, 57, 60, 68, 78, 79; *Decameron*, 54, 60, 62, 73. See also *Tre Corone*
body, 8, 34–35, 48, 50, 101, 103, 280n22
bon sens, 121–22, 136, 156
Borghini, Vincenzo: *Annotationi*, 62
Bouhours, Dominique, 124; *Entretiens d'Ariste et d'Eugène*, 122–24, 127, 136, 148
Boyle, Robert, 194; *Occasional Reflections*, 213
Bravo, Bartolomé, 100, 117; *Thesaurus verborum ac phrasium ad orationem ex hispana latinam*... , 100, 106, 117
Brunelleschi, Filippo, 70, 77, 261n49
Budé, Guillaume, 23, 29, 33
Bullokar, John: *An English Expositor*, 214, 231
Buti, Francesco da: *Commento*, 73–75

Calepino, Ambrogio: *Dictionarium*, 23–25, 30–35, 38–41, 44, 48, 53, 200; — 1502, 9, 32; — 1590, 12, *31*, 102; — 1718, 34; *Il Dittionario di Ambrogio Calepino dalla lingua latina nella volgare brevemente ridotto*, 70, 81
du Cange. See du Fresnes, Charles
capriccio (I.), 76, 175, 276n87
capricho (S.), 113–14, 116, 276n89
caractère (F.), 134, 138, 140
Cardano, Girolamo: *De subtilitate*, 238–40
Caro, Annibal: *Apologia*, 66, 67
Castiglione, Baldassare, 58; *Libro del cortegiano*, 77–78
Cawdrey, Robert: *A Table Alphabeticall*, 198, 199, 213, 231
Censorinus: *De die natali*, 30, 34, 112, 250n32

Cervantes, Miguel de, 90, 118, 119; *El ingenioso hidalgo Don Quijote de la Mancha*, 87–88, *89*, 92, 105
Chambers, Ephraim: *Cyclopaedia, or, An Universal Dictionary of Arts and Sciences*, 199, 204, 207–8; *Dizionario universale delle arti e delle scienze*, 57, 64, 67
character, 122, 134–35, 138–41, 183–84, 188, 229, 235
Charleton, Walter, 232–33, 301n95
Chauvin, Stéphane: *Lexicon rationale sive thesaurus philosophicus*, 28, 48–50, 51, 255n96
Christianization, 28, 33, 65, 91, 146
Cicero, 3, 20–21, 44, 69, 71, 189, 239–40; *De finibus bonorum et malorum*, 40–41, 49, 53
civility, 3, 127, 225. *See also* decorum, politeness
class. *See* social order/class
Cockeram, Henry: *The English Dictionarie*, 11, 223–24
cognitive axis, 7, 235, 238–39; French, 129, 132–33, 135, 146–49, 151; German & Dutch, 156–57, 159, 161, 168, 174–76, 182, 188–90; Italian, 69, 79; Latin, 21–22, 29, 37, 39, 41, 45–50
Coles, Elisha: *An English Dictionary*, 199, 205, 209, 220
comedy, 33, 41, 140, 149, 251n43
conceit (E.), 116–17, 195–97, 205, 219–20, 278n102, 294n10
conceptismo (S.), 117
concepto (S.), 93, 117–18, 276n87, 278n102
concettismo (I.), 58, 236
conduct, 44–45
congeniality, 67, 68, 205
Considine, John, 162, 168
conversation, 122, 130, 137
Cooper, Thomas: *Thesaurus linguae Romanae et Brittanicae*, 199, 206, 211
copia, 9–10, 200, 222, 245n32

Corneille, Thomas: *Dictionnaire des arts et des sciences*, 125–26, 129–30
Cotgrave, Randle: *A Dictionarie of the French and English Tongues*, 199, 206
courtiers, 3, 8–9, 71, 127, 213, 218, 240
courtliness, 21, 58, 78, 84, 236
courtly language, 60, 78, 127, 163
Covarrubias, Sebastián de: *Emblemas morales*, 106, *107*; *Tesoro de la lengua castellana, o española*, 12, 93, 95–96, 103–6, 109–10, 112–13, 114–17, 119
Cowley, Abraham, 221; *Ode: Of Wit*, 220
craft, 5, 9, 40, 160, 168, 173, 216, 229, 238; goods, 3, 218
craftiness, 79, 91, 106, 129, 173, 198, 213, 229, 231
craftsmanship, 22, 51, 70, 102, 215. *See also* workmanship
creativity, 1, 5, 236; English, 194, 205–6, 208, 211, 214, 216, 222; French, 148; German & Dutch, 157, 180; Italian, 55–56, 58–59, 72, 75–76; Latin, 30, 35
critic, 124, 138–40, 151
criticism, 10–11, 14, 17, 58, 148, 209, 220
cunning, 9, 21, 58, 79, 129, 159, 173; as devious, 46–47, 106, 174, 197–98. *See also cunning* (E.)
cunning (E.), 5, 195, 197–98, 206, 210, 213–15, 222, 224–26, 229–32 (*230*)
curiosity, 2, 80, 149, 218
curiosity (E.), 213, 224

daemon (Gk.), 20, 33, 181, 189, 203–4
Dante, 13, 57, 60, 76–77, *77*; *Divina Commedia*, 55, 73–75; *De vulgari eloquentia*, 59. *See also Tre Corone*
Dasypodius, Petrus: *Dictionarium latinogermanicum*, 13, 157–58, 159, 162, 173, 174, 186
decadence, 13–14, 118, 137
decorum, 8, 127, 149. *See also* civility, politeness

definition, 4, 6, 40, 94, 109, 119, 201; precision of, 15, 48–50, 63, 131
Defoe, Benjamin: *A New English Dictionary*, 218, 226, 232
Descartes, René, 150; *Regulae ad directionem ingenii*, 6, 133
Detienne, Marcel and Jean-Pierre Vernant, 6, 231
devices, 47, 58–59, 80–83, 91, 109–10, 196, 218, 236
dexterity, 88, 91, 105, 112, 229, 232
Dictionnaire de Trévoux, 97, 125, 126, 129–30, 132–33, 135–48
Diderot, Denis et al.: *Encyclopédie ou Dictionnaire raisonné des arts et des sciences*, 125, 126–27, 129–30, *130*, 138, 145–46, 151
discourse, 39, 71–72, 78, 104, 117, 193–94, 224
display, 8–9; English, 195, 223; French, 134, 136–37, 140; German & Dutch, 156, 160–61, 168, 183–84; Italian, 78; Latin, 21, 25, 41, 44. *See also* presentation
Dryden, John, 224, 297n49, 302n99
Dürer, Albrecht, 157, 187–88, 238, 269n19, 304n3; *Vier bücher von menschlicher Proportion*, 187–88
Dutch, 17, 153–54, 157, 161–64, 169–70, 190, 236; as Adamic language, 14, 25; *gemeenlands*, 163
dynamism, 32, 33, 48, 74, 116, 143, 250n36

earth, 156, 165, 169–70, 204, 233
education, 62, 124, 140, 142, 214, 284n62
elements, the, 25, 66, 110, 250n37
Elyot, Thomas: *The Dictionary of Syr Thomas Elyot*, 199, 204–5, 207, 210, 222, 231
emblems, 25, *26–27*, 97, 106, *107*, 180, 250n30
embodiment, 50, 141, 146, 161, 189, 191, 247n6
emotion, 157, 176, 186
engaño (S.), 91, 104–6

engeño/engenio (S.), 90, 99–100, 102–3
engin (F.), 47, 102, 124–26, 128–31 (128–29), *130*, 216, 218, 250n30; Old French, 5, 197
engine, 41; mechanical, 82, 85, 103, 238; siege, 9, 39, 82. *See also* machines
engine (E.), 102–3, 195–96, 218
engineering, 22, 70, 91, 130
engineers, 58–59, 70, 76–77, 81, 84–85, 88, 104, 109, 218, 240
English, 13–14, 17, 151, 194, 199, 236
enthusiasm, 206
epigrams, 10, 137, 140, 223–24, 245n38, 282n45
Erasmus, 19, 23
espíritu (S.), 112, 116
esprit (F.), 13, 17, 102, 122, 124–28, 131–40 (131), 142, 147–51, 236; *esprit fort*, 132–33, 135, 139; and German & Dutch, 161, 181–85, 188. See also *bel esprit*
Estienne, Robert, 11, 13, 23–25, 30, 33, 35, 38–45, *42, 43*, 48, 158, 161, 168, 181–82, 200, 249n18
etymology, 95–96, 104, 125–26, 159, 161–65, 170
examples, 40–41, 55–56, 63, 132
exceptionality, 1, 147, 189, 190, 235, 237

faculty psychology, 8, 48, 49, 88, 133, 198
fancy (E.), 195, 205, 216, 221–22, 224–27, 297n46
fantasy, 75, 114, 174
feeling, 187, 188, 236
Fernández de Palencia, Alfonso: *Universal vocabulario en latín y en romance*, 93–94, 98–99, 104, 106, 115
Fernández de Santaella, Rodrigo: *Vocabularium ecclesiasticum*, 98, 100
fertility, 8, 20, 156, 165. *See also* generative power; pregnancy
Florio, John: *A Worlde of Wordes*, 5, 63, 79, 85, 196, 199, 205–6, 213, 216, 223, 225

food and drink, 25, 26, 27, 30, 33, 67, 238–39
foreign languages, 12, 14, 64, 101, 164, 187
Franciosini, Lorenzo: *Vocabolario español e italiano*, 106, 115
frankness, 197, 209
freeborn, 37, 44, 71, 159–61, 209, 214–15
freedom, 58, 67, 71, 92, 110, 205, 214–15
free-thinking, 35, 147
French, 12, 14–15, 17, 236; and Dutch, 161–63, 169, 182; *génie de la langue*, 10n31, 14, 122, 126–28, 131, 151
du Fresne, Charles, seigneur du Cange: *Glossarium mediae et infimae latinitatis*, 25, 47
Frisius, Johannes, 161; *Dictionarium latino-germanicum*, 158, 159
Furetière, Antoine: *Dictionnaire universel*, 15, 97, 125–27, 129–30, 141–43, 146–47
furor (poeticus), 75, 111, 124, 206, 297n52, 298n53, 262n62

Galen, 20, 48, 49, 255n95, 304n9
geest (D.), 13, 155, 156–57, 161, 181–85 (181), 236
geestig (D.), 157, 184–85, 291n77
Geist (G.), 155–57, 181, 184
Gemüt (G.), 155, 156, 157, 160, 185–88 (185–86), 236
Gemüthsfähigkeit (G.), 153–54, 185
gender, 58, 135, 213, 225, 232–33
generative power, 28, 32, 56. *See also* fertility; pregnancy
genial (E.), 205, 220
geniale (I.), 67, 79, 205
génie (F.), 13, 122–28, 138, 134, 141, 143–51 (144–45), 188
génie de la langue. See under French
Genie (G.), 3, 153–54, 156, 185, 190
genio (I.), 13, 56, 57, 64–68 (64–65), 80, 205
genio (S.), 91, 110–12
genius: modern, 1, 3, 36; Romantic, 7, 8, 56, 91, 156, 185–86, 190, 194, 208

genius (E.), 6, 10, 17, 194–95, 197, 200–9 (202–3), 220, 233, 236–37
genius (L.), 2–3, 5, 13, 20, 28–36 (28–29), 39, 46–47, 51, 57, 64–67, 164, 169
genius loci, 20, 35, 46, 169, 197, 207, 208
gens (L.), 36, 45–46, 235
gens d'esprit (F.), 134
gentlemen, 141, 209, 214–16
Geoffrey the Grammarian: *Promptorium parvulorum*, 197, 207
German, 17, 153–54, 157–59, 161–64, 167–69, 175, 236; *ablaut*, 163; origins, 154, 157, 161, 175; *Stammwörter*, 162, 164, 175
German wit, 121, 124, 135
Gessner, Conrad, 23, 158, 159, 160–61, 164, 168; *Mithridates, de differentis linguarum*, 158
ghiribizzare (I.), 55, 72, 75–76, 262n63
gifts, 7, 37, 68, 70–71, 113; divine, 44, 75, 205, 216
gigno (L.), 20, 30, 36–37, 40, 104, 130, 165, 159, 235
Glocenius, Rudolph, 25, 48, 50
gods/deities, 20, 32, 44, 57, 64–65, 91, 147. *See also genius loci*; tutelary spirit
Goethe, Johann Wolfgang von, 157, 185, 186
good usage, 125–26
Goropius Becanus, Johannes: *Origines Antwerpianae*, 14, 24–25, 161, 288n23
Gottsched, Johann Christoph, 190–91
grace, 35, 75, 206, 216, 244n26, 252n51
grace (E.), 200, 206
Gracián, Baltasar, 90, 92–93, 110, 118–19, 180, 255n95
grammars, 94, 122, 162, 237, 246n50, 287n21

habilidad (S.), 91, 100–1, 105–6, 275n84
habitus, 49, 50, 184
Haecht, Guillem van: *The Gallery of Cornelis van der Geest*, 182–83, *183*

heart, 134, 186, 188
Henisch, Georg: *Teütsche Sprach und Weißheit*, 162, 167–68, 186
heredity, 44, 171, 254n83
Herrera, Fernando de, 103, 111
Hexham, Henry: *Het Groot woorden-boek gestelt in 't Engelsch ende Nederduytsch*, 176, 184
Hilliard, Nicholas, 215–16, *217*
Hobbes, Thomas, 198, 206, 221–22, 225
Hofmann, Johann Jacob: *Lexicon Universale*, 30, 35
Hogarth, Richard: *Gazophylacium Anglicanum*, 213, 218, 225
Homer, 44, 149; *Iliad*, 138
honnêtes gens, 132, 134, 138, 140
Howell, James: *Lexicon tetraglotton*, 14, 115
Huarte de San Juan, Juan: *Examen de ingenios para las ciencias*, 6, 7, 49, 90, 91, 100, 103, 113, 116, 135, 189–91
Huloet, Richard: *ABCEdarium Anglico Latinum*, 222
humoral theory, 48, 49, 141, 145–46, 205, 219, 247n6
humour(s), 113, 195, 205, 220, 221, 225

identity, 236; French, 122, 124, 127, 132–35, 138, 141, 144–45, 147; Italian, 59; German & Dutch, 151, 160–62, 164, 188; Latin, 21, 28, 33, 35, 37, 39, 41, 46–47, 50; national, 122, 127, 151, 162, 237
images, 180, 186–88, 225, 238, 250n30
imagination, 8, 76, 148–50, 187, 198, 205, 221–22
inborn: ability, 21, 37, 41, 68, 78, 103, 127, 141, 148, 214, 229; nature, 21, 36–37, 39–40, 44, 46, 50, 103–4, 153, 159, 168, 210–11, 235; inclination, 3, 7; power, 1, 21, 48, 103, 159; quality, 1, 20, 37, 127, 141, 156, 160, 197
inclination, 5, 7; English, 195, 197, 201, 205, 207, 210–11, 222; French, 144, 148;
German & Dutch, 182, 188; Italian, 56–57, 66, 85; Latin, 33, 40–41; Spanish, 91, 100–101, 104, 110–12
individuation, 29, 34–35, 45, 47, 111, 124, 132, 186, 189, 235–36
índole (S.), 112
indoles (L.), 21, 49, 104, 143, 148, 164, 210
industria (I.), 69, 71–72, 77–78
industria (S.), 87–88, 91, 105, 106, 112, 280n17
industrius/a (L.), 11, 69, 148, 165n3, 174, 210, 222
industry, 1, 9, 88, 103, 142, 155, 194; *engin as*, 129
inganno (I.), 78–79, 106
ingegno (I.), 53–59, *61*, 68–85 (82–84), 102, 175, 213, 216, 218, 236
ingegnoso (I.), 5, 59, 70, 79–80, 213, 216–17, 260n40
ingenia (L.), 8, 21, 36–37, 44, 48–50, 175, 188; siege machines, 3
ingenio (S.), 90–93, 97–114 (98), *108*, 118–19, 185, 189, 236
ingenio del açucar/azucar, 104–5, 109, 130
ingenios (S.), 90, 109–11, 113, 116, 189, 277n96
ingeniosity (E.), 211, 217, 220
ingenioso (S.), 88, 92, 99–102, 105, 109, 115, 185
ingeniosus (L.), 21–22, 102, 156, 159, 161, 171, 173–74
ingenious (E.), 5, 194, 197, 209, 211–14, 216–18
ingenium (L.), 2–4, 7–10, 12, 19–22, 24–25, 29, 33, 36–51 (37–38), 238–40; English, 197, 210–11, 218, 222; French, 132–33, 148–49; German & Dutch, 155, 156, 159, 161, 164–65, 170–71, 182; Italian, 53, 55–58, 64, 67–71, 79; Spanish, 90, 92, 99–100, 102, 104, 106, *107*
ingenuatus (L.), 44–45
ingenuidad (S.), 92, 110

ingenuità (I.), 71, 214
ingenuitas (L.), 37, 45, 67, 92, 110, 159–60, 161
ingenuity (E.), 195, 209–20 (211–12), 222, 231
ingenuous (E.), 197, 209, 211, 214–15, 219
ingenuus (L.), 7, 21, 36–37, 45–46, 92, 142, 159, 175–76, 214, 252n55
ingeny (E.), 210, 211, 212, 220, 233
innate abilities/qualities. *See under* inborn
inspiration, 6, 56–57, 75, 124, 189, 205–6, 237
intellect, 6, 8, 235, 237; English, 194, 208, 215, 232; French, 122, 124, 149, 151; German & Dutch, 154–55, 182, 186, 188; Italian, 56–58, 74, 76; Latin, 21, 50–51; Spanish, 104, 111–12. *See also* mental capacity
intellectus (L.), 74, 101, 155, 159
intelletto (I.), 56, 69
intention, 9, 46–47, 129, 139
invention, 3, 5, 7–9; English, 194–98, 210, 216–18, 221–22, 224; French, 122, 133, 137, 143, 147, 150; German & Dutch, 175, 180, 187; Italian, 55, 57, 59, 71–72, 75–76, 78, 82, 84; Latin, 19, 21–22, 37, 39–40, 44, 46, 48–49; Spanish, 90, 99, 103–4, 109, 112–14
inzegno (I.), 69, 80, 260n43
Iperen, Josue van, 170
Isidore of Seville: *Etymologiae*, 22, 38, 39, 71, 95, 248n12
Italian, 13, 16, 57, 62–63, 236; *questione della lingua*, 16, 55, 59–60, 78, 84

jargon, 126, 127, 130–31
Jaucourt, Chevalier de, 127, 145–47
je-ne-sais-quoi, 119, 220, 244n31
Johnson, Samuel, 224; *Dictionary*, 8, 10, 13–14, 17, 194, 197–99, 201, 204, 208–10, 213, 216, 221–22, 226, 229, 232–33
Jonson, Ben: *Entertainment for Britain's Burse*, 195–96; *Every Man out of his Humour*, 218–19

judgment, 8, 10; English, 194, 198, 208, 221–24, 226; French, 122, 133, 137, 140; German & Dutch, 171; Italian, 70; Latin, 45, 48–49

Kahl, Johann: *Lexicon juridicum*, 25, 29, 30, 33, 35, 45–46
Kant, Immanuel, 3, 155–56, 171
Kate, Lambertten: *Aenleiding tot de Kennisse van het verhevene deel der nederduitsche sprake*, 162–64, 170
Keil, Cornelis. *See* Kiliaan, Cornelis
Kenny, Neil, 2, 224
Kersey, John: *New English Dictionary*, 199, 213, 232, 233; *New World of English Worlds*, 199, 209, 218, 220, 225
Kiliaan, Cornelius, 24, 154; *Dictionarium tetraglotton*, 45, 30, 160–62, 170; *Etymologicum teutonicae linguae*, 173–74
Kopf (G.), 154, 190–91
Kramer, Matthias, 180; *Il nuovo dizzionario delle due lingue italiana-tedesca et tedesca-italiana / Das neue Dictionarium oder Wort-Buch, in Italiänischer-Teutscher und Teutsch-Italiänischer Sprach*, 163, 168–69, 175; *Le vraiment parfait dictionnaire roial, radical, etimologique, sinonymique, phraseologique, & syntactique, françois-allemand / Das recht vollkommen-Königliche Dictionarium Radicale, Etymologicum, Synonymicum, Phraseologicum et Syntacticum, Frantzösisch-Teutsch*, 163, 169, 173, 188
Kunst (G.), 160, 161, 173

Landino, Cristoforo: *Comento . . . sopra la Comedia di Dante Alighieri*, 74–75
language, 3, 137, 235, 237; origins of, 95; purity, 126, 127–28, 163. *See also* individual languages
Latin, 12–13, 16, 19–20, 22, 25, 125, 132, 161–62, 164, 167, 236

law, 3, 9, 22, 25, 36–39, 45–47, 110, 139, 203
learning. *See* ability: to learn; education; study; teachability; training
Le Blond, 127, 145
Lessing, Gotthold Ephraim, 190–91
Lewis, C. S., 196, 197, 200, 214
liberty, 71, 246n46. *See also* freedom
literary canon, 16, 17, 57–58, 62, 63, 194, 238
literary criticism. *See* criticism
literature, 17, 117, 127, 128, 137, 138, 148; literary theory, 118; the novel, 106, 218
Lloyd, William and John Wilkins: *An Alphabetical Dictionary*, 206, 216
loan words, 17, 199, 224
Locke, John, 149–50, 198, 221, 225
logodaedalus/i, 10, 95, 233, 238
Low Countries, the, 163, 236

Maaler, Josua, 154; *Die Teütsch Spraach*, 158, 160, 161–62, 164–65, 168, 174
machina (I./L/.S.), 47, 60, 69, 99–100, 102, 109
machination, 3, 22, 59, 180, 197
machines, 236, 239; French, 124, 129–31; Italian, 58–59, 70–71, 80; Latin, 22, 24–25, 36, 47; Spanish, 91, 99, 102, 104–5, 109; war, 3, 47, 58, 91, 104, 129. *See also* engine
Madoets, André: *Thesaurus theutonicae linguae*, 23, 24, 161, 169, 174, 181, 184
madrigals, 137, 138, 282n45
magic, 230, 231–32
maña (S.), 91, 105–6, 109, 112
manners, 129, 140, 142, 145
Marinelli, Giovanni: *Dittionario di tutte le voci italiane usate da migliori scrittori àntichi, et moderni*, 69, 79, 83, 93
Martial, 10, 13, 22, 35–36, 90, 110, 116
Martin, Benjamin: *Lingua britannica reformata*, 204, 213, 224
mechanical arts, 63, 70, 84–85, 104, 126, 130, 194, 214, 218

medicine, 3, 48–50, 56, 88, 189, 215, 239; writings on, 6, 20, 48, 57, 90, 238
melancholy, 6, 49, 57, 176, 205, 255n95
memory, 40, 48–49, 69, 97, 133, 149–50, 188, 238
Ménage, Gilles: *Dictionnaire étymologique de la langue françoise*, 125; *Les origines de la langue françoise*, 125, 129; *Le origini della lingua italiana*, 64, 79
mens (L.), 131, 132, 189, 197
mental capacity, 3, 57, 104, 160, 174, 186, 201, 207–8, 210, 229. *See also* intellect
Merlin-Kajman, Hélène, 127
métis (Gk.), 5–6, 79, 231
metonymic uses, 40, 91, 109, 133, 135, 147–48, 190
Michelangelo, 58, 63
Micraelius, Johannes: *Lexicon philosophicum*, 28, 49
Miège, Guy: *A New Dictionary French and English*, 206, 217
mimesis, 68, 141, 142–44
mind, 17, 49, 57, 72, 136–37, 186, 194, 221, 225–26; -body relation, 48, 50, 82
Minsheu, John: *A Dictionarie in Spanish and English*, 102–03, 116; *Ductor in linguas / Guide into Tongues*, 25, 106, 113, 199, 231
moral axis: French, 133–34, 143–47; German & Dutch, 155–56, 159, 161, 168, 174, 176, 184, 188, 190; Italian, 69, 92; Latin, 28, 29–30, 32–34, 37, 39, 41, 45, 40; moral ambivalence, 79, 105–6, 112, 174, 211, 213, 225
morphology, 39, 132, 159

Nashe, Thomas, 219–20
nationalism/national identity, 14, 237; French, 122, 127–28, 151; German & Dutch, 158, 161–62; Spanish, 96–97
national languages. *See under* French; Italian
Natur (G.), 153, 156, 164

natura (I.), 56, 64, 168
natura (L.), 21, 148, 159, 189
natural (S.), 91, 99, 100, 102, 104
naturaleza (S.), 99, 100, 102
naturalism, 4, 20, 30, 144, 147, 250n37
naturalization, 28, 30, 32–35, 125, 146–47, 250n37, 251n39
natural philosophy, 32, 34, 48, 55, 57, 64, 84, 133, 190, 215, 220, 238–40. *See also* new science, the
nature, 40, 41, 54–55, 75; art and, 3, 9, 80, 76, 140; gifts of, 39, 53, 75, 80, 113; god/spirit of, 30, 32, 34, 145, 146–47; good, 20, 49, 141–42; itself, 9, 32, 34, 66, 112, 147, 187; one's, 7, 10, 40–42, 45, 47, 49, 66–68, 71, 85, 91, 99, 107, 134, 146–47, 161, 164, 168, 184, 188, 205, 208, 222, 239; one's *inborn*, 21, 36–37, 39–40, 44–46, 50, 104, 141, 153, 159, 197, 210–11, 235; of things, 103, 137, 197. See also *nature* (E.); *nature* (F.); *naturel* (F.); *natuur* (D.)
nature (E.), 167, 195, 200, 207, 210
nature (F.), 134, 140–44, 161, 167, 280n21
naturel (F.), 122, 124, 127–28, 134–44 (141), 147, 151, 169, 188
natuur (D.), 156, 161, 164
Nebrija, Antonio de, 95, 106; *Diccionari latino-español / Dictionarium Latino-Hispanicum*, 13, 24, 93–94, 98–100, 115, 237; *Grammatica*, 94, 237; *Vocabulario español-latino*, 93–94, 98, 99, 115, 237
Neoclassicism, 97, 118
neologisms, 198, 199, 215, 219
Neoplatonism, 74, 124, 145, 147
new science, the, 51, 149, 194, 240
New World, the, 90, 92, 105, 237–38, 304n3
Nicot, Jean: *Thrésor de la langue françoyse*, 23, 124–25, 129, 131–33, 139, 140–41, 144–47
nobility, 7, 45, 92, 159, 161, 186, 214–15

objects, 9, 238; English, 195, 208, 217, 222, 224; French, 129; German & Dutch, 176, 180, 186; Italian, 79–81; Latin, 41, 46–47; Spanish, 92, 114
Ortus vocabulorum, 204, 229–30
Oudin, Cesar: *Tesoro de las dos lenguas francesa y española*, 102, 106, 113

paganism, 20, 30, 33–35, 65, 144–47, 170, 204
Papias, 73; *Elementarium doctrinae rudimentum*, 93–94, 99
Pascal, Blaise, 135, 137
passions, 146, 149, 251n43
pedagogy, 21, 37, 48, 50, 141, 164, 191
Percivale, Richard: *Bibliotheca hispanica*, 102
Pergamino, Giacomo: *Memoriale*, 72
Perotti, Niccolò: *Cornucopiae*, 10, 22–24, 32, 36–40, 44, 170
Persio, Antonio, 57, 67; *Trattato dell'ingegno dell'huomo*, 53–55
Petrarch, 21, 44, 55, 60, 79. See also *Tre Corone*
Phillips, Edward: *The New World of English Words*, 199, 209, 214–15, 218, 220, 225
philosophy, 127, 138, 150, 155–56, 198, 209; debates, 50, 51, 225; scholastic, 48, 76, 238; treatises, 90
phusis/physis (Gk.), 20, 32, 161, 168, 210
Pictorius. *See* Maaler, Josua
Plantin, Christoph, 24, 45, 160–61, 162, 174
Plato, 33–34, 66–67, 111
Plautus, 11, 24, 33, 41, 44–45, 168
pleasantness, 35, 196, 205–6, 223, 225
poetics, 48, 55, 58, 91, 128, 143–44, 149, 206, 221, 223–24, 238
politeness, 127, 129, 134, 136–37, 139–40, 142–43
Pope, Alexander, 222, 224; *Essay on Criticism*, 10
Porcacchi, Tommaso: "Vocabulario nuovo del Porcacchi", 66–67

pregnancy, 11, 195, 197, 205, 210, 223. *See also* fertility; generative power
presentational axis, 7–8, 35, 44, 69, 124, 132, 134, 151, 157, 180. *See also* display; representation
proverbs, 90, 129, 140, 168, 186
psychology. *See* faculty psychology
puns. *See* wordplay

Quevedo, 90, 118, 119
quickness: English, 209, 211, 213–14, 221, 225; French, 122, 137, 150; German & Dutch, 184; Italian, 79; Latin, 37; Spanish, 92, 109, 116–17
Quintilian, 3, 20, 21, 41, 68, 247n5, 248n12

Rabelais, 129, 206
Racine, Jean, 137–38, 149
Real Academia Española, 97; *Diccionario de la lengua castellana / Diccionario de autoridades*, 93, 96–97, *108*, 109–12, 116–18, 119
religion, 20, 28, 144, 145–47, 158, 188; fanaticism, 206; wars of, 127, 163
representation, 124, 143–44, 151, 239. *See also* mimesis; presentation
Reuchlin, Johannes, 22, 35, 39
rhetoric, 3, 9, 19, 68, 78, 137, 184, 191, 198, 220, 222, 239–40
Richelet, César-Pierre: *Dictionnaire des mots et des choses*, 14–15, 97, 126, 129–35, 139–40, 142, 145–47
Rider, John: *Bibliotheca scholastica*, 199, 200, 210, 222–23
Ripa, Cesare: *Iconologia*, 75, 216
Rosal, Francisco del: *Origen y etimología de todos los vocablos originales de la lengua castellana*, 95–96, 113
Royal Society, the, 194, 215, 233, 294n15

Saint Lambert, Jean François de, 127, 145, 148–50, 151
salon culture, 8, 9, 127, 136, 137
Sallust, 8, 46, 48, 99
salt, 117, 193, 195–96, *196*, 223
Salviati, Lionardo, 62–63
Sánchez de la Ballesta, Alonso: *Dictionario de vocablos castellanos*, 100, 117
satire, 33, 130, 135, 163
Savary des Bruslons, Jacques: *Dictionnaire universel de commerce*, 126, 130
Scaliger, Julius Caesar: *Exotericarum exercitationum*, 239–40
Scharfsinnig (G.), 156, 171, 173
scharfsinnigen Inschriften, 180
Scharfsinnigkeit, 174, 236
schemes, 109, 180
scherpsinnigh (D.), 176, 184
scholasticism. *See under* philosophy
Scoppa, Lucio Giovanni: *Spicilegium*, 60, 79
semiosis, 134, 138, 233
sense/senses, 171, 174–75, 184, 194, 222
sensibility, 125, 148, 150, 151, 171, 237
sentiment, 171, 222, 289n46
Sewel, Willem: *A Large Dictionary English and Dutch*, 174, 184
Shaftesbury, Anthony Ashley Cooper, Earl of, 149–50, 208, 285n77
Shakespeare, William, 14, 194; *Othello*, 226
sharpness: English, 197, 205, 210, 213, 219, 223, 225, 231; German & Dutch, 156, 165, 174–76, 184, 187; Italian, 55, 72, 78–79, 81; Latin, 41, 48; Spanish, 92, 109, 114–17
shine, 99, 136–37
sincerity, 110, 142, 209
Sinn/Sinn-/sin- (G. / D.), 153, 155–57, 159–61, 171–81 (171–73), *177–79*
Sinnlich (G.), 174–75
Sinnlichkeit (G.), 3, 13, 156, 168, 171, 176
Sinnreich (G.), 156, 173–75, 180, 187
sinrijk/sinrijck (D.), 155, 156, 158, 174
social order/class, 37, 44–45, 128, 133, 135–36, 140, 151, 214; disciplining, 140, 141, 151

society, 2, 6, 92, 111; good, 133, 134, 140
solertia (L.), 69, 79, 106, 197, 210, 231
sophistication, 106, 134–38
sophistry, 231, 240
soul, the, 7; English, 198, 208–9, 221, 224–26; French 133, 146–47; German & Dutch, 154, 185–89; Italian, 56, 73–75; Latin, 28, 32, 34–35, 39, 48, 50, 51; Spanish, 99, 103, 114
sovereignty, 8, 126, 127, 139, 237, 270n26
Spanish, 12–13, 16–17, 90, 95–97; linguistic purity, 119
spirit, 91, 111, 161, 195, 197, 201–8, 236. See also *esprit*; *geest*; *spirito*; *spiritus*; tutelary spirit
spirito (I.), 56–57, 66, 216
spirituality, 57, 157, 181–82, 184, 188, 197, 216
spiritus (L.), 131–32, 181, 197, 236
sprezzatura (I.), 58, 78, 244n26, 283n51
Stammwörter. See under German
Stevens, John: *A New Spanish and English Dictionary*, 111, 116, 117
Stieler, Kaspar: *Der Teutschen Sprache Stammbaum und Fortwachs*, 162, 163–64, 175–80 (*177–79*), 188
stuccio (I.), 81, *81*
study, 7, 67–70, 72, 82. See also acquired ability; training
sublime, the, 6, 125, 149, 222
subject (modern), 151
subtijl (D.), 157, 176
subtilitas (L.), 21, 79, 176, 210, 239
subtlety 6, 79, 100, 112, 115–16, 176, 195, 231, 238–40
synonyms, 5, 63, 94, 126
synonymy, 69, 197
systematicity, 16, 95, 97, 149, 150

Tacitus, 46, 154, 170
Takama, Henryk: *Onderzoek der byzondere vernuftens*, 189–90
talent, 9, 67–68, 70, 91, 110–13, 122, 194, 197, 216, 229, 237; god-given, 215; inborn, 21; natural, 68, 70, 78, 82, 100–101, 124, 142, 145, 148, 207
talento (S.), 113
taste, 17, 111, 114, 122, 132, 145, 149, 151, 223
teachability, 20, 40, 49, 69, 74, 168, 237
temperament, 7, 235, 240; English, 205; French, 122, 124, 128, 132, 134, 141, 144, 146–48, 150–51; German & Dutch, 164, 174, 176; Italian66; Latin, 20–21, 29–30, 32–33, 37, 40–42, 48–49; Spanish, 88, 103
Terence, 11, 24, 25, 30, 33, 41, 251n43
Tesauro, Emanuele, 8, 58, 76; *Cannocchiale Aristotelico*, 180
Thomas, Thomas: *Dictionarium linguae Latinae et Anglicanae*, 196–97, 199, 210, 223, 230–31
Tieck, Johann: *Franz Sternbalds Wanderungen*, 187
training, 1, 70, 103, 168. See also acquired ability; study
Tranchedini, Nicodemo, 84; *Vocabolario italiano-latino*, 60, 69–72, 79, 80
translatability, 4–5, 12, 16
translation, 11–14, 138, 139–40, 148, 161, 163
Tre Corone, 13, 16, 55, 57–58, 60, 62–64, 73
Trévoux. See *Dictionnaire de Trévoux*
trickery, 3, 7, 22, 24–25, 36, 46, 59, 79, 104–6, 157, 174, 180, 232
Turriano, Juanelo, 104, 112, 239
tutelary being, 20, 28–34, 51, 65–67, 112, 127, 144–47, 170
typography, 23, 41, 62, 94, 99, 117, 119

understanding: English, 221, 224–25; French, 132–33, 139; German & Dutch, 154–55, 161, 171, 174, 181, 186, 189; Italian, 55, 57; Latin, 45; Spanish, 90, 100, 104–5
unruliness, 116, 149

Varro, 35, 204
Vega, Garcilaso de la, 103, 111
Vega, Lope de, 90, 118, 119
Velázquez, Diego, 114, 268n16, 274n63
Venuti, Filippo: *Dittionario volgare, et latino*, 66, 71
Vergil, Polydore, 10, 11
vernacular languages, 13, 16, 19, 24, 51, 58–59, 94–95, 161–62
vernuft (D.), 155, 182, 184, 189
Vernunft (G.), 154, 155, 168, 174, 187
verstand (D.), 155, 161, 182, 184
Verstand (G.), 154, 155, 159–61, 168, 171, 174, 186
verstant (D.), 167
Virgil, 33, 46, 60, 149
virtù (I.), 69, 73–74, 76, 215
virtuoso (E.), 194, 214–15, 299n76
Vitruvianism, 70
Vives, Juan Luis, 7, 91
Volckmar, Henning: *Dictionarium Philosophicum*, 28, 48, 49–50
Voltaire, 127, 132–34, 136, 138, 140, 151
Vossius, Isaac, 34–35

Wachter, Johann Georg: *Glossarium germanicum*, 169–71, 180
wag (E.), 225, 232
Wiley, Margaret L., 201
wine, 25, 26, 27, 67
wit, 26, 27, 45, 78, 132–33, 137–40, 142, 184, 239; culture of, 157, 181. See also *wit* (E.)
wit (E.), 10–11, 17, 156, 167, 193–98, 208–10, 218–29 (226–28), 231–32, 236
wits, 135, 136, 139, 151, 224, 240
witticisms, 11, 93, 116–17, 223–24, 228–29
witty (E.), 5, 11, 115, 205, 217–18, 222–23, 228
witty sayings. *See* witticisms
Witz (G.), 155, 171
women. *See* gender
wordplay, 10, 19, 137, 140, 239
workmanship, 88, 106, 218, 230. *See also* craftsmanship
Wotton, Sir Henry, 205, 206
writer. *See* author